SOUTHERN LIMESTONES UNDER WESTERN EYES

THE MODERN WORLD EVOLVING IN SOUTHERN AUSTRALIA

SOUTHERN LIMESTONES UNDER WESTERN EYES

THE MODERN WORLD EVOLVING IN SOUTHERN AUSTRALIA

BRIAN McGOWRAN

Australian
National
University

ANU PRESS

For Susi, for believing in this project

Australian
National
University

ANU PRESS

Published by ANU Press
The Australian National University
Canberra ACT 2600, Australia
Email: anupress@anu.edu.au

Available to download for free at press.anu.edu.au

ISBN (print): 9781760465872
ISBN (online): 9781760465889

WorldCat (print): 1388684719
WorldCat (online): 1388684630

DOI: 10.22459/SLWE.2023

Cover design and layout by ANU Press

Cover image: *Acarinina mcgowrani* (Wade and Pearson, in Berggren et al. 2006). A planktonic foraminifer possessing a shell less than 1 millimetre in diameter, inhabiting the sunlit upper waters of the Eocene ocean in partnership (photosymbiosis) with a halo of photosynthesising algae outside the shell. *Acarinina mcgowrani* was painted in 2022 by Richard Bizley with scientific input from Paul Pearson.

This book is published under the aegis of the Science and Engineering editorial board of ANU Press.

Contents

List of figures and tables

Figures

Tables

Preface: Discovering biogeohistory

Homer. Moses. Socrates. Shakespeare. Newton. Beethoven. Einstein. Darwin. Most of these names cropped up at home and at school. Newton and Shakespeare had a good run; Homer and Moses and Beethoven but not Socrates were glimpsed; Einstein was both famous and still with us. But to the best of my dimming recollections the other name on this list was never heard or seen. Biology in junior high school was tepid stuff, briefly enlivened by an awkward account of how frogs perpetuate their kind; and in senior high school, physics and chemistry were the serious science for our class of boys, physiology and botany being mostly for girls. In physics I became aware of vast distances, to the moon and sun and far beyond, and of the unimaginable speed of light but not of concomitantly vast amounts of time. But trudging more or less competently through maths and physical science, I discovered geology.

The discovery might have been through growing up where we lived in the country, in valleys and on hills with rocky outcrops and running water, where the rising sun marches along the crest of the scarp from solstice to equinox and back to solstice and the moon and the stars were not contaminated by city light, and where exotic minerals were scattered around abandoned mine shafts. But it was not. The discovery occurred in picking up at home and casually flipping through WW Norton's *Elements of geology*, a 1921 (first published 1905) American textbook which my Aunt Alice had used when doing Geology I in 1928 with Sir Douglas Mawson at the University of Adelaide. Norton presented William Morris Davis's theory of landscape evolution in stages, from youth, through maturity, to old age, using examples from today's landscape. It had a photograph of the rugged and youthful Cascade Mountains in Washington with the remnants of an ancient land surface on top, implying massive uplift of the terrain, of rejuvenation, of the very old becoming young again. Norton's book illuminated the awesome vastness of geological time as it described marine and terrestrial life and times past and utterly unlike our own. It was magic, it was *romantic*. It brought on a tingling that I was not experiencing from physics or chemistry. I never met my Aunt Alice but I have owed her for seven decades because her book changed my life. The last 18 months of school were a blur of cricket and baseball and getting through the necessary academic subjects, but they were also a kind of spiritual marking of time until I could get to the university and seize geology. Or, be seized.

At university, the Geology I practical class was a cold shock. We were confronted with trays of wooden models of crystals. Following their disposal, in a manner of speaking, came a forbidding array of minerals, a parade of oxides and sulphides and silicates. And then trays of rocks, and then trays of fossils. There was geological mapping, too, the dreaded mapping. The objective of those mapping exercises was to learn to make the perceptual jump from a two-dimensional plan, or map, to its three-dimensional underground structure. (For those of us exposed to solid geometry in senior school, the three-dimensional thinking was no great problem. For some of the others, first-year mapping was the most frustrating item in the entire geological curriculum.) But they weren't real maps, they were dreary exercises; real maps would have recaptured the romance, for South Australia in the 1950s was leading in the great flurry to geologically map the entire continent. Meanwhile, the lectures were pretty dry; the day lecturer, the professor, had gravitas and a soul somewhat more chemical than geological, but he was not a showman (unlike the night lecturer); the textbook was Arthur Holmes's outstanding *Principles of physical geology*—but being 'physical' it did not emphasise earth and life history.

Several years' hindsight clarified for me the age-old tensions in first-year education in science (as distinct from medicine, law, engineering). On the one hand it is not so difficult to build a course when future professionals are the outcome, and here we see the old philosophy of induction at work. It seems reasonable to begin at the smallest scale and most basic data, the crystals and all those economic minerals from the mighty Broken Hill and the mines in the Copper Triangle, and work up to the grandest theories, via lots of rock-forming silicates and characteristic fossils—things you 'need to know' along the way. But Geology I was also to be the first and the last experience of the subject for most in the class, and it seemed to me that we were jibbing at an insistent double question. First, how do we entice some of these bright-eyed freshers to stick around in geology? Second, what should we be giving to all the others in one brief year, their first and last experience of geology? With the same lashings of hindsight I came to realise, along with Karl Popper,[1] that for both cohorts the theory is the thing: the theories, with their supporting evidence and with their emerging scientific problems, always provide more questions than answers.

Still I had not reached the eighth figure in the pantheon. I tried Zoology I in high hopes of some sort of intellectual fulfilment (not that such a phrase ever occurred to me). But the first lecture was on the chemistry of carbohydrates—worthy and necessary as building blocks of the biosphere, but about as exciting as the wooden models of crystals had been next door. Things did not improve—the textbook was early twentieth-century stodge; I found Darwin in a desultory section, 'evidence for evolution'. Evolution and Darwin were not front and centre, announcing what holds the biosphere and its science together: instead, they were tacked on as a late chapter. So it was during a state of some apathy towards life on earth in textbooks that I was reignited by another book, George Gaylord Simpson's *The meaning of evolution*. Simpson's books kept me going as I trudged through zoology, where ecology and evolution were sparse, and genetics, which was agricultural. Days before the final exams, I discovered in the library (for he was not mentioned in lectures) Theodosius Dobzhansky and *Genetics and the origin of species*. Graduation duly negotiated and research underway, I had discovered Ernst Mayr and *Systematics and the origin of species*, so I now had the central triad in the Darwinian Restoration: Simpson/Mayr/Dobzhansky. At the time, to be sure, I did not see things as crisply as all that; but I did suddenly acquire some clarity a couple of years later. When Mayr, pre-eminently persuasive Darwinian biologist of the twentieth century, visited us during the Darwin Centenary in 1959, he held court for a spread of biologists and a few geologists in the Mawson Building, hosted by Martin Glaessner. He gave no seminar and had no notes or props and merely invited questions, which he answered fluently for quite some time (for me, he could have continued far into the night). His central theme was the hope, forcefully expressed and defended, that all biological education be assembled under two great headings, functional biology and historical biology. That quite stirred things up, and I abruptly realised that my erstwhile teachers in biology (with whom I remained on very good terms) were rather reactionary. They did not get it. They believed in evolution, of course they did; but evolution was not central and critical, and systematics

---

1   Munz, *Beyond Wittgenstein's poker* (2004).

and taxonomy were low-profile endeavours. Instead, they thought not historically but structurally and functionally and experimentally, and *that* was their real science.

In the very year that I became a twofold-published author in geology and palaeontology, I awoke at last to what I had been wanting to become— a historian and more than that, simultaneously a biohistorian and a geohistorian.

It did not take seven decades to write this book, but the world view that I portray here has been almost that long in its making and maturing and expressing. The good people to whom I have owed so much cumulatively down the decades are overwhelmingly numerous and diverse. So I am taking the unusual step of thanking them all and presenting a very short list of some who especially expanded, challenged, upended or upset that world view along the way: Martin Glaessner (1906–1989), Al Fischer (1920–2017), Mary Wade (1928–2005), Bill Berggren (1931–), Lukas Hottinger (1933–2011), Wolfgang Berger (1937–2017), Qianyu Li (1956–) and Henk Brinkhuis (1960–).

For assessing the first draft constructively and encouragingly I am indebted to Henk Brinkhuis and Patrick DeDeckker. Brian Kennett, chair of the Science and Engineering editorial board at ANU Press, has been steadfastly positive and encouraging all the way through since the first tentative proposal. Beth Battrick did a marvellous job of copyediting and indexing and did it with patience and good humour. Likewise, Sarah Sky and Teresa Prowse and ANU Press did this book, and me, proud.

Introduction

The age of natural history: What did Joseph Banks know and how did he know it?

Australian plants and animals were exceedingly strange to Portuguese and Dutch eyes in the sixteenth and seventeenth centuries, and doubtless to the eyes of east Asian sailors long before then. The trees looked different, the animals looked different and for that matter so too did the people. Sustained European interest in Australia dates from the eighteenth century and especially from James Cook's voyage in the *Endeavour* (1768–1771). In Cook's day so-called true science, as distinct from geography and natural history, was causal and physical and dominated by mathematics, and exemplified by widespread interest in the impending transit of Venus across the sun, which event conveniently brought Cook to the south seas and Tahiti. (The word 'science' is anachronistic here, being coined in the next century. 'Natural philosophy' was the term at that time.)

It is the next phase of that voyage that interests us. Cook carried the so-called secret instructions to search out the great south land, *Terra australis* or *Terra incognita* hypothesised by Aristotle and by wise men who subscribed down the millennia to the hypothesis that there should be a large southern land balancing the land in the northern hemisphere. Whether that land actually existed was of great interest to the seafaring nations of Europe. It interested philosophers too, for a very early geophysical conjecture was that there should be some balance between land and sea in the two hemispheres. Accompanying Cook were the naturalist/botanists Joseph Banks and Daniel Solander and their support staff, all funded by Banks. They did not see southern Australia but the scientific impact of the visit to the east coast was profound, not least on the career and influence of Banks himself over the following decades.

Joseph Banks (1743–1820), gentleman commoner, educated at Eton and Oxford, became fascinated at an early age by botany and the diversity of the plant kingdom. He grew up in the heyday of natural history, exemplified by the parson-naturalists of rural England. Banks knew many of them, in the country or as university classmates. They observed the diversity of life and its exquisite designs fitting the plant or animal to the environment and its seasonal rhythms, from mountain, hedgerow and woodland to marginal sea, all celebrating the power and goodness of the Creator in a distinctively British intellectual atmosphere known as natural theology. They kept cabinets of specimens and they maintained diaries.

The cultural ferment known as the 'Enlightenment' of Western or European or Judaeo-Greek civilisation flared in the sixteenth–seventeenth centuries. The rise of what came to be known later as 'science' was broadly threefold. There was natural philosophy, especially mathematics and its applications in physics and astronomy. There was medicine and physiology. And there was natural history, pertaining to the diversity of living things, botany and zoology, and their substrate of rocks and muds and soils, called mineralogy. The three categories were by no means insulated from one another while the disciplines within them arose and took shape in the eighteenth century. Natural history is the central concern in our narrative.

The two words 'natural history' together were pregnant with meaning. 'Natural', pertaining to nature, faced 'human' and 'divine'. There still existed the hangover from medieval thought known as scholasticism. In scholasticism truth was arrived at by logic, not by observation or experiment. The oft-quoted example is a dispute over the number of teeth in a horse's mouth, the answer to be found in Aristotle's writings, not in the horse's mouth. This tradition was sustained most strongly in the overriding authority of the 'revealed word' of Christianity, and the revealed word was the route by which you connected with nature. Meanwhile 'history', then as now, meant the bleeding obvious, namely human history. We knew about the past just as we knew about distant lands—because ancient scribes, known as chroniclers, had written about it. The greatest event in the deep past was Noah's flood, the Noachian Deluge, and what was known of earlier times was in *Genesis*. If there was a deep past, there was no other way into it.

So! One basic problem for the natural historian was how to escape the trammels of ecclesiasticism (lovely Victorian phrase) in seeking explanation and meaning in the richness of nature, while at the same time sustaining their religious faith. Thus arose natural theology. Meanwhile physical scientists known as natural philosophers could live comfortably with a created world with natural laws that required minimal divine intervention. They could seek to explain the physical world and its universe in terms of a limited number of basic, divine laws, the laws of physics. Bishops and philosophers were also comfortable with the divinely inspired laws of physics. But physics itself was totally unsuited to the living world of diversity and complexity and multitudinous interactions with each other and with the environment. These interactions were known as adaptations. That is, the living world was inherently unpredictable. And so the devout natural theologian saw the hand of the Creator in the smallest and subtlest aspects of adaptation and diversity; and natural theology arose in parallel with natural philosophy.

The other central problem for the natural historian was of the day-to-day kind: how to manage the sheer variety, the exuberant diversity, of things in nature. It was exacerbated as global exploration expanded the geographic horizons of European societies,[1] and the bewildering array of new and strange organisms expanded accordingly. The big question got bigger: how do you cope with variety and diversity among the well-known categories of animal, vegetable and mineral? The early explorers coped via the eminently human instinct of imposing what they knew upon what was new. Rottnest Island off Western Australia was so named because Dutch sailors thought that the local quokkas were rats. Koalas seemed to be a small kind of bear. Tall straight trees eminently suitable for ships' masts looked like the pines of northern lands, so we still have the colloquial name 'Norfolk Island Pine'. Both birds and both black and white, the northern 'magpie' and the southern 'magpie' have little else to justify a shared name, but share it they still do.

Back in Europe, though, the familiar organisms were being scrutinised anew. What do you actually do as a serious naturalist? Your objective is to make sense of biodiversity by organising it into a classification. You first sort the objects, just as an interested child can nimbly sort a bucket of seashells, unjumble the jumbled kinds. You describe the grouped kinds and classify them, and you discover soon enough while learning how to identify them that you must grope for some system of organisation and retrieval. This sounds simple and straightforward, like an accountant purchasing a filing system, but how to do it depended upon why to do it, and the why had to entail some sort of philosophy of meaning. And

---

1   Glyn Williams, *Naturalists at sea* (2013).

scholars struggled for centuries with this framework, known as taxonomy, underpinning the study of diversity, known as systematics. Although we honour the philosophers who largely built Western civilisation, it was the practical people who were driving things at the practical level—the domesticators and breeders of animals; the herbalists and witch doctors and medicine men; the agriculturists domesticating and improving crops; those who exploited rocks in prospecting, mining and metallurgy and engineered them for roads, bridges, canals and harbours; and the astrologers mapping the rhythms of the heavens. They all had their names, identifications and categories and they ranged from pedestrian to competent to brilliant and visionary at what they did. And the accurate sorting and identifying of animals, vegetables and minerals could be utterly essential.

Take, for example, the trade in medicinal herbs. You had to know what you were buying or risk being seriously duped, to say nothing of poisoning someone. Here is the newly qualified apothecary Thomas Johnson (~1600–1644) on the perils of the London market for herbs:

> Almost every day in the herb market, one or other of the druggists, to the great peril of their patients, lays himself open to the mockery of the women who deal in roots. These women know only too well the unskilled and thrust upon them brazenly what they please for what you will ... Is not the fate of patients who rely upon the help of such doctors and druggists pitiable? For the doctor relies on the druggist and the druggist on a greedy and dirty old woman with the audacity and the capacity to impose anything on him. So it often happens that the patient's safety depends on the herbal knowledge of an ignorant and crafty woman.[2]

To us it is obvious that the ultimate basis for safety and progress in herbal medicine had to be in knowing the kinds of plants and how to identify them. But it is obvious only because of centuries of effort in disentangling solid evidence from folklore, hearsay and superstition, and concentrating on expanding, testing and improving that evidence. Only by such rigorous pathways could natural history be transformed into science. We can generalise from this and state that for all the thousands of years since our forebears accomplished the Neolithic agricultural revolution and began building civilisations, their pragmatic, suck-it-and-see, sophisticated technologies were underpinned by experience and tradition, but not by what the modern mind would identify as robust and reliable scientific knowledge.

---

2   Pavord (2005, p. 5).

Things changed in the seventeenth century and accelerated in the eighteenth. Think of these key words and prompts: Enlightenment emerging from Renaissance, spirit of enquiry, Industrial Revolution, imperial exploration, collecting, natural theology … natural history! If we were to come to grips with the sheer exuberance of different organic kinds, plants and animals, we had to find some organising principle, some way of detecting order in the seeming chaos of the organic world. The organising principle had to be found within the plant and animal kingdoms, not imposed from without or above by some clerk or cleric. In short, the classification had to be natural. The story of the centuries-long search in the plant kingdom is told splendidly in Anna Pavord's *The naming of names: The search for order in the world of plants* (2005).

We can pick up the prelude quite late in the seventeenth century with the botanists John Ray (1627–1705) in England and Joseph Pitton de Tournefort (1651–1708) in France. Ray, son of a village blacksmith and a herbalist, was a bright boy, able to win a scholarship to university. Trained to be a Christian minister, he produced as a systematist *The catalogue of Cambridge plants* in 1660. As a thinker about the problems of classification, believing strongly in seeking a natural classification bringing together related groups, he published (1682–1704) *New method of plants* and the grand *History of plants*. As if these accomplishments were not enough, Ray produced several authoritative volumes on zoology and is regarded as a pioneer of geology in Britain. And it was all done by a natural theologian and preacher inspired by religious faith, culminating in *The wisdom of God manifested in the works of the Creation* (1691). The wisdom was based in very sound natural history, containing a powerful argument for design and adaptation, thus becoming an early work on ecology as well as the founding document for eighteenth-century natural theology.

Ray and Tournefort argued about the basis for classifying plants, but more important for us than the rights and wrongs of the disputes is their influence in shifting study of the living world from mysticism and fable and hearsay towards evidence-based science. But the tension in organic classification inherited from the ancients was going to rise and fall as the exploration of organic richness picked up in the eighteenth century. It seems eminently sensible that a classification and system of identification be clear and unambiguous, straightforward and practical. But, should we not be searching out and exposing the deep structure of the living world, the blueprint in the mind of the Creator? Two clear and ambitious objectives. If they are compatible, so much the better. But what if they are not?

The towering figures in natural history were Carl Linnaeus (1707–1778) and Georges-Louis Leclerc, Comte de Buffon (1707–1788). Each has been described as resembling Aristotle in being more a committee than an individual. Taxonomy and classification reached a peak in Linnaeus's works as he achieved simplicity and consistency where there had been chaos, and his binomial nomenclature (the Genus including one or more species) and all the necessary techniques flourished in the science of biosystematics (Winsor, 2006b). Buffon was not so interested in all that detail. He was roaming in the big picture, where such resonant concepts as plenitude, continuity and scale of perfection drove his *Histoire Naturelle*, remaining uncompleted after four decades. He was at once the last serious, ambitious storyteller, the last cosmologist and the first explorer of deep time. He clearly endorsed the bottom-up strategies of assembling things into groups and groups into bigger groups, labours which would produce in due course the tree of life:

> It would seem to me that the only way to design an instructive and natural method is to group together things that resemble each other and to separate things that differ from each other.[3]

About this book: The watcher on the rock

YOU'D THINK BETTER IF YOU PUT SOME CLOTHES ON

Figure 0.1. The thinker on the rock. After the late Ron Tandberg and *The Age* (Melbourne).
Source: Glenys Tandberg.

Think of squatting on a rock in southern Australia, such as a limestone stuffed with fossils (Figure 0.1). If you aspired to emulate Rodin's naked and existentially challenged Thinker, you surmounted the scratchy bits of shell and our grandmothers' warnings on the risks of piles. But *Le Penseur* was sculpted as recently as the 1880s, and our thinker was there 50,000 years ago. On the cliffs facing the Southern Ocean she would have been aware of both winter storms roaring in from the south-west and summer storms

---

3   Mayr, *The growth of biological thought* (1982, p. 193).

from the north: in colder times than ours, more of the former; in warmer times, more of the latter. Perhaps the cliffs were those along the River Murray; if so, did our watcher realise that the abundant objects falling out of the rocks resembled shells on the sea shore, far to the south? Did her people have an explanatory narrative of floods, of ice caps on the mountains in the south-east, perhaps of hills arising out of the sea? Interesting and important as such questions are, they are beyond my time frame. Indeed, similar questions could be asked of thinkers possibly arriving in southern Australia only centuries ago, perhaps Chinese, perhaps Portuguese, and these too are outside the time frame.

Our beginnings are in the late 1700s, when the horizons of science, by then a couple of centuries old, expanded dramatically to include deep time. Natural history acquired the fourth dimension. Natural history also gained some of the respectability accorded natural philosophy, meaning physics, and natural theology, which contemplated the glory of the Creation.

Table 0.1 charts the development of biohistory and geohistory through almost a quarter of a millennium. Think of a river increasing in volume at the successive arrivals of its tributaries. The river is an accreting entity; and I am proposing that this story is like the river flowing down the decades: eight great surges, I–VIII, have accreted to form the discipline of biogeohistory.[4]

Table 0.1. The conceptual framework for this book: The rise and development of biogeohistory through a quarter of a millennium. Each surge in insight built upon its antecedents.

Fossils, strata and Cenozoic southern Australia: Eight accreting surges of insight into the history of the earth and its biosphere. This is biogeohistory!		
1980s–2000s	VIII	The shells of microbes hold simultaneously the signals of age, family tree, lifestyles ranging from symbiotic to low-oxygen, and environmental shifts at scales from local puddle to global ocean. The signals deliver high consilience in revealing the evolution of the biosphere and in describing the hothouse–icehouse transitions in ancient oceans on a dynamic earth during Cenozoic time.
1960s–1970s	VII	Continental drift is confirmed. Oceanfloor spreading and plate tectonics transformed the global environment and its history. After the India–Asia collision, Australia separated from Antarctica and Zealandia in Gondwanaland's disintegration. The Australo-Antarctic Gulf disappeared into the new Southern Ocean.

4   Implying that our river loses identity in the ocean of knowledge is misleading. Braiding or anastomosing the great themes like rivers in arid lands not reaching lake or sea is better but also misleading; best is to treat all naturalistic metaphors like supping with the devil: use a long spoon.

Fossils, strata and Cenozoic southern Australia: Eight accreting surges of insight into the history of the earth and its biosphere. This is biogeohistory!		
1930s–1940s	VI	In the Darwinian Restoration, the 'Evolutionary Synthesis' reconciled palaeontology, field biology and population (transmission) genetics, and natural selection roared back into favour. Goal-directed and internally driven theories of evolution were cast out.
1910s	V	The radiometric calibration of earth history confirmed deep time with years in the billions. The Cenozoic Era is 65 million years old.
1860s–1870s	IV	Oceanography was born of oceanic expeditions, especially the British *Challenger*. The highly informative microbes in the pelagic realm produced calcareous, opaline and acid-resistant skeletons in an enormously rich fossil record. Deep-ocean drilling and micropalaeontology triggered the rise of palaeoceanography and modern views of global climatic change.
1850s–1860s	III	In the Darwinian Revolution, organic evolution in deep time explained all of biology in historical terms. Genealogy, the 'tree of life', was accepted simultaneously as (i) a fact, (ii) a theory and (iii) a research program. Genealogy explained the success of the fossil-based geological time scale. (But Darwin's mechanism of natural selection remained controversial.)
1820s–1830s 1790s–1810s	II	Earth and its life had a discoverable global history (time's arrow). This was the Cuviero-Lyellian Revolution built on the Palaeontological Synthesis. Fossils in strata had an orderly pattern through time, thus biostratigraphy built the geological time scale of epochs, periods and eras. The world views of time's arrow and time's cycle have been waltzing ever since. Lamarck was a great taxonomist and the first evolutionist. Although his theory failed, his historical thinking flourished in due course.
1790s–1810s	I	Earth as machine was discovered in several rock relationships: (i) igneous intrusion and extrusion; (ii) erosion, deposition of sediment and unconformity; and (iii) deformation by folding, faulting and metamorphism. Mountain building (uplift) was the essential driver, but not understood. Immensely great stretches of time were analogous to immensely great cosmic distances. Deep time supplanted 'biblical time' in time's cycle.
Late 1600s–1700s		The Enlightenment project of describing nature's diversity got underway. Fossils and fossilisation were becoming understood.

Source: Author's summary.

I should note here the intertwining of topics known familiarly as geology and biology. Pulmonary physiology and gold prospecting (to take two rather disparate examples) would seem to be separate biological and geological fields of endeavour with no obvious symbiosis. But when we acknowledge the fundamental importance of rock relationships and earth history, we see where biology and geology are inextricable. Arduino, a minerals prospector in the mid-eighteenth century, envisioned a fossil succession up through strata accumulating kilometres in thickness, from the unfamiliar life forms below to the familiar above. William Smith, surveyor to the builders of the Industrial Revolution, used fossils to make the map that changed the world. Georges Cuvier, the biologist, shows us that pre-human historical sciences are discoverable. Most of an entire volume of Charles Lyell's *Principles of geology* is about animals and plants, their distribution in space and time, and the dangerous theory of their evolution. Generations of biology teachers have underappreciated Charles Darwin for the geologist that he was, most significantly. As our knowledge increases, our questions change and the role of the fossils changes—but fossils and life have remained intimately embedded in the science of rocks and deep time ever since.

Modern science is impatient with the silos of knowledge called chemistry, physics, geology and biology. Doublets such as biophysics, biochemistry, geophysics and geochemistry are commonplace, and triple hybrids such as biogeochemistry are found to be necessary. Geohistory and biohistory are excellent words but I frequently need to couple them. Thus *biogeohistory* is what this book is about, as it recounts the history of the history, the ever-changing perceptions of the rocks and fossils and landscapes, as chronicled in Table 0.1. And the broad-brush statements in Table 0.2 should keep the biohistorical side of the story in perspective.

Table 0.2. General statements about 'informed cultural' beliefs through the centuries about the 'bio-' side of biogeohistory.

Late 20th–21st century	Organic evolution transmutes into evolution of the biosphere on a mobile lithosphere.
Middle 20th century	The Darwinian Restoration (the Modern Synthesis).
Early 20th century	Variational evolution is rescued by naturalists and population genetics.
Late 19th century	Almost all believe in transformational evolution. Very few believe in variational evolution by natural selection.
Middle 19th century	Charles Darwin and AR Wallace: introducing variational evolution by natural selection as fact, as theory and as research program.
Early 19th century	Almost all believe in deep time and the global fossil succession. Only a minority believe in the evolution discovered by Lamarck.
Late 18th century	Deep time is here to stay; earth history and life history awaken.
Early 18th century	Almost all believe in shallow (biblical) time and in biblical creation. Pre-human history is almost impossible to imagine.

Source: Author's summary.

Why southern Australia? Why the Cenozoic Era?

South of Adelaide in central-southern Australia is Granite Island off Victor Harbor, a small lump of, well, mostly granite, shaped by the ice a quarter-billion years ago. On Granite Island there used to be a signpost pointing to various foreign places so-and-so many miles distant, and one such place was the South Pole, 3,850 miles. Standing on Granite Island, aware of northern sayings such as 'down under' and 'the ends of the earth', I would stare southwards at the ocean extending far beyond the horizon. Abundant evidence demonstrates that ice carved the landscape and seascape of the district, and scratches and grooves on the rock faces indicate that the ice came from the south. How then did the ice get to Victor Harbor? There was a theory that its traverse from Antarctica to Granite Island was entirely overland on the supercontinent Gondwanaland; but the theory was contested. In due course that theory won. The supercontinent indeed did split apart and a chunk in the east now known as Australia was facing a new seaway, the Australo-Antarctic Gulf. There was lush vegetation on both sides, rainforests actually, and tropical rainforests in the winter darkness. And once more there was ice, now on Antarctica, just across the water—from Granite Island to the other side of the gulf was about the distance from Adelaide to Melbourne. Once more? The aforementioned ice was on Gondwanaland; but Gondwanaland itself was preceded by the supercontinent called Rodinia, and Rodinia in its last days more than half a billion years ago experienced episodes of the climatic crisis known as snowball earth—the Sturtian, named after a small stream through Adelaide, and the Marinoan, after a seaside suburb of Adelaide. Most recently the Adelaide district, sandwiched as it is between the red desert and the deep blue sea, experienced the ice ages indirectly (the ice caps were on south-eastern Australia) when the Roaring Forties lashing the Southern Ocean became, as Reg Sprigg used to say, the Roaring Thirties.[5] Our first thinkers were there, hunting and foraging. Or squatting on the limestones and wondering.

So the biogeohistory in this book is biased cheerfully towards southern Australia, north shore of the dying Australo-Antarctic Gulf, north shore of the emerging Southern Ocean, just across the water from the Antarctic deep freeze.

---

5   Reg Sprigg was the most dynamic and productive contributor to the postwar expansion of Australian geology. See Note 10 in Chapter 9.

'Too beautiful for our ears, my dear Mozart, and monstrous many notes.' 'Exactly as many as are necessary, Your Majesty.'

My prose attracts no such accolade as 'too beautiful' and terminology is my problem analogous to Mozart's Emperor's problem with his notes, a problem not of one arcane discipline's jargon but of several, across the fields of geology and biology. My responses are a glossary and, most importantly, generous figures. Maps are basic to geology, but figures including or implying the dimension of time are utterly essential in biogeohistory. Such figures are stratigraphic, stratigraphy being the discovery and understanding of strata, the archives of things and events in space and time. Forbidding as they may be when flashed up in PowerPoint, charts and tables are better than words in showing stratigraphic relationships and carrying stratigraphic arguments. They are not for viewing for a minute or two before we all move on; they are for staring at, and pondering, and revisiting, repeatedly. We know that the elements of biogeohistory are easy to grasp piece by piece, item by item. Problems arise when, say, the word 'Eocene' is spoken, then 'Miocene'; and while you hesitate as to which comes first in the succession the speaker moves on and you are lost. But contemplate a chart or two and the names in their right order are soon engraved on your soul.

We can begin with a hypothetical cross-section through the uppermost part of the earth's crust in a district apparently deeply incised by a rushing river (Figure 0.2).

There is a stack of strata with fossils, different fossils in the upper from the lower parts of the stack. Immediately there are environmental implications—for there are marine shells high up. Did the land rise, or was it the biblical Flood? More subtly, the sea seems to have advanced and retreated, for some strata have nonmarine bones and leaves. Again, we see a break, an interruption, a caesura dividing the scene into an upper and a lower, implying a younger and an older, with profound earth movements in between—for the tilted strata originally had been horizontal. Thus it is possible already to deconstruct the scene into no less than nine items in a chronological succession—and with no jargon, no mention of time scales or organic evolution. But the scene and its deconstruction are pregnant with meaning. The nine items are altogether too orderly for the chaos of a biblical Flood to be plausible. There is evidence of repeated uplift, to say nothing of the tilting of the lower strata, all implying deep forces in the earth. Clearly the answers are not to be found within this regional scene but more broadly. Even so, our modestly deconstructed local scene whispers, 'Deep Time!' And perhaps even, 'Transmutation?'

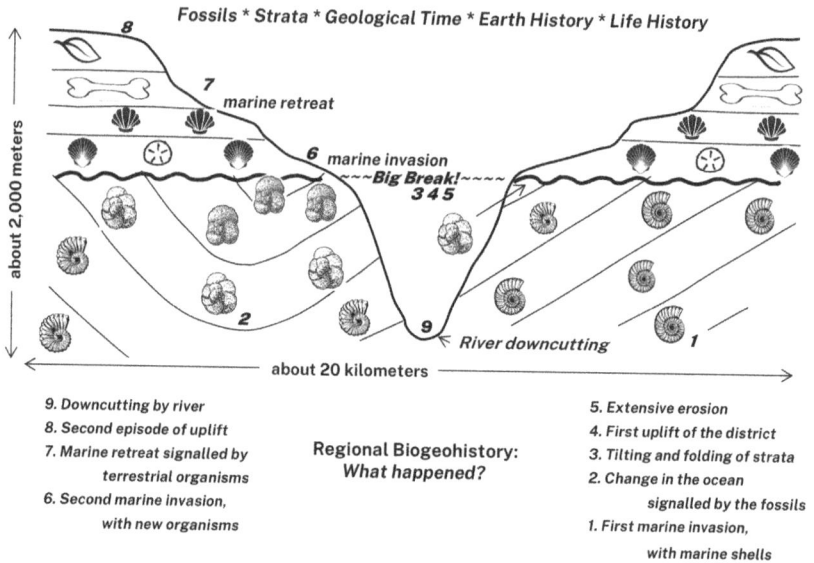

Figure 0.2. A dreamed-up cross-section of fossiliferous strata exposed in a mountainous terrain.

The fossils change through the succession from below (older) to above (younger), telling us something about the geological ages of the strata and the marine or terrestrial environments of their deposition. Numbered 1 to 9 is a succession of events teasing order out of what the innocent eye might perceive as chaos. But much happened within the 'Big break!' (the unconformity), implying a large local gap in time (a hiatus) and knowledge, which has to be filled in elsewhere. Note the huge exaggeration in the vertical scale (metres) over the horizontal (kilometres), making things legible.

Source: Author's depiction.

So we begin in Chapter 1 with rock relationships and how our stumbling enthusiast learned to 'get his eye in', the phrase which universally means reading the rocks, at your feet or on the horizon, or spotting and identifying the fossils. Tracing strata among scattered outcrops, asking 'which way is up?' is as essential today as a quarter-millennium ago.

In Chapter 2 our enthusiast engages with the familial lineage of the archetypal ancient mariner, the pearly nautilus, thence to how understanding fossils came to expose deep time and reveal the history written in the strata. It took time, did understanding the meaning of tens of thousands of shells. But in southern Australia we had a century of frustration as our fossils, rich and diverse though they were, failed to connect this region with the wider world. That's putting it too negatively, for this country is rich in the strange and the unique, and they have their roots in the fossil record.

Meanwhile, as geology and palaeontology went global, thanks largely to the reach of the Royal Navy, the pressures of explaining the patterns of life in deep time kept growing—the fact of extinction, the law of succession—until Charles Darwin broke through, if not completely. Referring to familiar kinds of fossils—shells and bones—Chapter 3 ventures some way into evolutionary biology in deep time.

Having met the broadly familiar, such as bones and shells and plants (*macrobes*), we now meet the unfamiliar, the marine microfossils, in Chapter 4. Several major groups of microorganisms (*microbes*) construct shells (or 'tests') which fossilise in their millions. Known for all of the nineteenth century, their biogeohistorical potential suppressed for decades somewhat unwittingly by Charles Darwin, the scientifically best-known group and the most informative, the Foraminifera, came into their own in the twentieth, at first in advancing the ages of strata and their ancient environments, then in the reconstruction of ancient oceans, no less, in the discipline known as palaeoceanography. Before the invention of the scanning electron microscope, our enthusiast had to learn how to sketch their tests under the stereo binocular light microscope.

Fossils gave us deep time, and fossils also have the dimension of space, biogeography. In Chapter 5 we come to a long-simmering question: while plants and animals can move, can continents and oceans also move? That the answer at last was a triumphant 'Yes!' is at the heart of continental drift and the dynamic earth's crust, known as plate tectonics. Just at the height of the revolution in the late 1960s–1970s, offshore petroleum drilling technology was successfully exported from the shallow seas to the deep oceans, and our enthusiast was there. And the narrative must span and reconcile evidence as disparate as the magnetic patterns in the oceans' basement, chronicling the birth and death of ocean basins and the making and breaking of supercontinents, and the family tree of the flightless birds on the southern continents.

In Chapter 6 we have to look up and think up to the global level. We build palaeoceanography on some generalisations about model oceans. Our enthusiast is comfortable with high-school-level equations wherein carbon dioxide moves busily from the land to the sea and the sky as it weathers continents, builds and dissolves skeletons (carbonate), and constructs 'carbohydrate' (organic carbon), representing the materials of the biosphere. We encounter some gentle geochemistry in the form of isotopes, especially the isotopes of oxygen and carbon. As the humble individual foraminifer grows, it is filing information about the global ocean in its shell. (And the forams' collective range of adaptations is huge—they live in, on and above the mud and in the upper waters of the open ocean at all latitudes and in a wide spectrum of oxygen and salinity; and extremely interesting are the cooperatives, in which various 'algae' live symbiotically with their hosts, the forams.) Having outlined methods of looking into oceans, we compare the results through deep time, for comparison and contrast are the most powerful methods at our disposal in this historical science. And now: our first glimpse of the differences between the greenhouse world of the Palaeogene Period and the icehouse world of the Neogene Period (and we live in the Neogene world).

Added recently to the neologisms 'icehouse' and 'greenhouse' are 'hothouse' and 'warmhouse'. Fifty million years ago when the world was in a hothouse state, the Australo-Antarctic Gulf, between 60 and 70°S latitude, was rimmed by tropical vegetation, and its planktonic microbes were tropical too. This is Chapter 7. Chapters 8 and 9 tell the environmental story of how we got from this extreme Palaeogene condition to our present, Neogene condition of a desert with damp fringes, facing the Southern Ocean and its Antarctic Circumpolar Current, the engine room of the modern ocean. It is a regional narrative, but always a part—crucial part!—of the global story.

Biogeohistory has always been on the edge and so too is Chapter 10. Biology or geology? Creation or evolution? By design or contingent? Historicist or mechanist? Catastrophist or uniformitarian? One of the more sustained and interesting are the two pillars of Darwinian evolution, natural selection and the tree of life, respectively variational and transformational evolution. They can be traced from the eighteenth century onwards, and out of science and into human history and the 'Humanities'. Biogeohistory is or should be embedded deeply in the grandly envisioned Big History. Likewise, biogeohistory is disgracefully missing from a 'Humanities' furore in Australian academia about Western civilisation. But a better place to sign-off is in the microfossils glimpsing the event that interests us all, the end of the Cretaceous Period and beginning of the Palaeogene Period, and 'what killed the dinosaurs?'

Précis: If you must leave early ...

This book tells a story in the history of earth and its life—in geohistory and biohistory, more conveniently called biogeohistory. It focuses on the Cenozoic Era, that slice of some 65 million years between two notorious catastrophes, the death of the dinosaurs on a greenhouse or Warmhouse Earth and the advent of humans on an Icehouse Earth. Needing a geographic anchor, the tale is centred in southern Australia, at one time the north shore of the Australo-Antarctic Gulf during the disintegration of eastern Gondwanaland, becoming the longest northern coast of the new Southern Ocean. Third, it is about the history of biogeohistory, about historical consciousness, about historicity. Biogeohistory erupted from natural history late in the eighteenth century, about two centuries after mechanical science emerged from natural philosophy.

Dutch and English explorers came to southern Australia while Europe was discovering that the biblical timescale obfuscated the geological past instead of revealing it. But it was the French explorers who discovered our natural history, for they were under instructions from the naturalists in Paris, from Lamarck who had discovered organic evolution and from Cuvier who had discovered organic extinction; and for all their different styles and opposing world views, Lamarck and Cuvier discovered biogeohistory and triggered the grand project of building the essential geological time scale. This was around 1800, and the Cuviero-Lyellian Revolution was underway. In Table 0.1 (see Introduction) it is surge #II of the eight (VIII) surges in the growth of reliable knowledge in the historical sciences of rocks, fossils and earth history.

In my notion of surges, evidence-based (reliable) knowledge accretes in highly episodic patterns. Insights, always a bit mysterious, arise from new discoveries in travel, technology, the demands of society or pipe dreams in the armchair. All cultures have had theories to explain their world and its components and how that world came to be. The only criterion for dealing with such theories scientifically is this: a theory is only as good as the research program it inspires. Creation myths don't satisfy that criterion. After thousands of years of human interaction with rocks and the environment, a few thousand years with theologians and philosophers, and two hundred years after the Enlightenment produced modern science, we inhabitants of 'the West' produced two kinds of theory in the late eighteenth century. In one mindset of the Enlightenment, earth's development was directional and historical (*time's arrow*). The other mindset was to be found in those thinkers of ahistorical temperament, perceiving a steady-state earth or an earth in cyclic equilibrium (*time's cycle*). In this category James Hutton's 'rock cycle' distinguished igneous, sedimentary and metamorphic rocks and rock processes. Better than 'cycle' is *Earth as machine* operating during vast but unknowable amounts of time—in the strikingly ungrammatical but deeply appropriate neologism, in deep time. Knowledge of rocks in their three dimensions became surge #I and geology's task of discovering earth's deep history was underway at last.

The phrase 'rocks in three dimensions' arouses questions about fossils in strata and about strata in deep time. Hutton's vision of a dynamic machine-like earth required uplift, that is, mountain building, whereupon gravity could act in rain, weathering and erosion, transportation and sedimentation. He had no mechanism for uplift but he had the results of uplift—such as rocks with fossils of indubitably marine organisms, 2 kilometres up in the sky! The seashell was indeed on the mountaintop. This way of thinking invoked fossils as ecological indicators of times past; Leonardo in the fifteenth century and Thomas Hooke in the sixteenth had similar thoughts about fossils. Great thinkers: they were never going to anticipate Lamarck and Darwin.

But what about fossils as chronological indicators, giving us entry into deep time? In outcrops of layered rocks (strata) lower implied older and upper implied younger. William Smith discovered that a body of rock, a limestone, say, or a claystone, could have its own characteristic collection of fossils, and that another seemingly identical rock had a distinctly different set of fossils. Thus the clays, the sands and the marls could come and go, as in time's cycle, going nowhere. But Smith's fossils were not jumbled: they formed an orderly succession, as in time's arrow, going somewhere. Smith's work was driven by the demands of surveying, mining and engineering in England's Industrial Revolution. His fossil-controlled stratigraphic succession worked so well that he could produce a marvellous and unique geological map. Meanwhile the French restored historical thinking in the natural science of biogeohistory by reconstructing ancient environments, marine and nonmarine, restlessly changing and populated by organisms now long extinct.

The geological time scale was built with fossils in about six decades between Cuvier's extinctions in the 1790s and Charles Darwin's organic evolution in the 1850s. The time scale is the grand centrepiece of surge #II, which once was called the heroic age of geology. The malacologists of Europe pieced together a succession of faunas (assemblages of shells) wherein the younger the fauna, the more it resembled the modern. Charles Lyell seized upon this accomplishment in the 1820s to erect three epochs in the Tertiary—the Eocene, Miocene and Pliocene epochs. Fossils as timekeepers worked in Europe, brilliantly—but also pragmatically, with no coherent theory as to why this species arrived in the shallow seas when it did and why that species departed forever when it did. Meanwhile the fossil discoveries tumbled out of strata of the neritic (shallow seas) and terrestrial environmental realms in lands distant from Europe, such as Australia and South America. Australia with its marsupials and lungfish was different, perhaps sheltered from global progress by its isolation? The plants were different too, and the marine shells. And most notably of all, the fossil bones and teeth of the big Pleistocene animals, the 'megafaunas' on the different continents, were like their respective living counterparts: they were not like each other. There never was a pre-biblical-Flood world fauna succeeded catastrophically by a post-Flood world fauna. Australia produced its own megafauna instead, with its own history extending back through deep time. And the disorderly litter of rocks and sands strewn over much of Europe and North America turned out to have been left not by the Flood but by the ice sheets of the Pleistocene ice age. Biogeohistorical awareness was encompassing fossils, strata, environment, geography and climate change.

A half century after Hutton there was another range of mindsets. We detect in those geologists and palaeontologists a spectrum from episodic or 'catastrophic' thinking to gradualist or 'uniformitarian' thinking. Some saw the jumps and gaps in the record of fossils and strata as meaningful, while other workers saw them as due mostly to destruction of the evidence. There was a range of views on organic evolution versus creation (and probably much reverential silence). The uniformitarian Lyell strongly advocated that known earthly processes then operating were sufficient to explain previous states of the earth. In this view, theories of vast catastrophes in ancient times were not required and not wanted in geology.

Charles Darwin the geologist absorbed everything geological, palaeontological and biological produced by the Cuviero-Lyellian Revolution as evidence for his theory of evolution. His ecological mindset was influenced by William Paley's *Natural theology* and his lateral thinking and geohistorical instincts came from Lyell, the prominent loner in British geology, and Lyell's *Principles of geology*. Darwin's *On the origin of species*, as significant a book as ever was written, is a super-theory in two parts: genealogical descent (the tree of life was and is being assembled by transformational evolution or macroevolution) and natural selection (the force driving variational evolution or microevolution). It was Darwin's genius to achieve the theory (and surge #III) while being fully aware of the four vast gaps in the potentially accessible knowledge of his times. There was no mechanism for inheritance. There was no mechanism for an organism's growth and development. There was no quantified geological time scale; no plausible numbers. And, fourth, there was virtually no understanding of environment and its propensity to change through geological time. In the major exception to that last point, the polar forests of the Eocene and the ice ages of the Pleistocene Epoch were discovered in Darwin's time. Although variational evolution did not prosper, geological and palaeontological knowledge grew apace in the later nineteenth century. Warm shallow seas spilled onto all the continents from time to time. Sediments shed from rising mountains in the Rockies–Andes chains in the Americas were rich in bones and teeth. Geological mapping spread around the world meeting the demands of economic exploration and exploitation in the European empires.

But by far the most important advance in the later nineteenth century was the discovery of a new environmental realm—new, that is, to biogeohistory—in the still new intellectual environment of organic evolution. Thus far it has been about the lands and bones and fruits of the terrestrial realm and the shallow seas and shells of the neritic realm. The global ocean, deep and dark and presumably primordial, was known more for its imaginative fiction than its science. The study of the pelagic realm, oceanography, celebrated in *On a piece of chalk* by evolution's celebrant TH Huxley,[1] came of age with the four-year global expedition of HMS *Challenger*. This was my surge #IV. Underappreciated for a century, the foraminifera, shelled microbes, came into their own, hundreds of species living on the ocean floor (the benthos) and some tens of species in the surface waters (the plankton). The shells of

---

1   See Eiseley (1967).

the planktonic foraminifera and the coccoliths carpeted vast areas of the deep sea floor with 'calcareous ooze', along with the 'siliceous ooze' (opaline radiolarians and diatoms) and the undersea 'desert', the brown clay.

The pelagic realm was sampled and revealed by dredging the bottom. The next step in the ensuing decades would be to probe the third dimension with coring, thereby entering the fourth dimension of time and foreshadowing the rise of palaeoceanography in discovering ancient oceans. But here we recall the triumphs and frustrations of biostratigraphy. The triumphs of the ammonites in the Mesozoic Era and the seashells of Lyell's Tertiary epochs occurred when *local* fossil succession reached out to *regional* correlation. But there were severe geographic limits to the spread of Lamarck's Parisian faunas of Eocene age. The rocks in the Alpine mountain belts were often dark, thick and uninformative in lacking the rich faunas of seashells that geology needed for classifying and dating—but containing several kinds of microfossil, especially the foraminifera. Intensive drilling was seeking reservoirs of petroleum, where 'knowing where you are' in the strata and their ages was essential. (Losing your way geologically can be ruinously expensive financially.) This demand invigorated micropalaeontology in the early twentieth century. The petroleum industry learned how to drill offshore and in due course it learned how to drill under waters kilometres deep and recover long, complete cores richly informative about ancient oceans, beginning with huge numbers of microfossils in mere cubic centimetres of mud.

The Pleistocene megafauna with its marsupials is of great intrinsic interest in Australia but, after its key role in Darwin's evolutionary biogeography it played no great part in growing the theory of organic evolution. Likewise the seashells, after Lamarck's early excitement and discovery of the strong provincialism in the South Seas. Shells collected by Charles Sturt in 1830 from the limestone cliffs of the River Murray were dated in London as Tertiary, the first biostratigraphic determination in southern Australia. But the energetic collecting and study of these rich marine faunas by such competent workers as Julian Tenison-Woods and Ralph Tate produced a double non-result for the next hundred years. They found no support for (or against) organic evolution in the fossil record and indeed no clear evidence for a succession of fossil faunas. There was no recognising Lyell's epochs here or even the Palaeogene and Neogene periods. And hardly a productive Darwinian scientist was to be found in the country during the nineteenth century. Micropalaeontology solved those problems in southern Australia, beginning in the 1940s. By then there was emerging a credibly quantified Cenozoic time scale—the coming together of microfossil divisions of strata (zones) and radiometric dating, itself emerging in the 1910s not so long after the discovery of radioactive decay in minerals, triggering my surge #V. So, by the late 1960s we could construct our charts comparing marine with nonmarine strata, foraminifera with spores and pollens (sporomorphs), seashells with bones and teeth, and all within a credible and testable framework of millions of years.

In the Evolutionary Synthesis (aka the Darwinian Restoration), microevolution in the form of population (transmission) genetics and field biology filled one of Darwin's gaps and restored the mechanism of natural selection to its rightful place, and surge #VI. The rise of evolutionary and developmental biology (evo devo) in recent decades has filled another gap and reinvigorated the old and powerful notion of homology (e.g. the leg, the wing and the flipper built on the same basic body plan in bony animals). So, we now have three research styles in evolutionary biology, namely population thinking, tree thinking and homology thinking; and some of the old cultural differences between palaeontology and neontology have been healed.

But where is the environmental theatre hosting this evolutionary drama?

For the Darwin Restoration's biogeohistorians, global geography was stable and the global environment was largely unknown; both had changed little since Darwin's time. When earth's repeatedly reversing magnetic field was discovered, the regular and symmetrical magnetic stripes frozen into the ocean floor could be discovered too. The earth-as-machine was now perceived as driven thermally by convection in the mantle. Upwelling into the crust generated oceanfloor, spreading and a brand-new geochronology emerged from the stripes in the rocks as the geomagnetic polarity timescale. Continental drift was confirmed in the new earth-scientific discipline called plate tectonics (surge #VII). Our real interest in this purring machine is its interruptions necessitating resettings, for these interruptions came to explain why our fundamental biogeohistorical philosophy was changing in the 1960–1970s. Lyell the slick lawyer-advocate had managed to tar Cuvier and the other catastrophists with biblical creationism in constructing his brief for the uniformitarianism which became the paradigm for a century. However, diverse episodic patterns of biogeohistory were now pointing in the opposite direction. The aphorism 'the present is the key to the past' brandished by generations of sedimentologists and geology teachers should have been countered long ago by the at-least equally valid 'the past is the key to the present'.

For southern Australia's narrative, the plate tectonic interruption that really mattered was fast-moving India colliding with Asia about 50 million years ago (although the event is contested and not instantaneous, its critical point is pinned down in oceanic crust at 47.3 Ma). In response, the global reorganisation of oceanfloor spreading included reinvigorated spreading in the Australo-Antarctic Gulf as Australia broke free to move into lower latitudes. The north shore of the AAG became the longest coast facing the new Southern Ocean. There was literally a shakeup by faulting of our uplands and sedimentary basins. Oceanic waters were forced across continental margins in the extensive seas of the Khirthar Transgression. The vast river system draining more than half of Australia into the AAG during the Late Cretaceous was upwarped and diverted in large part into the newly subsiding Lake Eyre basin.

Local, regional or global; and tectonic, palaeoclimatic or biospheric—this geotectonic event within the Eocene Period triggered the critical interval in Cenozoic biogeohistory (Figure 0.3).

The strongest characteristic of biogeohistory is the consilience, meaning the mutual reinforcing of independent lines of evidence. To understand strata and geological time we have a toolkit drawn from across the sciences, as seen in the names of multiple chronologies—biostratigraphic, seismic-stratigraphic, radiometric, geomagnetic and cyclostratigraphic. Likewise with the geobiological and geochemical tools for environmental reconstruction. Consilience not only keeps each one honest: it promotes a higher order of insight and biogeohistorical progress. The modern surge #VIII began with stable isotopes from the ocean. Oxygen isotopes (^{16}O, ^{18}O) are incorporated in the carbonate minerals of shells in ratios signalling simultaneously the water temperature during calcification (i.e. growth) and global ice volume. The $^{16}O/^{18}O$ ratio in modern foraminifera differentiates species from the deep ocean, from those from the thermocline (at the base of the mixed ~200 m layer) and from the upper mixed layer. Carbon signals ($^{12}C/^{13}C$) in the same shells add vital information about nutrition and about the vast amounts of organic (photosynthesised) carbon shifting between ocean, atmosphere and rocks, including coals and oils and gas pools. These ratios within minute fossil shells have revolutionised our reconstructions of ancient oceans. There is a spectacular example in detecting photosymbiosis in lineages now long extinct—that is, with no living relatives to inform us directly. Like corals, numerous well-lit lineages of foraminifera acquired their own colonies of photosynthesising microbes.

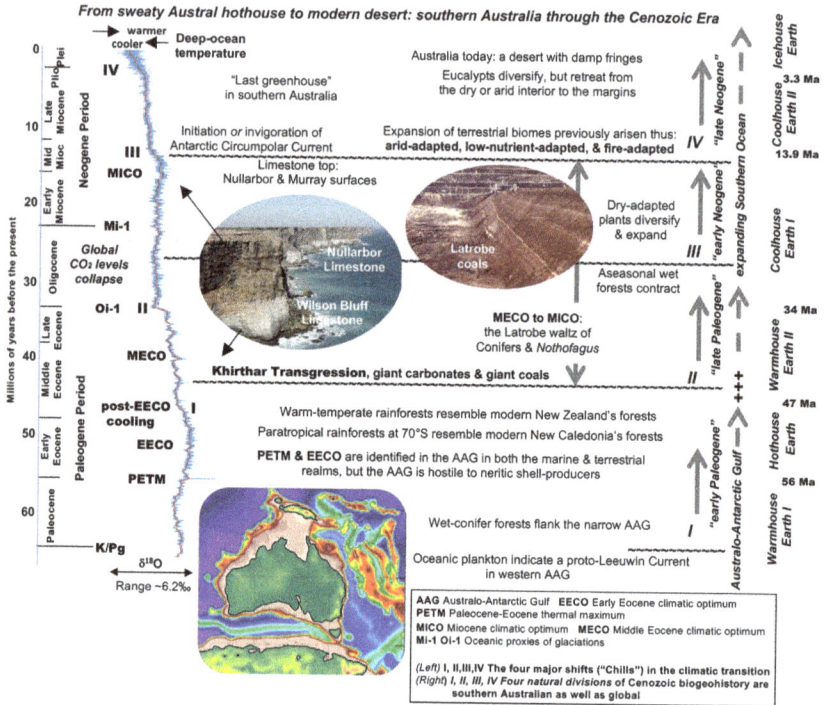

Figure 0.3. The Australo-Antarctic Gulf through the Cenozoic Era.

History of the Cenozoic Era in southern Australia in global context, crammed onto one page. To the left, the geological time scale with its epochs Palaeocene to Pleistocene in the Palaeogene and Neogene periods. The oxygen-isotopic compilation, proxy for global deep-ocean-bottom temperature, is from Westerhold and colleagues in 2020. The $^{16}O/^{18}O$ ratio expressed as $\delta^{18}O$ varies through the Cenozoic Era by only about six parts per thousand, variations physiologically irrelevant to the benthic foraminifera while being incorporated in their calcareous shells. To the right are the two major transitions: regionally, the Australo-Antarctic Gulf is swallowed in the Southern ocean; and globally, warmhouse and hothouse shift to coolhouse and icehouse climatic states (see Figure 0.5). Also at right are the four natural divisions I–IV of the record in the rocks and the highly convenient, informal employment of the terms Early and Late Palaeogene and Neogene biochrons. In the middle, at the appropriate level against the time scale, are the cryptic entries and bald statements, especially the 'Latrobe waltz' of the conifers and the southern beech, *Nothofagus*, meaning shifting balances of the dominating groups in the pollen ratios in the strata through 30 million years in the Gippsland coalfields. The map at lower left is for about 38 Ma in the Eocene, in the last days of the AAG, with Tasmania at about 60°S latitude. No apologies for the litter of acronyms.

Source: Westerhold et al. (2020); limestones, unknown; coals, Guy Holdgate (pers. comm.); 38 Ma reconstruction, extracted from a global reconstruction by Baatsen et al. (2020).

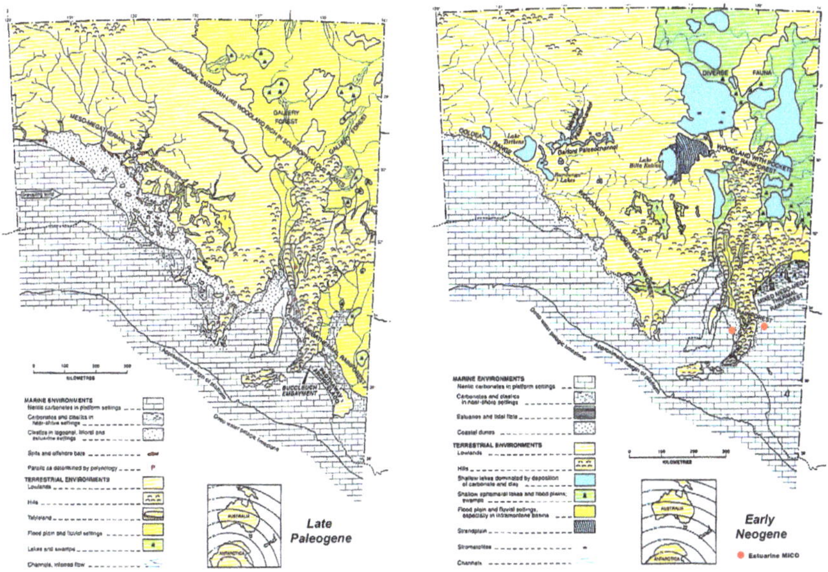

Figure 0.4. South Australia's Eocene and Miocene landscapes.

South Australia in Late Palaeogene and Early Neogene times, about 20 million years and 10 degrees latitude apart. The synthesis was by MC Benbow, NF Alley, RA Callen and DR Greenwood. At both times there was an extensive shallow limestone-forming sea in the region of the modern Nullarbor Plain (which indeed is mostly a solidified sea floor) and in the Neogene the Murravian Gulf stretched almost to Broken Hill. The rainforests are judged to be meso-megathermal, meaning mean annual temperatures of 20–24°C. There was more runoff from a larger drainage into the earlier Nullarbor sea. By Neogene times the bumpy ride of the Australian crust over the mantle was causing a tilt northwards and a sagging had formed the Lake Eyre Basin. But those lakes and wet areas were about to dry up as the aridification sets in. The plants which had adapted to aridity, fire and low soil nutrients had long been in place in the woodlands, perhaps originating as ghettoes in the Eocene rainforests, and were ready to expand, to radiate as the familiar dry-land floras of Australia. The two red dots identify the Munno Para Clay and Cadell Marl, respectively to the west and east of the rainforest-covered hills of the reincarnated Mt Lofty Ranges, and they mark pronounced wetness and brackishness at the peak of MICO (the Miocene climatic optimum).

Source: In Alley and Lindsay (1995); modified slightly, courtesy of Geological Survey of South Australia.

The oxygen-isotopic curves from the deep ocean give us an immediate sense of the global climatic changes on Cenozoic earth. We have a regional history in southern Australia that fits in with and contributes to global biogeohistory (Figure 0.4).

At the dawn of the Cenozoic Era and at 60–70°S the Australo-Antarctic Gulf (AAG) was in warm, wet, winter darkness. It was rimmed with rainforests becoming tropical as the AAG entered the Eocene hothouse; there were tropical mangrove palms in the swamps and estuaries and tropical dinoflagellates in the marine plankton. All this was highly consilient with the $^{16}O/^{18}O$ signal in the global deep ocean of bottom temperatures 10–15°C warmer than they are today. The wet-tropical climate at high southern latitudes deeply weathered the Australian continent. There were no limestones, no coral reefs, none of the banks and mounds of foraminifera flourishing in limestones at low latitudes, and the shelly fossils and calcareous foraminifera were found only sporadically. Why was this? Wetness could exceed levels beyond any terrestrial environment known today. The runoff of the great rivers formed brackish lids over the gulf which not only discouraged the calcifying organisms on the bottom and in the plankton, but also stifled CO_2/O_2 gas exchange with the atmosphere, causing extensive marine anoxia and excluding swathes of marine organisms (including the calcifiers). And all this in winter darkness. Anthropocentrically the north shore of the AAG in the hothouse would not have been a welcoming place.

As the AAG suddenly widened in the Middle Eocene and, not coincidentally, the global hothouse state ended, the low-lit forests changed too, from more like New Caledonia's to more like New Zealand's in modern, well-lit analogues. In the main bio-event on land, forests of southern beeches (*Nothofagus*) expanded in response to global cooling at the expense of the conifers. Here are two scenarios reconstructed for the state of South Australia for Late Palaeogene and Early Neogene times (the Middle Eocene climatic optimum to the Miocene climatic optimum: MECO to MICO)— limestone seas, flood plains, coal swamps (see Figure 0.4). In the middle was the oceanic breakthrough of the AAG to the south-west Pacific Ocean and the onset of modern times. Glaciations on Antarctica probably began in the Middle Eocene with the end of Hothouse Earth, and came and went modestly and episodically until the ice sheets began (and they too were episodic). Just across the water from Wilkes Land on the north shore beneath Adelaide, we see glacioeustatic canyon-cutting and backfilling at the level of the oceanic proxy for this glaciation, Oi-1.

The vast shallow seas returned and expanded as Oligocene time became Miocene in the Early Neogene. Both the shallow limestone seas and the rainforests, including the Gippsland coal forests, contracted abruptly at about 14 million years ago. This is a regional version of the thoroughly global pattern in which the Miocene climatic optimum was truncated by a major cooling as the Antarctic ice expanded. Indeed, the microfaunas in the limestones and clays of our shallow seas show deepenings and shallowings and interruptions of the strata which fit very well with the scenario being developed in the deep ocean. That is, they are basically glacioeustatic.

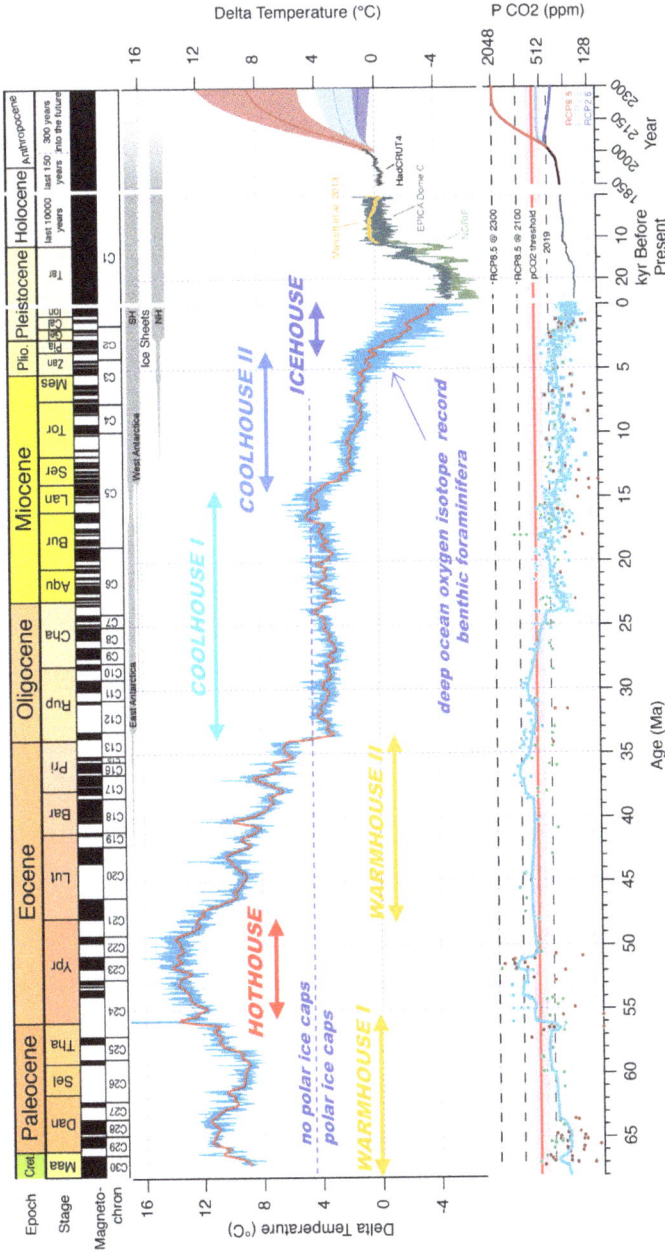

Figure 0.5. Climatic states on planet Earth for the Cenozoic Era and some forecasting.

Westerhold and colleagues have translated the deep-ocean $^{16}O/^{18}O$ compilation into a temperature curve and projected it to surface air temperature change, the average global surface value expressed as a difference from today's. The warmhouse, hothouse, coolhouse and icehouse states are sharply defined. Note the successive growth of icesheets, one then two in the south, and then the northern. Note the scale changes from millions of years to thousands to hundreds in the three time scales for comparing the future with the past. Cores drilled from the polar icecaps show the shift from the last glacial event in the Pleistocene to the current warm period, the Holocene and, from about 1850, the Anthropocene. Below is a CO_2 curve (ppm, parts per million). Three scenarios based on CO2 emissions and levels are 'Representative concentration pathways' (RCP). On our present trajectory the climate state by 2100 will resemble the Miocene Climatic Optimum of about 16 million years ago. The authors point out that if emissions are constant and are not stabilised before 2250, the globe will enter the hothouse state with no major ice sheets and prone to multiple heating events known as hyperthermals.

Source: From Westerhold et al. (2020), copyright © AAAS and courtesy of Tomas Westerhold.

Not so fast. There is also a structural-geological argument claiming that this pattern has partly tectonic causes. The most prominent claim is that the limey sea floors which now are the Nullarbor and Murray plains (also coeval is the top of the great Gippsland coals) were suddenly drained as the continent tilted northwards in its bumpy ride over the mantle. Did the land rise up, or did the sea drain away? After a quarter of a millennium of biogeohistorical science, that question is as alive as ever. This is the algebraic dilemma of one equation with two variables. In the Late Neogene of southern Australia, Reg Sprigg demonstrated that a series of stranded beach ridges recorded the rhythms of the Pleistocene ice ages. At the same time, the Flinders – Mt Lofty ranges are rising under compression.

Half a century of drilling the deep ocean basins has uncovered a marvellous archive of Cenozoic biogeohistory in the form of strata rich in microfossils, ever more finely resolved in time and ever more accurately dated—biochronologically, radiochronologically, geomagnetochronologically, chemochronologically and astrochronologically. In their synthesis of 2020, Westerhold and his 23 co-authors estimate their chronology to be accurate to ±100 kyrs (thousand years) in the Palaeocene and Eocene, ±50 kyrs in the Oligocene to Miocene, and ±10 kyrs in the Late Miocene to Pleistocene (see Figure 0.5). Research has extracted from the deep oceans a detailed trajectory of average global surface temperature. And we can see sharply delimited global climatic states of hothouse, warmhouse, coolhouse and icehouse. Memorise these numbers marking turning points: 56, 47, 34, 13.9 and 3.3 Ma (millions of years before the present). (And note further that we have seen that date 47 Ma as pivotal in plate-tectonic history.)

Although the temperature curve is continuous, the chunkiness of climatic states implicit in emphasising those numbers seems to be real. We have sensed for some time that the transition from the Eocene world to the Oligocene world was special. It now turns out to be the most prominent transition in the entire Cenozoic Era as the role of polar icecaps in modulating earth's climatic state is clarified.

1

What are rock relationships?

Solving a rock relationship

The southern Mount Lofty Ranges sweep down to the sea in several arcs around Adelaide. One arc, the Willunga Range, displays a belt of fossiliferous rocks from Sellick Hill to Carrickalinga Head via Myponga Beach. The belt was assigned to my honours degree colleague Chas Abele and me for mapping in 1957, my sector being the northern part. This project was to be our first actual mapping project—our three undergraduate years included field work, to be sure, but not producing a real map, from the first tentative identification of a rock and the first awkward reading of the dip-and-strike to the finished product, complete with plausible cross-sections in the third dimension. Sir Douglas Mawson had visited many outcrops and run traverses across country, but he did little actual mapping because he was not comfortable thinking in three dimensions, and so his students did cross-country hikes too. Producing a map of an area rich in outcrop and geological variety was introduced into Geology II only in 1962.[1]

The Willunga Range had been mapped recently by Bruno Campana and Bruce Wilson[2] following earlier studies by Walter Howchin and Cecil Madigan, but attempts to reconstruct plausible geological cross-sections revealed lingering problems. A good geological map is a wondrous creation. It is richly informative factually as to which rocks, accurately determined, are to be found and where; but it is simultaneously a theory, a theory of

---

1   Although this quirk of our most eminent geologist Sir Douglas Mawson (1880–1958) was well known among geologists, it is rediscovered, sceptically, from time to time.
2   Campana and Wilson (1953); Campana (1955).

the architecture of the landscape in the four dimensions of length, breadth, depth and time. One learns to perceive the solid structure from the lines, symbols and the colours on the sheet. Constructed under the pressures of budgets and deadlines, a published map will likely have patches of uncertainty, where coded colours and boundaries had to be interpolated more in hope than in confidence. And fair enough—as a good theory, it will contain the seeds of its own supersession.

Our strata had been formed as marine muds hardened over time; the strata were upended and folded, producing slopes (dips) between zero (flat) and 90 degrees (vertical); and the deformed strata were uplifted and exposed by erosion. My cross-sections at Sellick Hill and Myponga Beach show the beds dipping east (Figures 1.1 and 1.2).

Figure 1.1. Myponga Creek, looking upstream (east) from Myponga Beach, watercolour by Glenice Stacey.

This is the location of section D-D in Fig. 1.2. The rocks outcropping above the left bank (middle right) are the Precambrian thrust over the Cambrian at the Black Hill Fault. The Cambrian below the thrust is orange and yellow due to more intensive weathering; this is characteristic of faults being deeply receptive to groundwater.

Source: Author's collection.

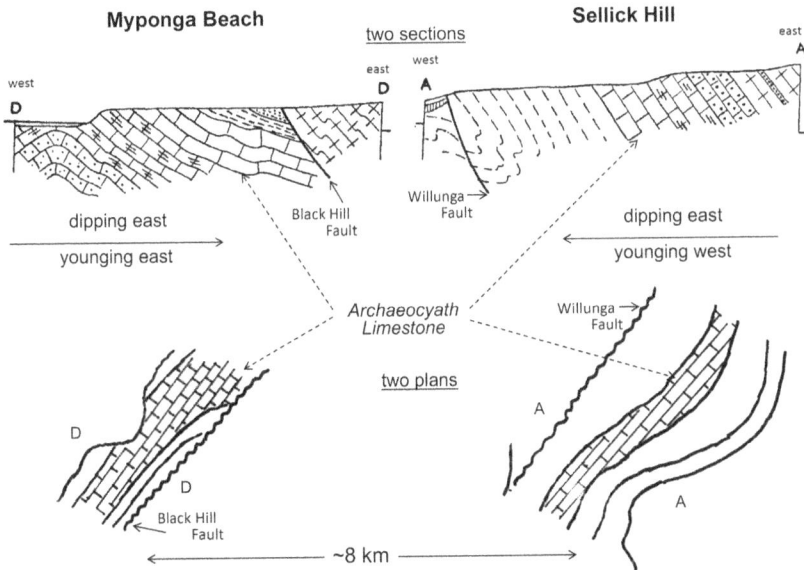

Figure 1.2. At Sellick Hill and Myponga Beach on the western flank of the Willunga Range, south of Adelaide, two geological cross-sections A-A and D-D show the strata dipping to the east.

The sketched plan views show only the mass of fossiliferous limestone. Howchin thought that the beds were also younging to the east. Madigan, Howchin after him, and Campana and Wilson, concluded that the beds were overturned, younging to the west. But we found that all were half right and half wrong. There had to be a fault at Myponga Beach, thrusting older strata over the fossiliferous limestones; we sought and found it, naming it the Black Hill Fault.

Source: From Abele and McGowran (1959).

Following the most fundamental tenet in geology, that in a stratigraphic succession the beds below are older and the beds above are younger, it was reasonable to postulate that the strata dipping to the east are younging to the east. ('Younging' may have been an abrasive neologism offending the linguistic purists, but it was too useful in geology not to survive.) Certainly Howchin thought that the strata were younging to the east; but Madigan saw on the contrary that the strata to the east were older, not younger. Thus the beds had been tilted through more than 90 degrees; that is, they were overturned. Howchin in turn accepted the overturn in both localities and this perception survived into the modern geological map.

The foundation stone, as it were, in this narrative is a limestone rich in the sponge-like fossils known as the Archaeocyathinae, whose securely identified presence confirmed the Cambrian age of the strata. By paying more attention to the details of the succession, I could show that the beds above but older than the Archaeocyath Limestone at Sellick Hill were actually below the same limestone at Myponga Beach—that section was 'the right way up'. That revision revealed a discontinuity to the east where the rocks were suddenly lower down—older—forcing us to predict and duly discover another fault, the Black Hill Fault. It all made sense and our paper with its map convinced Adelaide's geological elite[3].

Excepting my reference to fossils proving a Cambrian age (marginal to the argument), everything in this anecdote could well have been reconstructed during the late eighteenth century. Certainly the argument would have been understood, easily. The essential concepts handed down from those times are only six in number and easily assimilated, then and now. Here they are: (i) muds accumulating in the sea become hard strata, frequently entombing organic remains known as fossils; (ii) in a stack of strata the oldest are at the bottom and the youngest at the top (the law of superposition); (iii) strata can be folded and even inverted (upside down); (iv) rocks can be uplifted and eroded, producing discontinuous exposures (outcrops); (v) beds or strata with characteristics such as mineral composition or fossils can be traced to make a coherent pattern from outcrop to outcrop; and (vi) interruptions or discontinuities can indicate breaking (faulting).

During my three weeks' work experience in the northern Flinders Ranges with Campana's field-mapping party he had become a hero to me, a wide-eyed undergraduate. So, it was somewhat tentatively that I explained our revisions to him in 1958 when Glaessner, Campana and I co-led an ANZAAS (Australian and New Zealand Association for the Advancement of Science) excursion through the Fleurieu Peninsula. But I needn't have worried—Bruno was all too well aware of the problems inherent in the Campana-Wilson map and he was delighted with our solution. And there was another lesson for me in this research. At that time, at the height of modern geological mapping in Australia, there was simmering federal–state rivalry between the Bureau of Mineral Resources in Canberra and the Geological Survey of South Australia—the latter did not welcome the former on their territory. The rivalry manifested in rates of publishing geological mapping, rates that reached remarkable levels in the GSSA under the leadership and personal example of Reg Sprigg. Campana told me that

---

3   Abele and McGowran (1959).

he was pressured politically to publish, notwithstanding the shortcomings in their understanding of the Willunga Range. It was an example for a naive young enthusiast of the human context of science.

Rock relationships in the eighteenth century: Composition, classification, origin, succession

The Industrial Revolution demanded energy and minerals. Meeting that demand made the distribution of the rocks and minerals to be discovered and extracted critically important—meaning, the structure underlying the landscape. Imagine that you were a capitalist investing heavily in new coal mines, in terrain considered highly promising, but your technical consultant recommending the lease had confused two very similar sandstones, one lying below the target, the coal seams, and one lying above. The drillers were hired to drill through the overlying sandstone but it turned out to be the underlying sandstone … That mistake might well bankrupt the entire enterprise, exemplifying the economic motivation to analyse the structure of the country correctly.

But there was motivation other than the economic, and more was required for scientific understanding than the perception of economic opportunity. It came from natural history and the fashion for assembling mineral specimens in the specimen cabinets of gentlemen. The naming, identification and classification of minerals was called 'mineralogy'. Mineralogists looking inwards deciphered the inner structure of minerals, namely 'crystallography'. Outwardly, minerals were assembled as rocks, composite entities posing questions of composition and origin, the subject later named 'petrology'. The rocks making up the earth's crust were sorted on the basis of their origin into three great groups, named respectively and in due course as: igneous, crystallising from molten material at high temperatures; sedimentary, derived from erosion, transport by water or wind or ice, and deposition, mostly in the sea but also in nonmarine environments; and metamorphic rocks, in which heat and pressure have transformed one rock type and its constituent minerals into another. (A common binary of the times was crystalline rocks vis-à-vis earthy or muddy rocks. The sparkling of the crystallines comes from crystals forming and growing under heat, pressure and fluids in veins.)

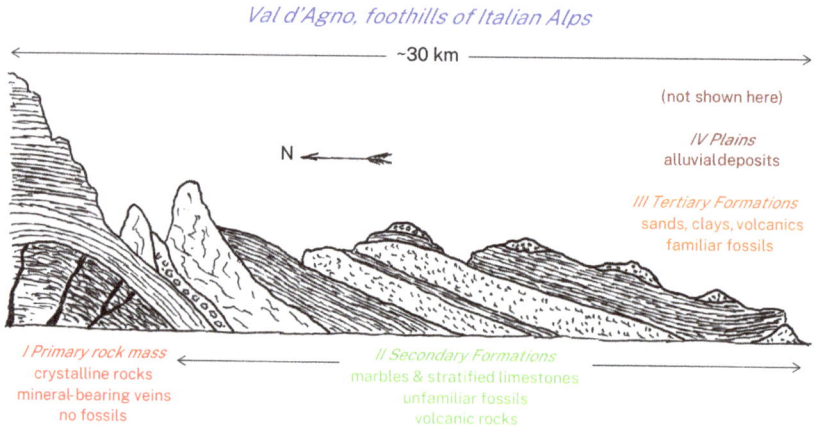

Figure 1.3. Arduino's 1758 section, Alpine foothills.

Giovanni Arduino, geologist (as they became known subsequently) and economic minerals prospector, constructed a geological section which became widely known although he never published it. However, he seems to have displayed his battered copy to colleagues and guests in the field and his four-part stratigraphic succession became widely known and published in due course (see e.g. Rudwick [2014, Fig. 4.3]). Such a great pile of rocks, thousands of metres thick and spanning 30 kilometres of country, implied vast and unmeasurable spans of time. Arduino distinguished and named the Primary and Secondary Formations, adding the Tertiary and Plains from the same region to his sequence.

Source: Originally unpublished, widely published since; here redrawn and relabelled by author.

This is all very neat, but the eighteenth-century story of scientific discovery and clarification was not so neat. Consider the first of two rock diagrams from the eighteenth century (Figure 1.3).

Giovanni Arduino (1714–1795), a Venetian mining specialist, compiled a famous (but unpublished) outcropping section through the southern Alps illustrating the notion of rocks making up a succession. In 1759 he classified his rocks in a sequential series: Primary, the hard crystalline cores of mountain ranges (schist, gneiss, granite, mineral veins); Secondary, the hard sedimentary rocks on the mountain flanks, often with fossils of unfamiliar and strange appearance and challengeable as the remains of real organisms; Tertiary, the hardened sedimentary rocks of the foothills, often with fossils of familiar appearance, resembling organisms in the modern ocean. This was a major step forward in bringing order and clarity into the jumble of materials beneath our feet, but still we feel that rock sequence somehow is quite different from rock composition.

Now inspect the second rock diagram from the eighteenth century (Figure 1.4).

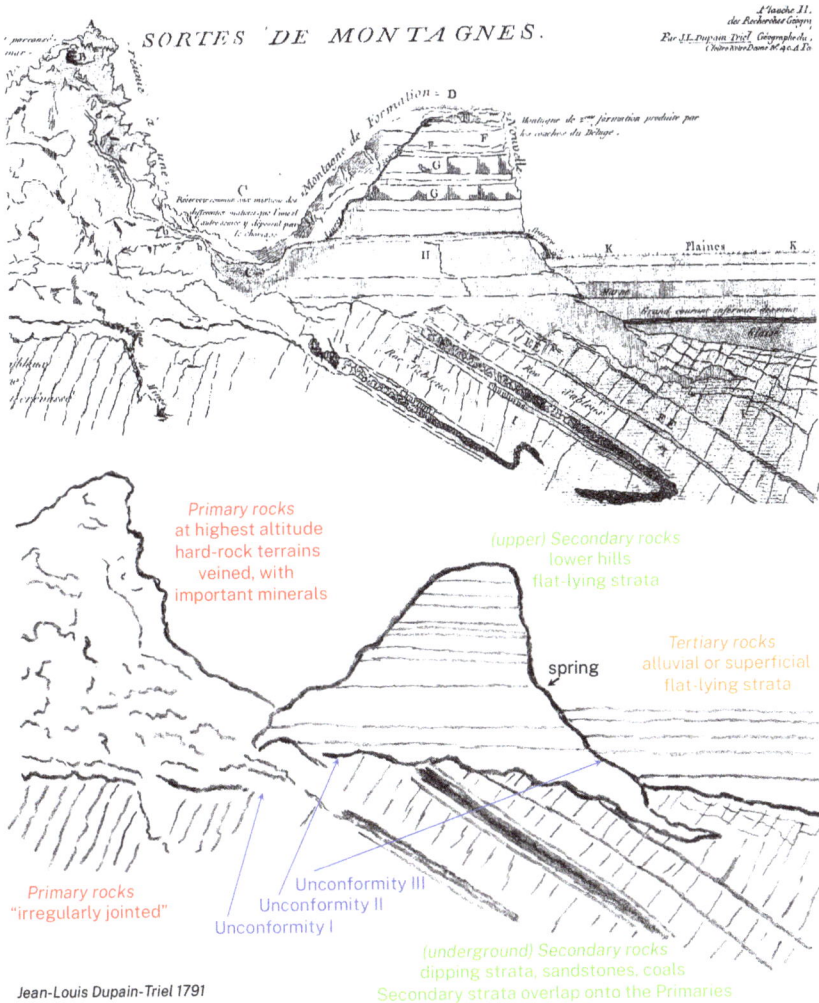

Jean-Louis Dupain-Triel 1791

Figure 1.4. Dupain-Triel 1791, *Primary Secondary Tertiary*.

Three decades after Arduino erected his fourfold chronological succession, this textbook figure displays the profoundly significant interruptions to the succession subsequently called unconformities. The upper-right to lower-left texture is jointing, due to the release of stresses, as in the physics of extension and compression in a bending beam. Unconformities imply a gap, a hiatus, in the rock record. In this example, evidently a lot happened by way of folding, uplift, erosion, and a new episode of sedimentation in the hiatus implied by Unconformity II, between the Secondaries underground and the upper Secondaries. The Secondary/Tertiary Unconformity III looks understandable, but Unconformity I not so. But this figure was constructed during the times of igneous clarifying, when basalts embedded in fossiliferous strata would be found to have been molten lavas rapidly cooled, and granites to have been intruded at various times into rocks of various ages and cooled slowly.

Source: Original from Rudwick (2005, Fig 2.18); author's sketch and labels.

33

Figure 1.4 is a simplification of a diagram drawn by Jean-Louis Dupain-Triel the younger, the royal geographer in Paris at the time of the French Revolution, for a booklet explaining the sciences of the earth for the general reader (1791). There are two masses of Primary rock, the upper mass quite different in aspect from the lower. An underground packet of Secondaries overlies the Primaries with a sharp discontinuity (Unconformity I) and these sandstones and coals are tilted, uplifted and eroded, forming a sharp discontinuity with the upper Secondaries (Unconformity II). The latter are flat-lying but also sharply discontinuous with the Tertiaries (Unconformity III).

These diagrams tell us several important things about the earth sciences in later eighteenth-century Europe. One is the cultural factor, such as the habit of invoking the biblical account of Noah's Flood as the universal explanation for the distribution of rocks on the planet. Arduino's demonstration of a pile of sedimentary strata to be measured in kilometres' thickness made that explanation highly implausible to all but the most myopically religiose zealot. Another is the economic factor. Economic (metallic) minerals are to be sought first in the somewhat chaotic but crystalline rocks in the cores of the mountains; the coals are in the Secondaries; the positioning of springs indicating groundwater can be explained. In short, knowing your rocks makes exploration and production easier and cheaper and more successful.

Third, there is the notion of a succession which, in broad terms, runs from higher, crystalline and older rocks to lower, unconsolidated rocks and younger rocks. This implies that granites, say, were older than the rocks immediately above them, which brings us to a much-publicised controversy of the late eighteenth century concerning the origin of dark, hard and dense, fine-grained rocks such as basalt. In the mining school of Freiberg in Saxony, Abraham Werner (1749–1817) argued that basalts sandwiched between rocks clearly of sedimentary origin (because the latter contained fossils such as ammonites) must themselves have been sedimentary in origin. Basalts such as the Giant's Causeway in Ireland and Fingal's Cave in Scotland, not connected to an apparent volcanic source, can have hexagon-shaped jointing resembling huge crystals. In Werner's opinion, in the 'neptunist' theory, they must have precipitated from water. In central France, however, the French naturalist Nicolas Desmarest (1725–1815) could demonstrate a volcanic connection in the field, corroborating the 'vulcanist' theory that basalts were once molten lavas.

James Hutton (1726–1797), the Scottish farmer, medico and philosopher,[4] confirmed that granites too were not aqueous but igneous, having crystallised from molten material forced into the country rock at depth, the critical field relationship being intrusion into that country rock. Hutton was interested in the rhythms and cycles of nature and in whether they might be applied to this fundamental question: given the observable, modern processes of the weathering of rocks and the erosion and transporting of the products, how come the continents were not all worn down to sea level? Hence the notion of perpetual change and renovation, especially uplift, driven by Earth's internal heat and operating as dynamic but steady-state systems at vast time scales. More on this later; our concern here is the addition of intrusion to the catalogue of eighteenth-century rock relationships. European geologists found that molten material had been forced into rocks identified as Secondary and even Tertiary to form granites and mineralised veins, and they found too that the same kinds of rock had been metamorphosed to crystalline—schists, gneisses—all of which destroyed the chronological significance of the class of rocks known as 'Primary'.

So, by late in the eighteenth century we had a comprehensive working theory, known as geognosy, of the earth's structure. Becoming understood were igneous, sedimentary and metamorphic rocks and the processes generating them, and the kinds of contact or junction between bodies of rock—conformity, unconformity, intrusion and fault. The idea of succession, exemplified pre-eminently by Arduino's section, was appreciated; and so too was the significance of deformation and unconformity as indicating profound and repeated upheavals in Earth's crust. But still in the future was a workable time scale for the vast stretches of time past when all these things were happening.

Geological cross-sections, real and ideal

Although Sir Charles Lyell is prominent in the next chapter, we can usefully peruse here his ideal section of part of the earth's crust explaining the theory of the contemporaneous origin of the four great classes of rocks (Figure 1.5).

---

4   Hutton's *Theory of the Earth* (1788, 1795; Playfair, 1802) tied together the processes of erosion and sedimentation, of granitic intrusion and volcanic extrusion, of faulting and folding, of uplift and metamorphism. Hutton's theory was an extreme example of time's cycle, 'a machine without a history'. See Chapter 10.

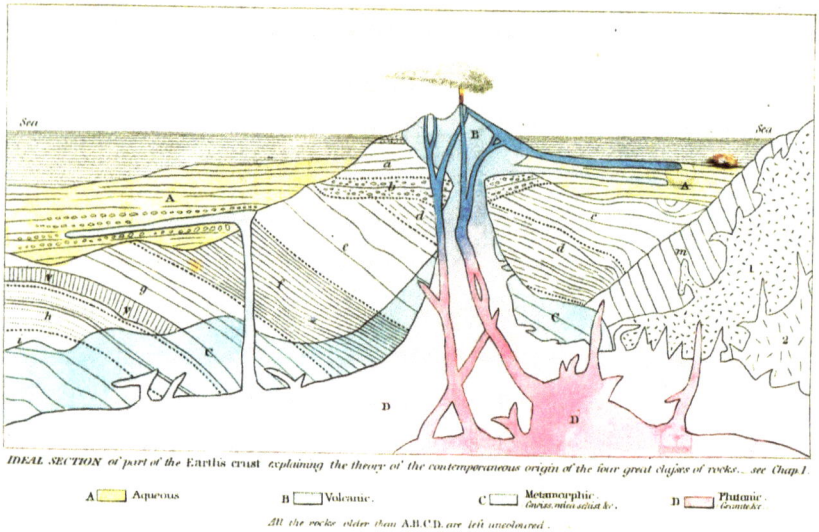

IDEAL SECTION of part of the Earth's crust explaining the theory of the contemporaneous origin of the four great classes of rocks... see Chap.1.

A ☐ Aqueous.   B ☐ Volcanic.   C ☐ Metamorphic *Gneiss, mica schist &c*.   D ☐ Plutonic. *Granite &c*.

All the rocks older than A.B.C.D. are left uncoloured.

Figure 1.5. Lyell's ideal section with four classes of rocks.

Lyell's title says it fulsomely: *Principles of geology, being an attempt to explain the former changes of the Earth's surface, by reference to causes now in operation*. His frontispiece packs in a compendium of geological concepts. We see these: intrusion by granites (plutonic); metamorphism by cooking and deforming; extrusion as volcanoes; sediment accumulating stratum by stratum (aqueous environments); unconformities implying vast hiatuses of time, during which mountain building and mountain erosion can be accomplished. Lyell's central and driving theme was that, given vast geological time, all things are possible without the need to invoke catastrophes of earthly violence such as Noah's flood.

Source: Frontispiece to Lyell, *Principles of geology* (Second American Edition, 1857), Public Domain via Wikimedia Commons.

Layered rocks (m) have been intruded repeatedly by magmas 1 and 2. The entire mass, worn down long since, was overlain unconformably by layered sediments i–c in that order; then a second lot b–a, after another angular unconformity implying uplift and erosion and subsidence; then a third lot, the modern sediments (aqueous A). And yet again there are multiple igneous activities—at depth (plutonic) and breaking into and over shallow strata (volcanic B), and cooking of the host rocks (metamorphism C). To modern eyes there is a lot that is naive about this diagram from almost two centuries ago, but that need not detain us. Lyell's point is Hutton's point. It is the uniformitarian thesis, that the formation of the most ancient and worked-over rocks can be understood in terms of earthly processes visibly operating today—processes ranging from the cold and the shallow to the hot and the deep.

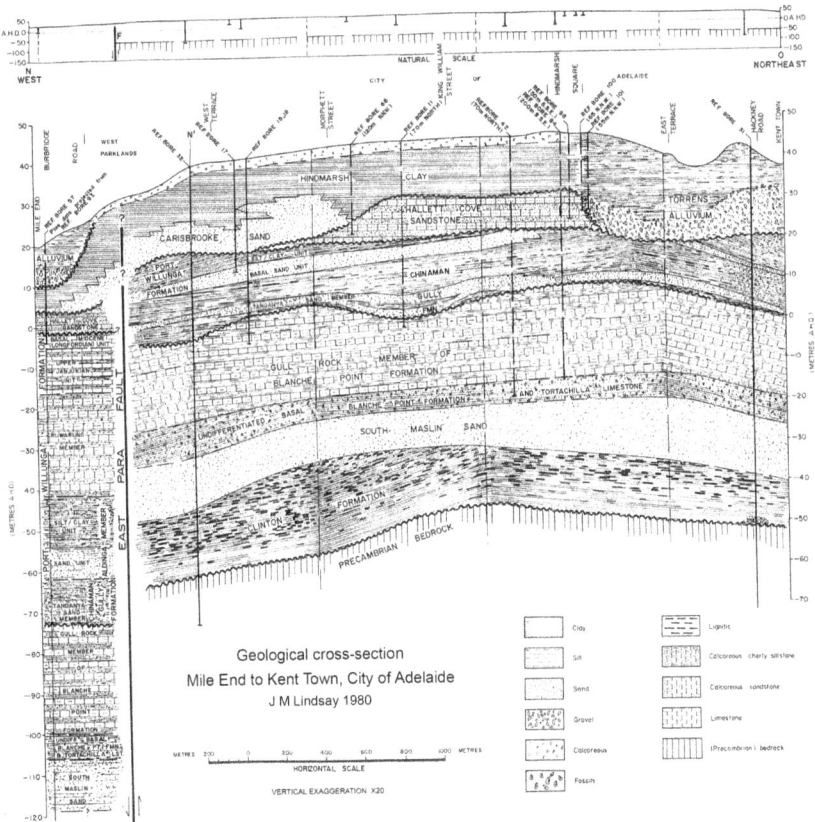

Figure 1.6. Lindsay's geological cross-section under Adelaide.

This compilation of strata is shown at 20 times' vertical exaggeration for clarity (the natural scale is at the top). Drillholes are shown; there are many more in the Adelaide district, primarily for mapping the water-bearing sediments (aquifers). The emphasised wriggly lines are four unconformities. The lowest unconformity on top of the Precambrian bedrock marks a hiatus of more than half a billion years (it is seen in outcrop in Figure 4.18). The next, below the Chinaman Gully Formation, marks the onset of Icehouse Earth, when the rapidly expanding ice sheet on Antarctica lowered sea levels (Chapter 8). The Hindmarsh Clay is a sheet shed by erosion from the modern, rising Mt Lofty Ranges.

Source: Plan 12 in Murray Lindsay's unpublished MSc thesis (1981, University of Adelaide). Four unconformities emphasised by me.

Hutton was a gentleman farmer, at ease in the intellectual ferment in the Edinburgh of his time, the 'Scottish Enlightenment'. Lyell was a lawyer with the wherewithal to travel extensively. I finish here with the other side to intellectual curiosity, the economic imperative, exemplified by a geological cross-section of the city of Adelaide (Fig. 1.6).

It is almost all underground, accessible by excavating and extensive drilling. Drilling costs money but the money was there. Engineers and builders discovered to their cost that they needed real knowledge of the rocks and soils. Thus a large block of solid rock, the Hallett Cove Sandstone, excellent for building foundations, looks like an island surrounded by sands and clays, much less stable and requiring extensive preparation of foundations. Then too is the need for water for the market gardens supplying Adelaide's burgeoning population. There was water in all the units labelled as 'sand', and driller contractors needed to know which was which; and market gardeners had to be warned that extensive pumping was causing drawdown, for this, together with careless techniques, raised the disastrous possibility of seawater coming sideways into the aquifers from beneath the Gulf St Vincent. The South Maslin Sand is a particularly important aquifer and found to be at −30 to −40 metres depth—but not to the west, where it is at −110 to −120 metres, thanks to displacement across the South Para Fault. Finally, fossil fuels: did the early discoveries of the underdone coal known as lignite (in the Clinton Formation) raise economically viable possibilities? (As it turned out: no.)

Soils, foundations, building stones and sands, water, coal and oil ... abundant reasons for investing in the detailed drilling, logging and mapping of the rock units, the lithostratigraphy of Adelaide and its surrounds. Beginning in the nineteenth century, the work reached its zenith in Murray Lindsay's career in the Geological Survey in 1964–1989.[5] But Lindsay needed much more than the identification and logging of sands and clays and limestones: he needed the power and precision of microfossils, which we defer until Chapter 4.

---

5   Murray Lindsay (1928–2004) analysed the stratigraphy and micropalaeontology of the Adelaide region in painstaking accuracy and detail unrivalled in Australian hydrogeology, in numerous reports and publications and an enormous MSc thesis, much of which was never published. This reconstructed cross-section and another, crucial to the story of Icehouse Earth (Figure 8.14), are from that thesis (1981, Plans 12 and 14).

2

Discovering earth history and calibrating deep time

Nautilus ancestors: Mariners in ancient southern seas

First, a confession. Drawing and labelling fossils in undergraduate classes was the surest way to understand their structure and preservation, but for me it could be tedious. Thanks especially to the writings of George Gaylord Simpson, evolutionary theory and earth history and life history were exciting. But to really experience that excitement we had to understand the essential documents, the fossils, and the best way by far to see them was to draw them, and that was unarguable—but less than exciting as an exercise, nonetheless. My attitude reversed dramatically when Martin Glaessner hired me as a vacation assistant late in 1956 to work on fossil nautiloids, forerunners of *Nautilus*, the pearly nautilus of the South Seas. A somewhat dull undergraduate exercise in procedure was transformed by the prospect of real research and perhaps real discovery. Glaessner had recently written on the nautiloid genus *Aturia*, one of the two groups of nautiloids found in the shallow-marine limestones and clays in southern Australia, and he set me to work up the other group, known as *Eutrephocera*s and *Cimomia* (Figure 2.1).

The project developed into an honours thesis submitted at the same time as the Cambrian thesis of the preceding chapter. Both handwritten theses were thrust into an academic pigeonhole at 4 am on Christmas Eve, 1957.

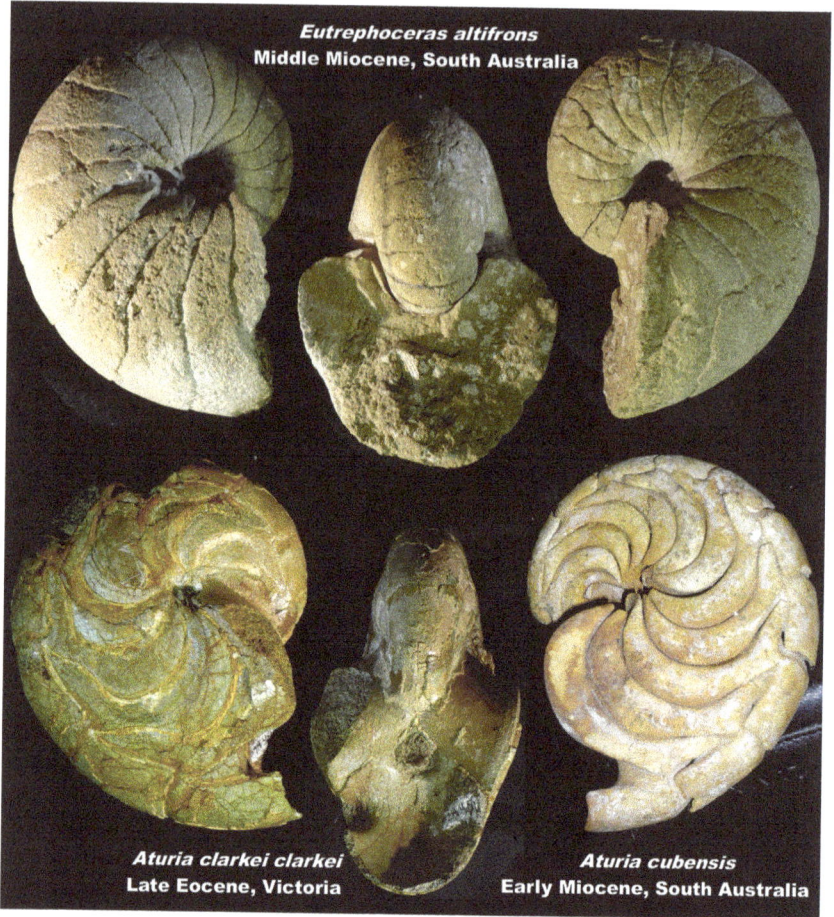

Figure 2.1. Nautiloids from the shallow seas of Cenozoic southern Australia.
These are internal moulds displaying the sutures, the junctions of the septa with the outer shell which, made of geologically unstable aragonite, is usually missing in porous limestones (some shell is retained on this *Aturia clarkei clarkei*). Each septum began life as the back wall of the living chamber and became the wall between two float chambers (camerae) as the animal grew. The body chamber also is usually missing but preserved independently as an internal mould (*upper left*). *Aturia cubensis* entered the literature as *Nautilus ziczac* (as it is in Figure 2.12) and became widely known as *Aturia australis*. So now the evidence of the fossils is that there was but one species of *Aturia* navigating the Miocene global ocean.

Source: Photographs by the author.

Represented by today's solitary genus *Nautilus*, the nautiloids once were the dominant members of the Cephalopoda, the group of highly mobile, predatory molluscs including the octopus, squid and cuttlefish, and the extinct legions, the famous ammonites. Like the ammonites, *Nautilus* has an external shell with chambers added in a spiral of growth. My fossil nautiloids typically show the spiral exposed because the shell composed of aragonite (the geologically unstable form of calcium carbonate) did not survive after the chambers were filled with mud. Thus the fossils shown here are internal moulds. Compared to the clams and snails they are not common, but they are of striking appearance resembling large lobsters (as indeed farmers and others tend to identify them). They had been described and named hastily and illustrated poorly in 1915. Full of evolutionary biology (Chapter 3) and modern taxonomy, the science of taxa (singular: taxon), and with the brash confidence of youth, I could see that nautiloid taxonomy was in an archaic and reactionary state, exemplifying the ultimate putdown known as 'stamp-collecting' among the physicists and the chemists.

The shell of *Nautilus* and its relatives varies little through their fossil record, the main varying character being the junction of the chamber wall, the septum, with the outer shell, giving a wavy line, the suture (Figure 2.2). Figure 2.2 shows suture lines arranged in a series, a typical reconstruction of transformational evolution through geological time (see Chapter 4), each stage being given the name of a genus.

The diagram looks neat and impressive, but it does not really inform us much as to the actual taxa. It seemed to me (it still does) that we had, in the shallow seas spread over large tracts of southern Australia, two clades, meaning two branches within this little sector of the great tree of life on earth. One clade is characterised by *Aturoidea–Aturia* sutures and one with *Eutrephoceras–Cimomia* sutures. Both clades survived for a long time; *Aturia* went extinct but the other clade, as the genus *Nautilus*, is still with us in the warm oceanic waters to our north. How many actual taxa, species, were coexisting in southern Australia was quite unclear. I concluded that I had two species in the Eocene and two in the Miocene but I cast a cold eye on the very existence of three distinct genera under the names *Eutrephoceras*, *Cimomia* and *Hercoglossa*.[1]

---

1   The studies referred to are by Chapman (1915), Glaessner (1955), McGowran (1959) and Miller (1947). The suture lines for *Nautilus* are from Ward et al. (2016), who regard *Nautilus praepompilius* as the direct ancestor of *Nautilus pompilius* (and of the several other living species of *Nautilus*), even though there is a gap in the fossil record of 50 million years between the two species. I lean to the alternative hypothesis, namely that *Eutrephoceras* did not just go extinct along with *Aturia* when the shallow Miocene seas in our region were drained; instead, a remnant population of *Eutrephoceras altifrons* in our local seas survived by making an adaptive jump into the oceanic waters to our north and have formed several new species in the genus *Nautilus*. *Nautilus* is often cited as a living fossil, a hangover from the great days of the nautiloids hundreds of millions of years ago; but its modern blooming (speciation) implies the opposite. The main threats to this ancient and noble lineage now are (i) human—our non-scientific, unscholarly and all-too-human urge to possess these lovely shells as ornaments on the mantelpiece; and (ii) environmental constraints imposed by deep-water-pressure upon the chambers.

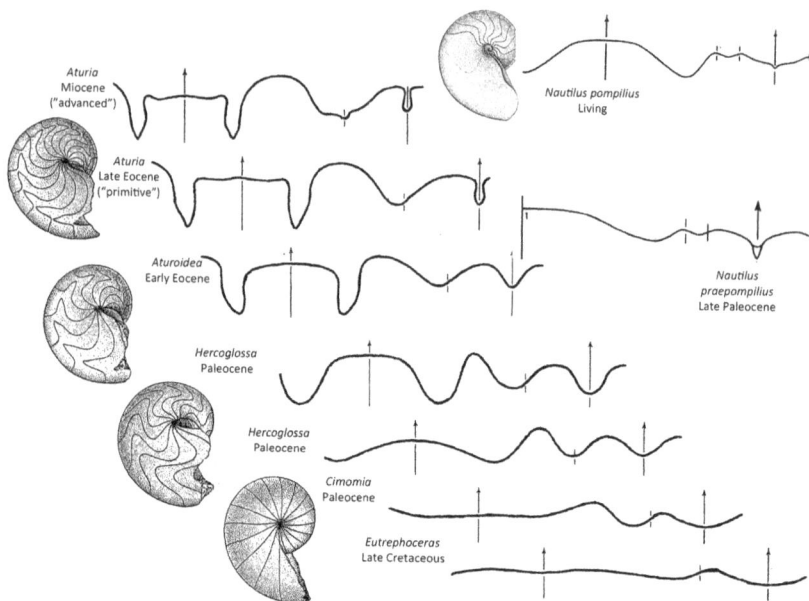

Figure 2.2. Evolutionary succession in nautiloids.

The septa in nautiloid shells are corrugated, an adaptation both for strengthening as the holdfast for the strongly muscular body and for the overall strength of the external shell. The corrugations show in the suture, which is unwrapped onto a flat page from outer midline (left arrow pointing forward) to inner midline (right arrow) on the bilaterally symmetrical shell. The small line or lines indicate the suture passing through the umbilicus, the dimple on the axis of coiling. This reconstruction is adapted from AK Miller, who presented it as an eye-catchingly grand sweep in nautiloid evolution; a morphological series like this became common in palaeontology after Darwin, but it says little about the evolutionary tree of the actual taxa, species and genera comprising the Cenozoic nautiloids. I thought this keenly in 1958 but lacked the spirit to say so clearly in print.

Source: Succession of sutures from Miller (1947) and Ward et al. (2016).

Meanwhile, this palaeontological project became an excellent lead into the bigger and broader panorama of fossils and strata of southern Australia, all displaying abundant evidence for extensive shallow seas spilling across the continental margin during the Cenozoic Era. The limestone cliffs looming 400–600 feet above the sea in the Great Australian Bight (Figure 2.3) were painted by William Westall, Flinders's landscape artist on the *Investigator*. Two units are shown clearly: the almost white lower part later named the Wilson Bluff Limestone and the brown upper part mostly comprising the Nullarbor Limestone. This 'extraordinary bank' (wrote Flinders) extending as uniform cliffs for many leagues and evidently made up of limestone, 'would bespeak to have been the *exterior line of some vast coral reef*' which might separate a shallow 'inland' sea from the Southern Ocean.[2]

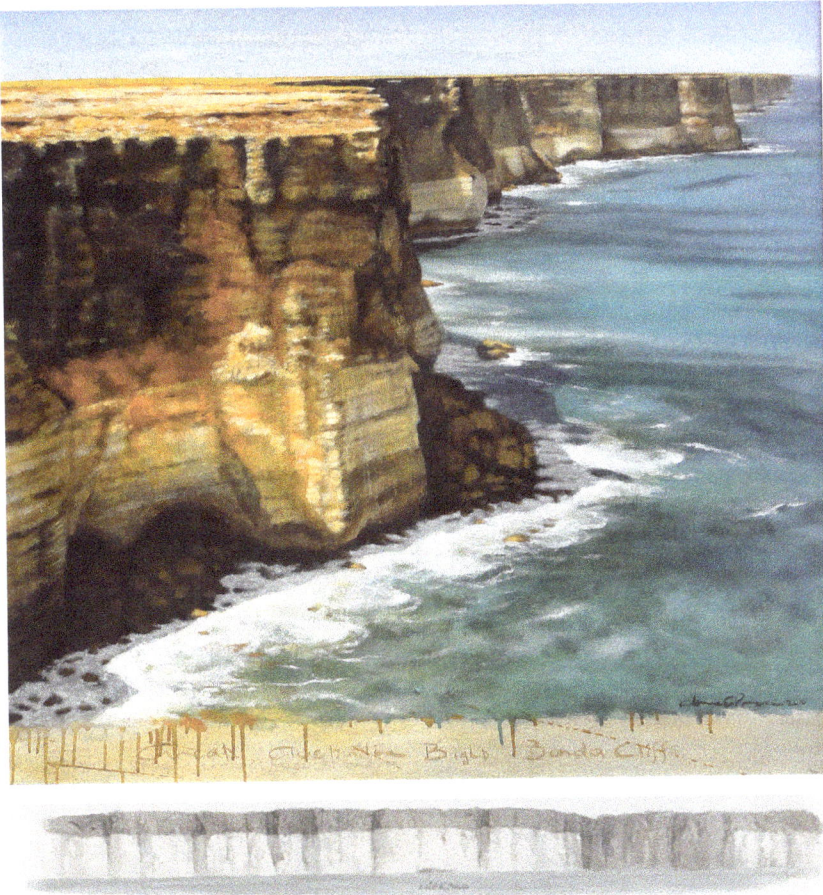

Figure 2.3. Limestone cliffs in the Great Australian Bight, painted by William Westall in 1802 (*below*) and by Alana Preece in 2010 (*above*).

The pale lower unit is the Wilson Bluff Limestone, of late Middle Eocene age. Above it are the Abrakurrie Limestone of Oligocene-Miocene age and the Nullarbor Limestone of Middle Miocene age. The surface, the Nullarbor Plain, represents the sudden withdrawal of the sea as the Antarctic ice sheets suddenly expanded and the basin tilted (Chapter 9).

Source: Westall's painting (lower) is from *Arcadian Quest* (Findlay, 1998), and is one of the 'Views on the South Coast of Terra Australis' reproduced from *A Voyage to Terra Australis* by Matthew Flinders (1814, Plate XVII). 'Bunda Cliffs, Great Australian Bight' (upper) was painted by Alana Preece in 2010, and shown here with her kind permission.

---

2   The quote is taken from Woods (1862, pp. 114–115: italics in original).

Figure 2.4. Charles Sturt's plate of Fossils of the Tertiary Formation, the limestone cliffs of the lower River Murray, as prepared and identified by James Sowerby in London.

Figures 1–15 display assorted bryozoans, echinoids, brachiopods and Bivalve molluscs (scallops), all retaining their shells of calcite. Figures 16–28 in contrast are all internal moulds of Bivalve and Gastropod molluscs, because their shells of aragonite were long dissolved.

Source: From *Two expeditions into the interior of Southern Australia* by Charles Sturt (1999).

In 1829–1830 Charles Sturt and his crew voyaged down the Murray River and recorded the limestone cliffs (the 'great fossil bank') and the oyster beds above. After his later expedition, having failed to discover the inland sea postulated by Flinders (and dreamed of by himself), Sturt dismissed the speculated coral reef as being no more than fossiliferous limestones 'similar in substance and formation to the fossil beds of the Murray, but differing in colour'. Edward John Eyre inspected the fossil cliffs of the Bight in 1840, describing the Nullarbor as a hard limestone with some shells, and the Wilson Bluff as a gritty chalk, 'full of broken shells and marine productions'.[3]

Sturt's voyage between the Murray's limestone cliffs exposing abundant marine fossils demonstrated that there had indeed been an extensive sea there in more ancient times. James Sowerby in London determined a collection of fossils submitted by Sturt to be of Tertiary age. Sowerby's plate prepared for Sturt illustrated Tunicata (i.e. Bryozoa), Radiata (i.e. echinoids, or sea urchins and sand dollars), Conchifera (i.e. Bivalvia, clams) and univalved Molusca [sic] (Gastropoda, snails), referring to species known from England, Westphalia and the Paris Basin (Figure 2.4). On the basis of the specimens identified, Sowerby concluded that Sturt's fossils were of Tertiary age.

These were the first published illustrations of marine fossils and the first identification of Tertiary strata in southern Australia, appearing in print during the decade of Lyell's division of the Cenozoic into its epochs on the basis of fossil contents, of which, more below.

William Smith and fossils and geological maps

This was the first time that strata in southern Australia were dated by using fossils (biostratigraphy!). The event occurred during a time of great ferment in palaeontology, a time bracketed by Georges Cuvier demonstrating extinction in 1796 and John Phillips erecting the Palaeozoic, Mesozoic and Cenozoic eras —the fossil-based time scale—in 1844.

We begin with the English surveyor William Smith, at work when England was changing rapidly in the Industrial Revolution. Smith needed accuracy and precision in his work in coal-prospecting and canal-building, where small errors in surveying meant big problems in development and construction.

---

3   Taken from Woods (1862, p. 386).

He needed geological structure in the broad eighteenth-century sense of the word: that is, three-dimensional structural analysis in large tracts of the country, where it was all too easy to confuse one low-dipping clay with another, one low-dipping limestone with another, and with displacement by faulting complicating the pattern still further. But as he went about his field work, miles and miles of it on foot, visiting natural outcrops, cuttings and other excavations, the fossils in the strata began to talk to him. First, Smith came to recognise a regular—an orderly and meaningful—succession of fossils through the stack of strata exposed, tunnelled or excavated, initially in a district in south-west England and then across most of the country. Second, he discovered that individual beds or formations of repeated, seemingly indistinguishable and easily confused lithologies (rock types) had its own characteristic assemblage of fossils. The fossil succession held true regardless of repetition of sediment types (clays, sands, limestones). Third, he could use those discoveries of his to compile a large-scale geological map of the whole country, the first such of any country and covering about 65,000 square miles. Large (measuring 8 feet by 6 feet), the printed map was complex, accurate and beautiful. The achievement was essentially solitary (but he shared the results widely) and it was truly immense in its magnitude and originality.

Smith discovered his generalisations in the 1790s and produced manuscript maps in 1801 and 1802, which he discussed and demonstrated widely through the years until the map was published at last in 1815. Thus, well before the big map appeared, his insights and methods had become by word of mouth general knowledge among a large body of English geologists, thus contributing to the progress of the science.

But what was Smith actually doing when he so impacted the culture of English geology? When he was tracing the Secondary formations and their individual beds across country, was he using his fossils to identify often confusingly similar units, or was he correlating? Correlation has a special meaning in stratigraphy, the science of strata, as in this textbook definition:

> Two units, belonging to two different local sections, are said to be *correlative* if they are judged to be time equivalents of each other, and *correlation* is the process by which stratigraphers attempt to determine the mutual time relations of local sections. Thus correlation is concerned with the synthesis of the data of established local sections into a composite time scheme applicable to a whole region.[4]

---

4   Dunbar and Rodgers (1957, p. 271).

Correlation means chronocorrelation and time-equivalence and coevality. Smith was not concerned so much with the temporal fourth dimension during his pioneering three-dimensional structural analysis and mapping. He was identifying, not correlating.

Biostratigraphy investigates the distribution of fossils in strata in space and time. As it invented and constructed the geological time scale, the discipline of biostratigraphy arose and grew (and still it grows) in three steps. The first step was recognising a succession of fossils. The fossils in higher strata outcropping in a district were found to differ from those lower down. Careful collecting, describing and identifying of the fossils produced a coherent pattern—it was not all jumble and chaos. In the second step, that pattern could be tested in another district. This enterprise was rather like solving a communal jigsaw pattern—everyone had something to contribute, some more than others, but very rarely did one practitioner hold all the pieces. All power and honour are due to William Smith for collecting so many of the pieces very largely by himself, achieving a composite succession of fossils and strata, and basing his magnificent map upon it. It was a huge accomplishment.[5]

Fossils and earth history: The spark that ignited global geology

Smith was less interested in the third step (publicly, at any rate), which was the perception that similarity among assemblages of fossils indicates similarity in geological age. ('Step' is rather insipid; the spark that ignited global geology is better.) We turn to continental Europe, to the great Parisians, Georges Cuvier and his colleague Alexandre Brongniart, and their advancing the building of a robust geological time scale. They constructed the succession of vertebrate fossil assemblages in the Paris Basin (Figure 2.5).[6]

---

5   Simon Winchester's popular *The map that changed the world: William Smith and the birth of modern geology* (Harper Collins, 2001) tells Smith's story of struggle, pain and setback, and dogged persistence, but it does not place Smith accurately in the development of biogeohistory in the late nineteenth century and his grand accomplishment was not the birth of modern geology. SJ Gould spells out these matters persuasively in his review, 'The man who set the clock back' (*New York Review of Books*, 4 October 2001). Rudwick's 2005 *Bursting the limits of time* treats Smith clearly and fairly.
6   See Martin Rudwick's *Georges Cuvier, fossil bones, and geological catastrophes* (1997) and *Bursting the limits of time* (2005).

Cuvier & Brongniart:

vertebrate fossil succession in the Paris Basin

Modern fauna with human remains

‐ ‐ ‐ ‐ ‐ ‐ ‐ ‐ ‐ ‐ ‐ ‐ ‐ ‐ ‐ ‐ ◄——— *revolution*

| Superficial deposits | Whole fauna of fossil mammals; no human remains. | Equivalent to the "megafauna", known genera, unknown species. |

‐ ‐ ‐ ‐ ‐ ‐ ‐ ‐ ‐ ‐ ‐ ‐ ‐ ‐ ‐ ‐ ◄——— *revolution*

| Parisian Tertiaries | Mammal fauna consists wholly of unknown genera. |

‐ ‐ ‐ ‐ ‐ ‐ ‐ ‐ ‐ ‐ ‐ ‐ ‐ ‐ ‐ ‐ ◄——— *revolution*

| Chalk & older strata | No mammals but diverse reptitles, some of the strange aspect: marine lizards [mosasaurs], flying reptiles [pterodactyls]. |

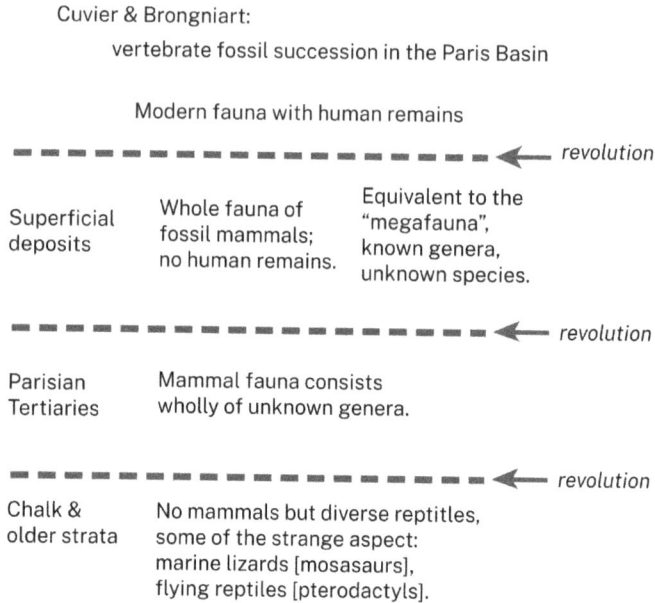

Figure 2.5. Paris district, succession of fossil vertebrates.
Cuvier and Brongniart assembled the faunal succession in the Paris Basin, the solidly evidence-based, bottom-up guts of biohistory and geohistory overthrowing the old, speculative geotheories, which used to be known as 'cosmologies'. The succession is punctuated by turnover or 'revolutions' (a stronger term then than now), because each fauna was found to be so distinct and different from the others. This implied wholesale extinctions and restocking from some vague source (but not from organic evolution). Older, below; younger, above.
Source: From McGowran (2013b, Figure 5).

This sounds simple when you say it quickly, but much effort was expended in discovering and excavating the bones and teeth, always with great care and diligence in recording precise levels in the strata and matching finds from different parts of the district; in reconstructing the skeleton, the animal and its affinities; and in defending the clear statement of a succession punctuated by 'revolutions' marking extinctions. Extinction, meaning the final disappearance of a species or higher taxon, had long been suspected, but people also suspected and hoped that ammonites or belemnites still lurked in uncharted antipodean oceans. President Thomas Jefferson cherished the hope that the Lewis-Clarke expedition would find terrestrial counterparts in the 'pristine' North American wilderness. But Cuvier once and for all established the reality of extinction and, with Brongniart, established the robustness of fossil-faunal succession. The Paris Basin as their canvas was local, and much more restricted than Smith's canvas, which covered much of England, and their research was correspondingly deeper. They interrogated their sedimentary rocks as to the ancient environments of deposition and what that might mean for the strange animals—the discipline that would be named ecology in due course. Where this particular line of enquiry might lead is best illustrated by the French geologist and palaeontologist Constant Prévost (Figure 2.6). His stratigraphic diagram for the Paris Basin is called 'theoretical' in the true sense of that often-pejorative word, for it is a coming-together, a synthesis, of an impressive array of observations and correlations expressing a dynamic balance in space and time. The sea comes in, the sea goes out, the climate changes; and the sediments and their fossils reflect all this (and the revolutions have been omitted).

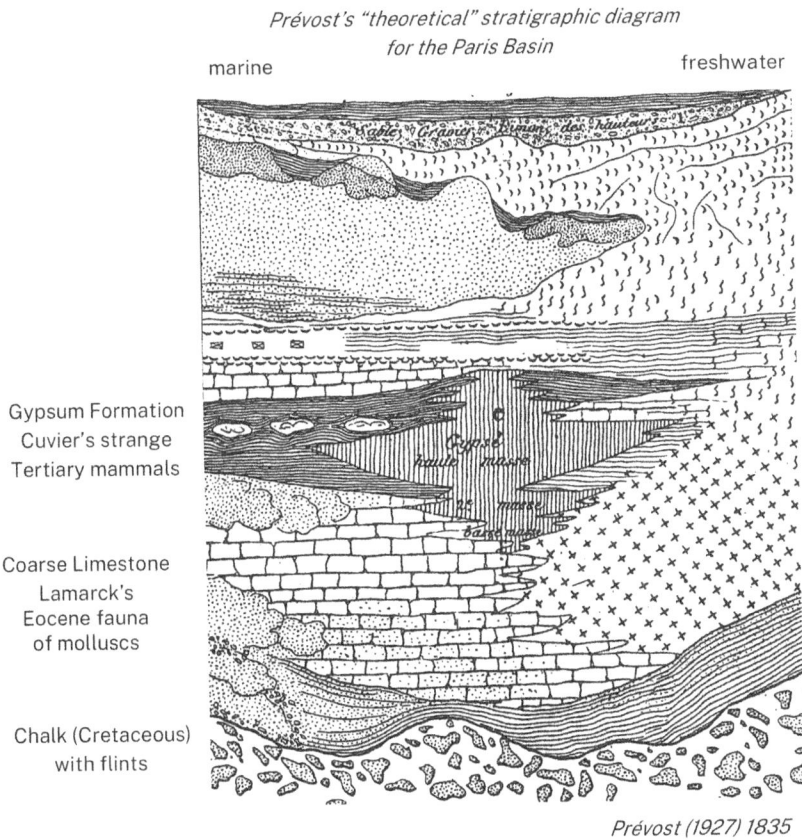

Prévost's "theoretical" stratigraphic diagram
for the Paris Basin

marine   freshwater

Gypsum Formation
Cuvier's strange
Tertiary mammals

Coarse Limestone
Lamarck's
Eocene fauna
of molluscs

Chalk (Cretaceous)
with flints

Prévost (1927) 1835

Figure 2.6. Prévost's stratigraphic diagram for the Paris Basin.
Constant Prévost reconstructed this theoretical section (or synthesis, or stratigraphic diagram) of the Paris Basin in the 1820s, building on the Cuvier-Brongniart efforts. Time runs up the page and the horizontal dimension spans the basin. Observe how the Coarse Limestone (left, marine) is replaced by sands (crosses) sideways and shoreward (right, freshwater). Note too that Cuvier's new, soon-to-be-famous genera of mammals were

entombed in a marginal-marine lagoon. The overall effect is of environments changing dynamically in space and time. The geological structure of the late 1700s is becoming the geological history of the early 1800s!

Source: Prévost's *Thesis* (1835), taken from Rudwick's (2014) publication of the plate.

Cuvier returns in Chapter 3; for now, we focus on how fossils built the geological time scale, with the Cenozoic Era as the prime example. The story begins with the detailed study of excellently preserved shells of molluscs from the limestones under and around Paris, by the evolutionary biologist, Lamarck (also in Chapter 3). These shells had been collected for many years; I write 'begins' because their taxonomic monographing by Lamarck in the years 1802–1809 became the touchstone for biostratigraphic progress. The stratigraphers of Europe advanced Tertiary studies on two fronts in the 1810s–1820s. First, they did the hard slog of detail and accuracy in describing the local successions of strata and meticulously locating their contained fossils. That work, second, permitted comparing and correlating the various scattered stratigraphic sections in the other sedimentary basins. Alexandre Brongniart pioneered this enterprise of reconstructing the Tertiary with his extensive travelling, visiting, fieldwork and collecting, less difficult after the Napoleonic wars had ended.

The power of fossils brings with it the problem of competing variables—how do we discriminate the age signal from the environmental signal and from the biogeographic signal in a fossil assemblage? This problem could be compared to solving a single equation containing more than one variable. The Italian Giambattista Brocchi found that fossil shell faunas in northern Italy were more similar to those far away in the Paris Basin than to those just across the Po Plain in the Sub-Apennines, his explanation for this anomaly being biogeographic, namely, a lateral or spatial climatic gradient. The above-mentioned Prévost proposed instead that the difference was not environmental but temporal, the Sub-Apennine faunas being the younger, like the faunas he studied in the Vienna Basin, both having more species in common with the modern seas than did Lamarck's faunas from the Paris Basin. And then Gérard-Paul Deshayes, successor to Lamarck as the world's leading conchologist, distinguished three natural assemblages of molluscs in succession, that is, through time, before the present day. This was the big breakthrough.

It was the breakthrough needed for subdividing Tertiary time using the fossils of shallow-marine organisms. While visiting field sites and museums in western Europe in the 1820s, the lawyer-turned-geologist Charles Lyell was privately nurturing a theory of species entering the stratigraphic record and departing that record. He knew not whence they came or where they went, but he sensed some regularity in the pattern of their geological record that might become a kind of chronometer of geological time, whereby the percentage of species in a fossil assemblage that are still living in European seas today could locate that assemblage on a time scale. The stranger the assemblage of species, the older it was; the more familiar, the younger. It was this glimpse of a possibly cyclical pattern that caught Lyell's imagination, not whence the species came and where they went. The enormous efforts by Deshayes, building on his antecedents' work, maintaining networks of conchological colleagues across Europe, and handling more than 40,000 specimens of molluscs, living and fossil, enabled the quantitative estimates in this method, the percentage method, and pioneered the fossil-based geological time scale—interpreting faunas assembled in stratigraphic order. Lyell from about 1829 adopted and actively supported (including financially) this Parisian conchological research program and he employed Deshayes's faunal succession to erect Eocene, Miocene, Pliocene and in due course Pleistocene faunal units as the proportion of still-living species present increased (Figures 2.7 and 2.8). (The Oligocene and Palaeocene divisions came later.)

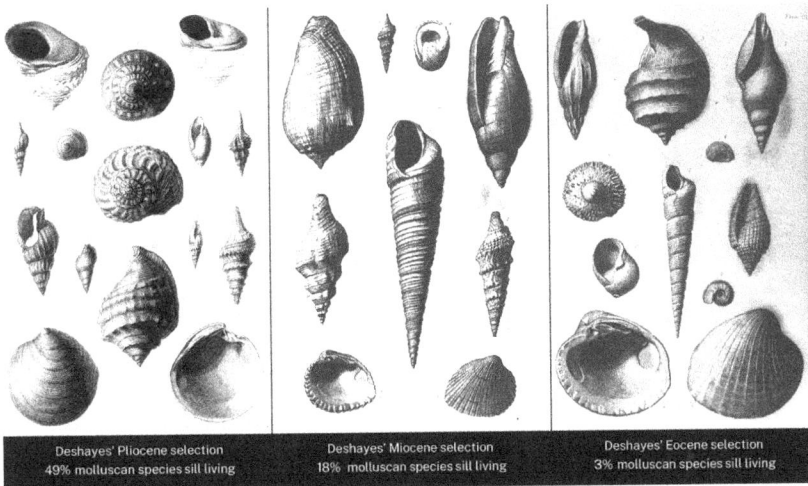

Deshayes' Pliocene selection
49% molluscan species sill living

Deshayes' Miocene selection
18% molluscan species sill living

Deshayes' Eocene selection
3% molluscan species sill living

Figure 2.7. Deshayes's molluscan assemblages for Eocene, Miocene, Pliocene.

By the late 1820s, after some decades of scientific description and classification by the conchologists of Europe, Gérard-Paul Deshayes found that the records of tens of thousands of molluscan fossils in the Tertiary strata could be sorted into a temporal pattern of three successive assemblages. Lyell seized upon this discovery as the key to subdividing the recalcitrant Tertiary Period (and paid Deshayes to round out his research). To illustrate the succession in Lyell's *Principles* (Vol. 3, 1833), Deshayes prepared these three plates of the few species that he selected as characteristic of the respective assemblage. Lyell believed that the assemblages — named by him Eocene, Miocene and Pliocene — were arbitrary slices of a smoothly continuous pattern of species coming into existence and species departing this vale of tears (by natural processes entirely unknown). But Deshayes himself saw the assemblages of species as real entities in nature and not merely artefacts of preservation; he would have appreciated the twentieth-century notion of chronofaunas.

Source: Lyell's *Principles of geology* (1833, 1990–91), Volume 3, plates 1–3.

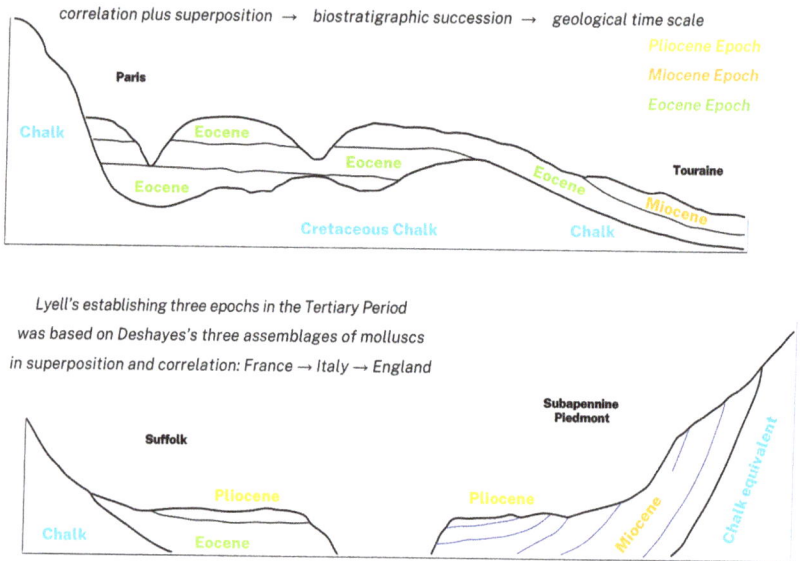

Figure 2.8. Deshayes and Lyell assemblage, Tertiary succession.

Fossils and strata are where you find them. This is the classical example of construction piecemeal by superposition and correlation, whence emerges biostratigraphic succession, thence a geological time scale. Lyell's sketches, here labelled more fulsomely than nineteenth-century sketches ever were, trace the strata containing Deshayes's three assemblages from Paris to the south-west, jumping across to Italy, and leapfrogging back to England. Superposition and correlation! The fossils missing from southern England give some perspective to the unconformity and hiatus in that district.

Source: Lyell's *Principles of geology* (1833, 1990–91), Volume 3, redrawn and labelled.

Lyell had the wherewithal to visit the crucial field sites across Europe, charm the local experts in museum and field, and put in the long hours at the writing desk. Lyell the generalist never did the sheer hard footwork that Smith pre-eminently accomplished in the field and Deshayes pre-eminently accomplished in the museum or laboratory. But Lyell was what nowadays might be called a lateral thinker and risk-taker. We discuss such matters in the final chapter.[7]

---

7 This was 'Charles Lyell's dream of a statistical palaeontology' (Rudwick, 1978).

So, well underway was the task of turning the more parochial rock (lithostratigraphic) succession of Primary, Secondary, Tertiary, Quaternary into the fossil-based (biostratigraphic) and hopefully international succession. The young-looking fossil shells of western Europe led the way. There was intellectual ferment in museums all over Christendom, but it is fair to say that the Paris of Lamarck and Cuvier, recovering after the Revolution, led the way in the first three decades of the nineteenth century, culminating in the palaeontological revolution of about 1830.

The British were lagging somewhat but they came into their own when focusing their attention in the 1830s on the two big systems below the Tertiary and the Secondary—the coal measures (Carboniferous) and the thick, dark and contorted rocks further down still, bearing the names Silurian System and Cambrian System. Insights into the Tertiary and Secondary fossils and strata were brought to bear upon their antecedents down below. There arose two famous controversies concerning succession and correlation—the Cambrian-Silurian, with the interpolation in due course of the Ordovician Period, and the Silurian-Carboniferous, with the interpolation of the Devonian Period. The two-highest-profile figures, the Cambridge divine and academic Adam Sedgwick, and the fox-hunting ex-army playboy Roderick Murchison, were colleagues and friends until all was undone in bitter disagreement and prolonged falling-out. The stories are well told,[8] but here I want to hammer the essential message that biostratigraphy was decisive in resolving both cases, for only the biostratigraphers had the keys (the fossils!) to the chronicles of the problem. Indeed, Sedgwick was one of the sadder figures on the geologic stage, for he, a structural geologist with only belated and tepid appreciation of biostratigraphy, never could see that his descriptive Cambrian, a pile of rocks, was profoundly inferior to Murchison's interpretive Silurian, another pile of rocks but with a characteristic assemblage of fossils demonstrably and uniquely powerful in achieving far-reaching correlations. Murchison, well-financed, army career and all, quickly grasped that difference while Sedgwick was stranded in structure and sedimentation.

The appropriate pinnacle of this narrative was John Phillips's vision of the history of life (Figure 2.9). Boots on the ground, Phillips came to understand the geology of the English countryside with his uncle, William Smith. He accumulated along the way unrivalled experience of the marine

---

8   Rudwick (1985); Secord (1986).

fossils distributed through the known fossil record. His own biostratigraphic research contributed resoundingly to resolving the controversy which gave birth to the Devonian System. Realising that a geological time scale applicable worldwide could only be based on the fossil succession, he proposed in 1841 the three great eras, Palaeozoic, Mesozoic and Cenozoic. His synthesis of the fossil record in *Life on the Earth* (1860), almost coeval with *On the origin of species*, included a compiled curve of fossil taxa through the three eras that markedly resembles modern Phanerozoic curves.

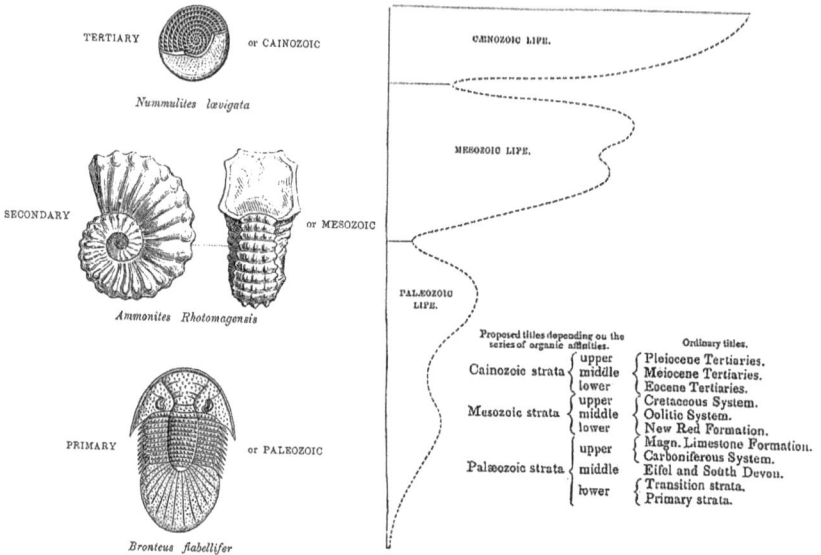

Figure 2.9. Phillips and Lyell: The three fossil-based eras.

Almost every geological problem perceived in more than half a century either was solved by fossils or had fossils in the thick of it. That applied especially to the grappling with deep time and the building of the geological time scale. But the old scale — Primary, Secondary, Tertiary, Quaternary — lingered on in the eighteenth-century minds of men until John Phillips could see (by the 1840s) a natural succession, three series of organic affinities, Palaeozoic, Mesozoic and Cenozoic life. Phillips had no chronometer; he knew the numbers of fossil kinds described by then, he knew the thicknesses of strata that had been measured by then, and he had vast experience of fossils and strata in the field and powerful geological instincts. The outcome in Phillips's (1860) *Life on the Earth* is this unscaled but remarkably modern-looking curve. The inset is a brief tabulation of where some well-known British strata sit in the fossil-based divisions. To the left is the frontispiece of Lyell's 1871 edition of his *Students' elements of geology*.

Source: *Left,* frontispiece from Lyell's *Students' elements of geology* (1871) and also McGowran's *Biostratigraphy* (2005a); *right,* from Phillips's *Life on the earth* (1860) and McGowran's *Organic evolution and deep time* (2013b).

The Australian problem

But Sedgwick did one very good thing, palaeontologically speaking. He hired Frederick McCoy, a highly productive Irish palaeontologist, as his assistant, and McCoy could establish that Sedgwick's pile of Cambrian rocks did indeed contain its own distinctive assemblage of fossils. Furthermore, McCoy could identify the level in the succession of strata where resided the great break, the major disruption in the faunal succession. McCoy emigrated to Melbourne in 1854 to become the first professor of natural history in this country; he came with compellingly strong references from both Sedgwick, who was forever grateful, and Murchison, who wanted him gone. McCoy soon recognised the Cambrian and the Silurian biostratigraphic ages of strata in this country and went on to do likewise with other European systems such as the Devonian and Cretaceous, thereby confirming unequivocally that the fossil-based geological column was a functionally coherent global entity. Put less magisterially: it worked, worldwide!

Anticipating the next chapter somewhat, I include here a tabulating of the development of natural history in this country (Table 2.1) and a gallery of four figures who struggled to make sense of the Tertiary record (Figure 2.10).

Table 2.1. A sweeping overview of natural history developing in Australia — geology, palaeontology, botany, zoology.

Early twentieth century	Variational evolution revived	Evolution in deep time widely accepted in Australia but variational evolution not comprehended or accepted	
Late nineteenth century	Anti-Darwinian decades	Tate and Etheridge, then Howchin: first real leaders of Australian palaeontology	Leading anti-evolutionists and last hold-outs against Darwinism (all respected natural historians): Tenison-Woods, McCoy, von Mueller, Halford, Macleay
Mid-nineteenth century	Darwin-Wallace variational evolution	Arrival of McCoy and von Mueller Establishment of museums, surveys and universities	
Early nineteenth century	Deep-time fossil-based time scale	Pioneering stage, dominated by European thought and opinion, materials shipped to England and Europe	1839: Rev William Clarke, our first geologist, and William Sharp Macleay, our first zoologist
Late eighteenth century	Biogeohistory and biogeography perceived dimly	The age of Lamarck and Cuvier Australian materials collected especially by the French and taken to Europe	First notions of Australia as a distinct biotic creation, a repository of living fossils or an evolutionary backwater

Source: Author's summary.

Frederick McCoy 1817–1889 | Julian Tenison-Woods 1832–1889 | Ralph Tate 1840–1901 | Walter Howchin 1845–1937

Figure 2.10. McCoy, Tenison-Woods, Tate, Howchin: Australian Tertiaries.

Sources: Respectively, from: *The Victorian Naturalist* (October 2001; McCoy); Sisters of St Joseph Congregational Archives, with permission of the Trustees (Tenison-Woods, B4086); University of Adelaide (Tate, Howchin).

We left the limestones of southern Australia with the date of Tertiary, as determined on Sturt's fossils by Sowerby in London right at the time that Lyell was erecting the subdivisions of the Tertiary Period—Eocene, Miocene, Pliocene epochs. Where were these divisions in this southern land? Did they even exist down here? Or was it a step too far, attempting to apply the global fossil-based column this far away at this level of resolution or refinement? Perhaps there were real limits to 'global'? (After all, there are real constraints to the distribution of all living plants and animals.)

Many Australian Tertiary fossils, especially the molluscs, were described between 1865 and 1899 by several energetic palaeontologists including McCoy himself, Father Julian Tenison-Woods SJ and Ralph Tate. Beginning with McCoy, they attempted to recognise Lyell's epochs in the fossiliferous strata, but it turns out that the biostratigraphers down under were still getting their Miocene confused with their Eocene, a full century after the conchologists of Europe took the lead in the palaeontological revolution.

Julian Tenison-Woods wrote *Geological observations in South Australia: Principally in the district south-east of Adelaide* (1862) (Figures 2.11 and 2.12) while based in Penola on missionary duties to a district of some 22,000 square miles (~57,000 km²) from the Murray to the sea, pastoral duties leaving him 'but little spare time' (but much solitary and contemplative time on horseback) and no access to library or museum, nor 'the aid of any scientific men nearer than England whose advice would have been most useful'.[9]

---

9   Woods (1862, p. viii). Note: Father Julian Woods subsequently changed his name to Tenison-Woods.

CAVES, MOSQUITO PLAINS THIRD CHAMBER

Figure 2.11. Limestone cave at Mosquito Plains (Naracoorte).

The South East district of South Australia, now known touristically, speleologically and viticulturally as the Limestone Coast, is dominated by the sheet of Gambier Limestone, of Oligocene and Miocene age. This was the pastoral territory of Father Julian Tenison-Woods (1862). The caves of Mosquito Plains, now Naracoorte, have developed in the Naracoorte Limestone Member of the Gambier Limestone, of Early Miocene age, and are famous for the accumulation of Pleistocene fossils.

Source: From Woods (1862), *Geological observations in South Australia*.

Tenison-Woods grasped the notion of a vast package of limestone strata deposited along virtually the full length of our long southern continental margin, the limestones and fossils remaining essentially similar throughout in the lithology and palaeontology of the strata, perhaps a little more 'tropical' in some districts than others. Woods believed strongly in getting the fossils described because they were so important to geology and the economic health of the colony of South Australia. Two themes recur in Woods's arguments on the Cenozoic limestones. One is their general similarity to the 'crag', referring to the provincial English term for the calcareous Pliocene in south-east England, intermediate in age and biogeography between the warmer Eocene and Miocene and the cooler Pleistocene. Woods could see variation in fossil content within the South Australian limestones but no fossil succession; he diagnosed the age of the Gambier Limestone as undoubtedly being Tertiary because it contained species known elsewhere to be variously Eocene, Miocene, Pliocene and younger, up to the Recent. In the end he plumped for a Pliocene age, believing that the similarities to the crag were more than merely environmental. His second theme came from Charles Darwin's book on the geology of coral reefs. One senses in Woods's writing, in the recurring 'coral reef', 'coralline rock' and 'coral seas', some real regret that Flinders's speculative reef rimming Sturt's hoped-for inland sea was never to be. He used the coral-words freely while warning the reader that 'true corals' are rare and coral structures absent. Instead, Woods suspected that the Bryozoa, especially the relatively massive *Cellepora gambierensis* (now *Celleporaria*; Figure 2.13), were capable of building reefs and atolls. He pointed to the close geological relationships in South Australia between granites, limestones and volcanic rocks and structures, and he drew parallels with Darwin's theory of atolls based on volcanic pedestals. It was a stretch too far.

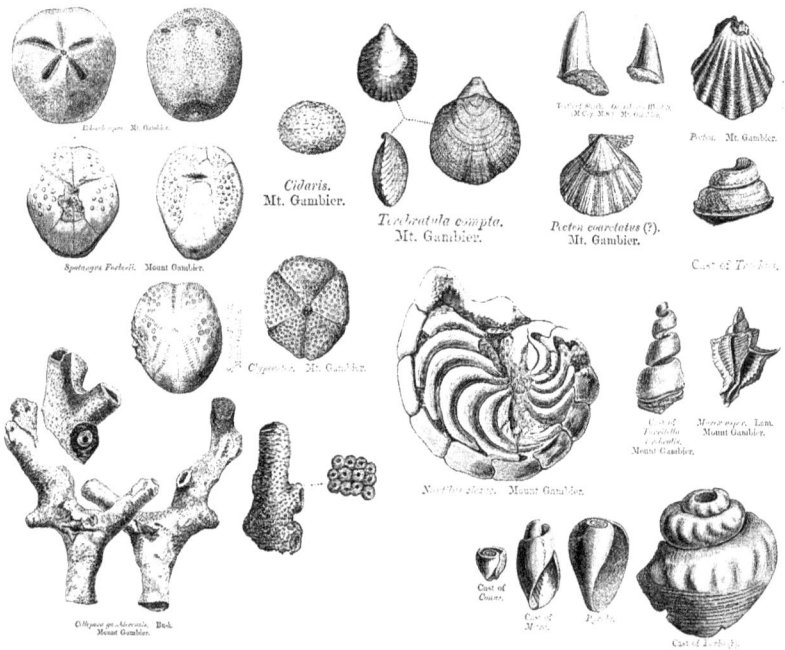

Figure 2.12. Tenison-Woods's assemblage of fossils.

Julian Woods (1862) illustrated echinoids, brachiopods, bryozoans, shark teeth and molluscs from the Gambier Limestone. As for Sturt's collection, the calcitic shells are intact but the molluscan aragonite has almost all disappeared (the 'casts' are actually internal moulds).

Source: From Woods (1862), *Geological observations in South Australia*.

Figure 2.13. A thicket of the bryozoan *Celleporaria gambierensis* from the Middle Miocene shallow sea in the Murravian Gulf (Chapter 9).

This species grows very quickly on 'finger sponges' in waters too rich in nutrient and too muddy for the comfort of most bryozoans. Here, the sponge has long since gone; the tubular form of the bryozoan colony remains. The thicket is about 35 centimetres in width.

Source: Author's photograph.

Two decades later, Woods could note with some satisfaction that the description of several fossil groups was making progress. But there was little progress in dating the rocks. It was a double problem, of regional sequence and of global dating. There were lots of limestones to be found and observed and collected, from river cliffs and sea cliffs, from caves and quarries and small outcrops. But, first, how to put all these strata and their abundant fossils into order, from lower to higher, from older to younger? How to translate the spatial to the temporal? And, second, how to relate (correlate) the local and the regional with the international and the global—that is, with Lyell's epochs? A well-known 1902 paper by Hall and Pritchard illustrated the uncertainties we were in by end-century, with this tabulation of estimated dates for a single little outcrop at Beaumaris on Port Phillip Bay, rich in significant fossils (see Table 2.2).

Table 2.2. Estimated dates for Beaumaris outcrop.

Older Pliocene	by McCoy	1875
Miocene	by Hall and Pritchard	1897
Miocene (?)	by Tate	1888
Oligocene (?)	by Tate	1899
Eocene	by Tate and Dennant	1893
Eocene	by Pritchard	1892

Source: Author's adaptation from Hall and Pritchard (1902).

We seem not to have made progress here, in addressing the scientific question of the age of the innocent little outcrop at Beaumaris. We have confusion instead, as Hall and Pritchard well knew. Outcrops with similar faunas could be clumped as fossil-based regional stages, a sort of halfway house en route to attaining the international holy grail of Lyell's epochs. But the outcrops with their fossils were scattered across southern Australia, as I keep saying; there was very little superposition to behold. So there arose differences of opinion as to the sequence of these units when isolated outcrops in southern Australia were brought together, as shown in Table 2.3, which is from the same paper, where stages are shown in descending order from younger to older, in three attempts at ordering.

Table 2.3. Outcrop stages ordered by age.

McCoy	Tate and Dennant	Hall and Pritchard
--	Werrikooian	Werrikooian
Kalimnan	Kalimnan	Kalimnan
Jan Jucian	Jan Jucian	Balcombian
Balcombian	Balcombian	Jan Jucian and Aldingan (in part)
	Aldingan (in part)	

Source: Author's adaptation from Hall and Pritchard (1902).

By now, the reader would be justified in wondering just how fossils are being used to sort and date strata in southern Australia. The clearest statement is in a textbook by Walter Howchin, *The geology of South Australia* (1929; first published 1918), who outlined the two approaches of the later nineteenth century. He identified Tate, Hall, Dennant and Pritchard as following the Lyellian method of classifying the marine strata on the percentage of still-living molluscs to be found as fossils. Howchin severely criticised this method on two grounds. First, the data base was quantitatively too thin—

the modern molluscan faunas off southern Australia were insufficiently sampled and too little known to be a reliable basis. Second, our knowledge of the fossil species was still too much a work in progress.

Howchin listed McCoy, Chapman and Gregory, to whom can be added Tenison-Woods and himself, as taking the other approach, which was to search for and identify certain fossil forms known to characterise a definite geological horizon in other parts of the world. But this too was very thin, and Howchin's clear summary of what we knew of southern Australian palaeontology shows how thin, and so too does his conclusion that we could not recognise Lyell's epochs in this part of the world. It was all very frustrating and the frustration boiled down to biogeography. Just as Lamarck recognised that our living molluscs had many counterparts in northern seas but not the same species, so too did the toilers among the fossil faunas know that there were simply not the species in common with the faunas of Deshayes and colleagues in Europe to recognise the epochs of the Tertiary Period in Australia.

A radically new approach was needed, but we are reminded that Darwin's theory of evolution was several decades old. Where was Darwin, in this story of the building of the geological time scale and its application in southern Australia?

3

Exploration and organic evolution

Rise and fall of a fossil kangaroo

Sorting the succession of Cambrian strata on the Willunga Range, I found a sandstone at the bottom with numerous specimens of the small cone-shaped shell known as *Hyolithes*, and we called this stratum informally the *Hyolithes*-Sandstone. People also noticed the distinct possibility that I had the oldest shelly fossil in Australia. Basking in this accomplishment was fine, but I had become more involved in my nautiloids and in the rocks and fossils of the Cenozoic Era, so I did not mind Brian Daily, curator at the museum and the authority on the South Australian Cambrian, taking over the *Hyolithes*-Sandstone and doing the thorough job on it that had not been possible in a crowded year. Even so, bragging rights were reserved to be paired with another newsworthy discovery, the oldest modern kangaroo in Australia; and this find ended rather differently. We—Mary Wade, Chas Abele and I—were in western Victoria in 1959, crawling over a section of richly fossiliferous strata of Miocene age. Sandwiched between a relatively hard limestone and a clayey, limey sediment known as a marl, was a sand with pebbles, worn marine shells and fragments of whale bone; and I found a more solid bone looking like part of a leg. Collected and labelled, in due course the bone was shown to Martin Glaessner, who saw its very close resemblance to the left leg (femur) of a large modern kangaroo—but what was this terrestrial creature doing in sediments millions of years older than the known, modern-type roos and among abundant marine fossils? Glaessner and Wade promptly returned to the site, checked and sampled and double-checked, and showed that the microfossils below and above the bone bed tightly bracketed its age as Middle Miocene (see the next chapter on microfossils). Even in retrospect, the age-bracketing could hardly have been tighter.[1]

So there! I had the oldest kangaroo in the land. But it was a fleeting triumph. Vertebrate palaeontologists took over the locality and collected an assemblage of fossils of land animals which turned out to be not Miocene but Pleistocene in age. The weight of the evidence was not that the fauna was much older than had been known; the evidence was that a young assemblage of fossils had somehow been emplaced within older rocks. Misleading rock relationships are common. Imagine a wombat bone among ammonites, meaning marine fossils of uncontested Mesozoic age. This configuration could be dismissed easily as a modern wombat burrow. Also well known is the neptunian dyke, where an advancing sea inserts sediment metres or tens of metres deeply into joints (cracks) in the old landscape. And then there is the redepositing of older fossils upwards into younger strata, having survived erosion and transport. (Things can be obscure to the uninitiated. At a soil scientist's seminar on the development of a 'soil profile', my fellow stratigrapher Brian Daily nudged me in the ribs and hissed, 'can you see what I am seeing?' Indeed I could: the slide was displaying a pronounced unconformity indicating that, seemingly unknown to him, the distinguished chemist's soil profile was missing millions of years from its middle.) Our painstaking logging, collecting and analysis had covered all eventualities, or so we had thought. Our southern limestones are prone to extensive dissolving by groundwater, leaving caves and sinkholes, well known to accumulate bones and teeth, and frequently infilled with almost-modern clay; but in our case groundwater had seeped through the sand between the Miocene limestone and the Miocene marl, slowly, ever so slowly, nudging young terrestrial bones into intimate association with old marine bones and marine shells. No older than the Pleistocene, the kangaroo was special no longer.

---

1   Glaessner et al. (1960).

Earth history written on a humble clam

Still, it is an ill wind … Schoolchildren like hearing about scientific research—the activity, not just the achievement and the progress—and they like it more to hear an elderly scientist confess his misfiring. That gets their attention. But the mystery of the misplaced kangaroo was also my modest introduction to the world of bones and their significance in earth and life history, which was immense, and in the story of Australian discovery too.

| James Cook 1728–1779 | Joseph Banks 1743–1820 | Jean-François de Galaup, Comte de Lapérouse 1741–1788 |

| Nicolas Baudin 1754–1803 | François Péron 1775–1810 *Voyage de Découvertes aux Terres Australes* | Matthew Flinders 1774–1814 | Robert Brown 1773–1858 *Prodromus Florae Novae Hollandiae* |

Figure 3.1. French and British explorers in the south seas.
Source: All portraits from Wikimedia Commons (commons.wikimedia.org).

We return to the South Seas (Figure 3.1) and to the French, to the shells of Lamarck and the bones of Cuvier triggering that revolutionary awakening in our culture, round about the turn of the eighteenth to the nineteenth centuries. A famous quote is a good resumption:

> 'Captain, if we had not been kept so long picking up shells and catching butterflies in Van Diemen's Land, you would not have discovered the south coast before us.' Lieutenant Henri de Freycinet (*Le Géographe*) to Captain Matthew Flinders (*Investigator*), Port Jackson, 1802.

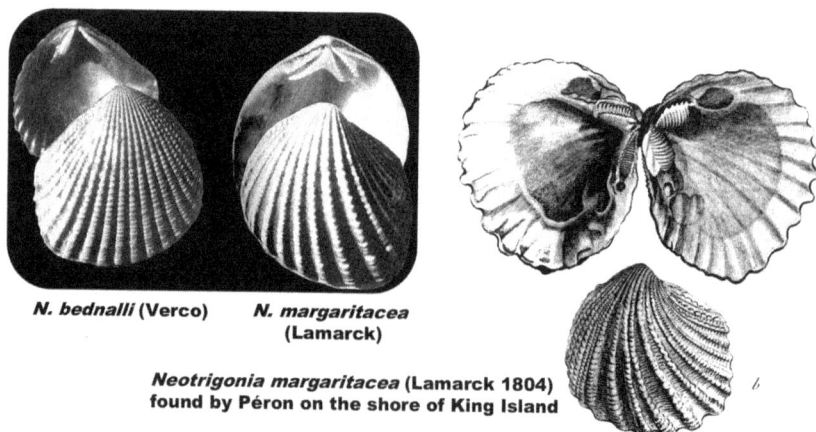

N. bednalli (Verco)   **N. margaritacea (Lamarck)**

**Neotrigonia margaritacea (Lamarck 1804)
found by Péron on the shore of King Island**

Figure 3.2. *Trigonia* and *Neotrigonia*: Lamarck's excitement.

Trigonia was well known in the Secondaries in Europe (the Mesozoic) but conspicuously missing from the Tertiaries. Its living presence in southern Australia contributed to the notion of this place as a repository for ancient survivors, the notion established by the strange fauna and flora. For Lamarck, *Trigonia* was evidence for lineages evolving but not extinguishing. From Lamarck, to McCoy, to Darwin, to Gould — the story of *Trigonia* as a celebrity in organic evolution is well told in Danielle Clode's *Continent of curiosities* (2006).

Source: *Left*, two species from Ludbrook's monograph of shells, courtesy of Geological Survey of South Australia; *right*, original illustrations of *Neotrigonia margaritacea* by Lamarck in 1804, here from Rudwick's (2005) reproduction.

The contrast between Baudin's expedition and Flinders's to the South Seas reflected the contrast between the scientific culture of Paris—fermenting, thriving, competing, expanding—and London, somewhat less of the above.[2] That the savants in Paris, led intellectually by Lamarck and Cuvier, were awake to the scientific possibilities is nicely illustrated by the story of *Trigonia*, a clam (Figure 3.2).

Trigonia had been named recently by Lamarck's friend and predecessor in malacology, the late Bruguière, and was familiar to students of the fossiliferous rocks of Europe known as the Secondaries. They were the cockles of the Mesozoic Era.[3] But *Trigonia* seemed to have gone extinct. Lamarck was finding no *Trigonia* in his great study of the rich molluscan assemblages around Paris, those later to be designated as Eocene in age; but

---

2   Danielle Clode tells the stories beautifully in *Voyages to the South Seas: In search of Terres Australes* (2007). From the blurb: 'It is the story of the scientists, collectors, savants and sailors who risked their lives in order to bring back untold riches, not of spices and gold, but of knowledge, for a fascinated public to devour.' Clode records (p. 278) de Freycinet's complaint, taking it from Flinders himself (1814).
3   Stanley (1978).

he was also nurturing a theory of organic evolution wherein species were phantoms, organic entities perpetually in transition as lineages meandered through geological time.

The thousands of marine specimens brought back to France from New Holland came primarily to Lamarck. In a modern handbook for shell collectors, *Molluscs of South Australia* (Hunt, 2011), I find these numbers: 420 taxa (kinds) are listed as having been named and illustrated in the past two centuries or so; almost all are molluscs, including snails, cockles and clams, cephalopods and chitons. Of the 420 taxa, 65 or almost one in six were described, named and classified by Lamarck in five works between 1811 and 1822. Lamarck was long convinced of the transformational evolution ('transmutation') of species and in working up his theory he was documenting close similarities between his Paris Basin faunas and modern molluscan assemblages. In the previous chapter I hail the diligence of the malacologists in dividing up the Tertiary Period using fossils, my emphasis being on fossil succession and difference leading to the fossil-based erection of Lyell's epochs; here, in contrast, the emphasis is on fossil similarity implying relationship. There were many new taxa to be discovered among that great pile of shells, numerous shells from the southern shores reminding Lamarck of shells long gone from the Northern Hemisphere; but the most exciting and to become the most famous was a small clam collected by Péron on Bruny Island off Tasmania, which Lamarck published promptly as *Trigonia margaritacea*, in 1804.

Think of it: *Trigonia* has never been sighted in any of the intensely prospected strata of the European Tertiaries and it must be extinct—yet up it pops in the Antipodes and is not extinct after all! In due course five more species of *Trigonia* were discovered, the six species being strewn around maritime Australia and nowhere else. And in due course the missing Tertiary species were found too, recognised first by McCoy in Melbourne, now amounting to 11 species and reclassified as the genera *Eotrigonia* and *Neotrigonia*, direct descendants of *Trigonia*, only in Australia (Figure 3.3).

There is yet more to the *Trigonia* story, but first we need to move from the sea into the terrestrial realm and on to the next episode in the saga of biohistory culminating in organic evolution.

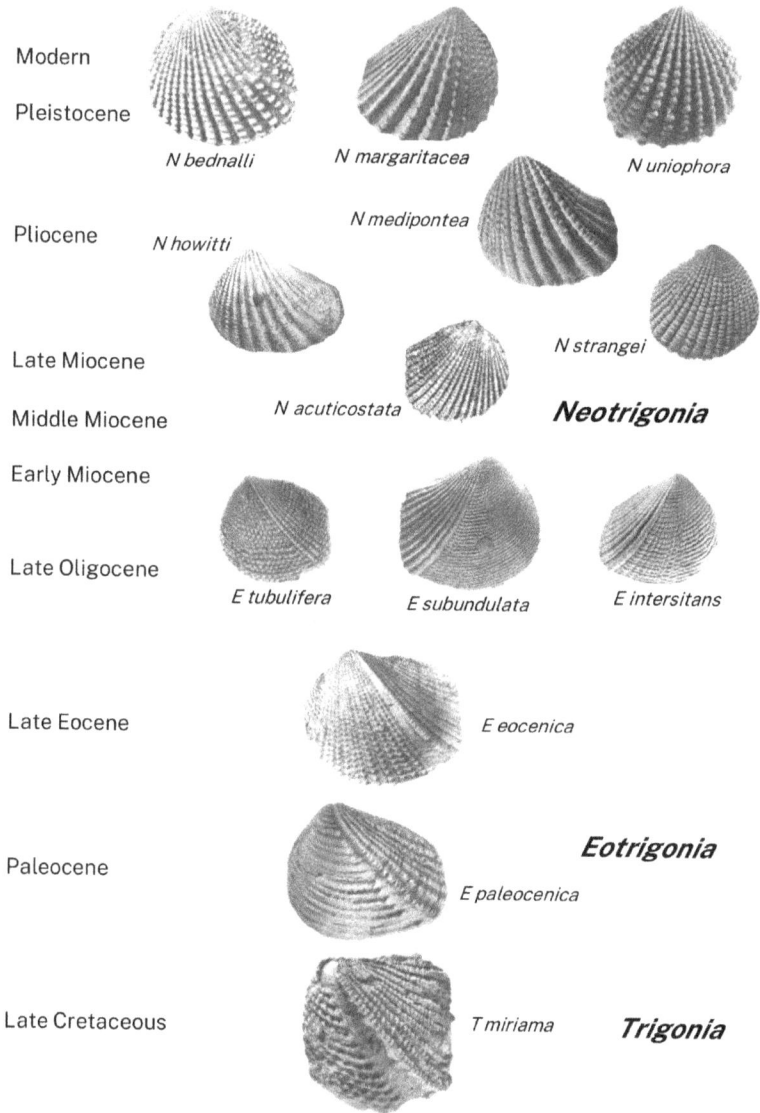

Figure 3.3. *Trigonia, Eotrigonia* and *Neotrigonia*: Darragh's array in time.

Bridging the yawning gap between *Trigonia* ancient (Mesozoic) and modern (living today) began with Richard Daintree's discovering a fossil at Bird Rock in Victoria in 1861, the fossil now known as *Eotrigonia subundulata*, of Oligocene age. The images telling that strictly Australian story are arranged here from the splendid review and synthesis of subsequent discoveries by Tom Darragh (1986). Just as *Nautilus* is not a 'living fossil' (Chapter 2), so too has *Trigonia* evolved and speciated to form groups now known as *Eotrigonia* and *Neotrigonia*, even though *Trigonia*'s status as the cockles of the Mesozoic seas has been appropriated long since by, well, cockles.

Source: Redrawn with the approval of Tom Darragh.

Of bones and time and Georges Cuvier

Here is another famous quote, eminently worth its reiteration down the decades:

> How was it not seen that the birth of the theory of the earth is due to fossils alone; and that without them we would perhaps never have dreamt that there had been successive epochs, and a series of different operations, in the formation of the globe?[4]

I have mentioned Cuvier's achievements with Brongniart, putting ancient strata with their fossils in the sequence of time, and demonstrating that the fossil faunas have changed markedly in the course of geological time (see Figure 2.5). This was the first of Cuvier's three big steps towards a history of life on earth. Second, he showed us that the patterns of earth history and life history are multiple, not linear, as illustrated by his classification of all animals into four great streams, or phyla (French, *embranchements*) which he named Radiata, Mollusca, Articulata and Vertebrata. This was the final death-blow to the Great Chain of Being (*scala naturae*), an ancient notion periodically rejuvenated, that all earthly life can be arranged in a linear hierarchy from the lowliest organisms all the way up to the angels. The *scala naturae* was the ultimate politicised rationale justifying stratified hierarchies of power and privilege in human society. Cuvier, more than anyone before or since, progressed the scientific discipline known as comparative anatomy or zoology and showed conclusively that organisms in no way fall into any single linear pattern. Third, Cuvier showed once and for all that extinction was a fundamental fact of life on this earth. He demonstrated that the mammoths of Europe and Siberia were different from the mastodons of North America, that neither was merely an African or Indian elephant in warm clothing and out of place, and that both were extinct. The older Tertiary fossils in the Paris Basin were of animals long extinct. The great swimming reptiles in the Secondary seas and the flying reptiles above the seas were extinct beyond argument. And during Cuvier's time and shortly thereafter people came to understand that the trilobites, the ammonites and belemnites, and multitudes of other animals preserved as fossils in the muds of ancient seas, had gone extinct.

---

4   Georges Cuvier's 1812 essay in *Preliminary discourse on the revolutions of the globe*, as newly translated in Rudwick (1997; quote on p. 205). Rudwick's book is the best place to appreciate Cuvier's place in founding the sciences of biogeohistory.

George Gaylord Simpson, the most incisive scrutineer of fossils and palaeontologists, observed that Cuvier, long acknowledged as father of vertebrate palaeontology, did not actually originate any of the fundamental principles of the discipline. Instead, he developed, exemplified and systematised them and brought together a mass of hitherto scattered observations and insights—achievements that assure his pre-eminent place in the history of science.

> It was Cuvier's good fortune, [wrote Simpson in 1942, p. 151] that he had the material and it was the world's good fortune that he had the intelligence to discriminate the true and the false in all the preceding work of the eighteenth century and to unify and amend these facts and inferences into a statement of the case that has proved to be permanently valid, aside from unimportant details.

Simpson's verdict still holds on Cuvier's towering achievements in instilling historicity and triggering the rise of geohistory and biohistory, the new historical sciences.[5]

Well aware of New Holland's extraordinary fauna of egg-laying mammals and pouched marsupials (this was three decades after Captain Cook), Cuvier knew too that there was an opossum in North America and that nothing at all was known of marsupials in the Old World. Receiving a new specimen of a vertebrate animal from the gypsum quarries in the Paris district, with his theatrical flair he made the prediction from its teeth that exhumation from the matrix would reveal the characteristic marsupial bones; and he performed the extraction as a dramatic confirmation in front of competent witnesses. A risky prediction but, yes, it was indeed a marsupial, a single specimen in the Old World Eocene, related in Cuvier's opinion to the American opossum but isolated in space and in time.[6] But for progress in the Australian story we move on a decade, adding a framework of time (i.e. back to the Pleistocene Epoch) to the spatial or biogeographic dimension (i.e. Australia vis-à-vis the other continents).

---

5   Simpson (1942).
6   Rudwick (2005, Fig. 7.20).

Owen and Darwin: Bones and extinction — biogeography and the law of succession

The Wellington caves in east–central New South Wales[7] expanded through the ages by extensive solution of Palaeozoic (Devonian) limestones. The action of percolating water enlarges cracks and fissures and develops caves and lets in dust, all good for accumulating fossils. Animals fall in and they defecate in, bones are dropped in, carcases are dragged in, pellets are vomited in (by owls); and chances of preservation are relatively good. The discovery of bones within the Wellington cave system caught the attention in 1830 of the surveyor-general of NSW, Major Thomas Mitchell, well-educated and well-connected, and friend of the most prominent theologian in the colony, the Reverend John Dunmore Lang. Mitchell was keenly aware of the work of Cuvier, including his theory of successive catastrophes based on his reconstructed fossil succession in the district of Paris, and of William Buckland, who took up the notion of a worldwide Biblical Deluge more seriously than Cuvier ever did, and who sought an 'antediluvian fauna'. Lang was interested in the implications of and for the Deluge. There was extensive collecting and the collections were shipped to Edinburgh (to Robert Jameson at the university), to London (to William Clift at the Royal College of Surgeons), and to Cuvier in Paris. Cuvier died in 1832 and the bones in Paris were studied by his assistant Joseph Pentland.[8]

Clift took on as his assistant the young Richard Owen. Owen had been apprenticed to a surgeon doing post-mortems, learning his bones and sinews and organs the rigorous way and very well, but making his name in zoological circles with a monograph on the pearly *Nautilus* in 1828. Owen was to become the most significant practitioner of vertebrate palaeontology in the ensuing decades; his published research on the Wellington Cave fossils spanned half a century, 1838–1888. In his appendix to Mitchell's 1838 account, Owen reported modern genera of 'possum', potoroo, wombat, 'native cat', thylacine and kangaroo, but in each case the species was different and several were bigger than the modern counterparts. There was also the huge animal to be called the Diprotodon. (Two decades later, Owen, now superintendent in natural history at the British Museum, successfully applied for NSW governmental funding to sharply expand the

---

7   Dawson (1985).
8   Pentland reported in 1831 that Cuvier (like Clift) had identified kangaroo, wombat, koala and other marsupials similar to the living fauna, plus a large and unfamiliar bone, which was a marsupial later to be named *Diprotodon* (Rudwick, 2008).

excavating and collecting under the direction of Gerard Krefft, curator at the Australian Museum. The Wellington Caves fossil assemblages, the first to be found on this continent, are of Late Pliocene to Late Pleistocene age; they now comprise about 60 species of marsupials, half of which are extinct; and they include reptiles, birds, bats, rodents and monotremes.)

Word of the rich bone hauls spread quickly in the halls and crypts of natural history and anatomy. These animals were marsupial, like the strange modern Australian fauna and unlike the Pleistocene fossils on the other continents. But they were extinct. Other continents had their extinct giant animals, later known as the megafauna. And now Australia had its own extinct giants (Figure 3.4).

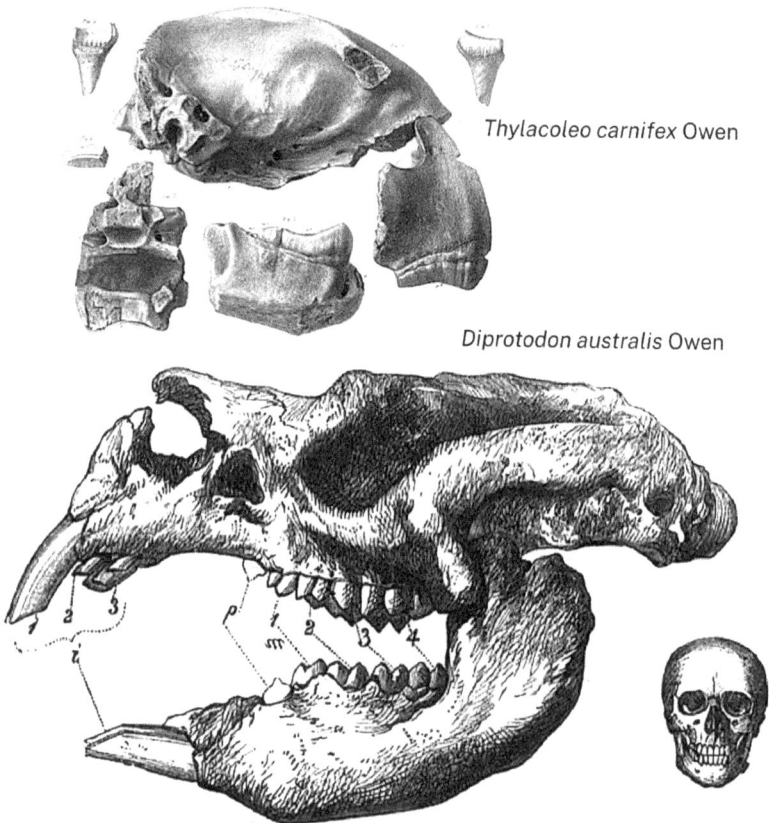

Thylacoleo carnifex Owen

Diprotodon australis Owen

Figure 3.4. Owen's big Pleistocene marsupials.

Two marsupials described by Richard Owen became the best known members of the Pleistocene megafauna in Australia: the large-leopard-sized carnivore *Thylacoleo* and the massive herbivore *Diprotodon* (human skull is for scale).

Source: Both illustrations are from Owen's 1860 textbook *Palæontology*.

Already the well-networked Charles Lyell could write in 1832, in *Principles of geology*:

> These facts [on the large marsupials] are full of interest, for they prove that the peculiar type of organization which now characterizes the marsupial tribes has prevailed from a remote period in Australia, and that in that continent … many species of mammalia have become extinct. It also appears … that land quadrupeds, far exceeding in magnitude the wild species now inhabiting New Holland, have, at some former period, existed in that country.[9]

Richard Owen was excellently placed to access and compare the various Pleistocene megafaunas of the world. He found a close correspondence between the faunas of the Pliocene and Pleistocene epochs in continental Europe and Asia and the modern faunas in the same large region—most of the genera but not the species were represented among the still living. Darwin's collections of mammal fossils from South America went to Owen. On South America, Owen opined: 'most of the fossil Mammalia … are as distinct from the Europaeo-Asiatic forms as they are closely allied to the peculiarly South American existing genera'. And now the 'ossiferous caves of Australia' were confirming the law, namely that 'with extinct as with existing Mammalia, particular forms were assigned to particular provinces', and that '*the same forms were restricted to the same provinces at a former geological period as they are at the present day*' (emphasis added). In his 1860 textbook *Palaeontology*,[10] Owen makes much of the pattern repeated around the world, wherein an extinct fauna is 'matched only by species now peculiar to that continent'. In Australia, the one great natural group, the Marsupialia, includes counterparts of the more diverse placentals on the larger continents. Thus the dasyures (quolls etc.) play the parts of the Carnivora, the bandicoots of the Insectivora, the phalangers of the Quadrumana (four-handed primates), the wombat of the Rodentia, and the kangaroos 'in a remoter degree' of the Ruminantia (grazing mammals). Owen is making an important ecological generalisation here: the stages are different, the players are local, but the drama, the play and even the times of the performance are pretty much the same.

---

9   Lyell (1990).
10   The quotes and discussion in this paragraph are taken from Owen's section headed 'Geographical distribution of Pleistocene mammals' (Owen, 1860, pp. 387–397).

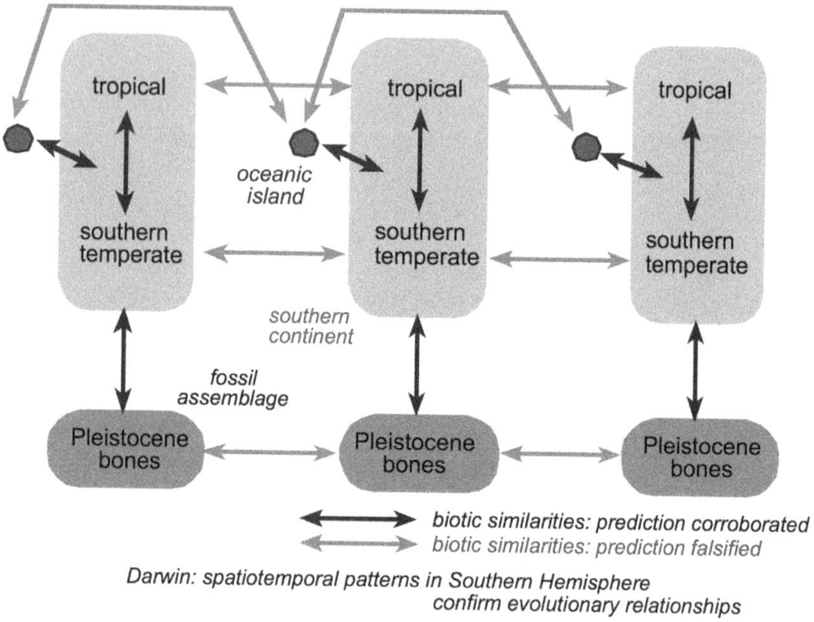

Darwin: spatiotemporal patterns in Southern Hemisphere
confirm evolutionary relationships

Figure 3.5. Darwin's argument from southern biogeography.

This is a model of southern biogeography, of three southern continents such as South America, Africa and Australia, and including the all-important fourth dimension of time. Two explanations are presented. A biblical creationist in evolution-denying mode, conscious of the catastrophic Deluge on the very young earth and the subsequent repopulating of the earth, would predict similarities (or affinities, or relationships; these words were used widely and loosely) within the realms of tropical–continental, temperate–continental and oceanic–island habitats, respectively. That prediction is shown with pale arrows. Several biogeographers observed the patterns that falsified the prediction, a falsification compounded by adding in the Pleistocene megafaunas (as they were later called) observed acutely by Clift, Lyell, Darwin and Owen. That interpretation is shown with the dark arrows. But only Darwin made the inspired cognitive jump from geographic similarity to evolutionary centres of origin.

Source: From McGowran (2013a).

But there is murkiness here too, as to who realised what and when they realised it. Writing his Chapter XI on geological succession in *On the origin* …, Charles Darwin begins his section, 'On the succession of the same types within the same areas, during the later Tertiary periods', with the clear statement that William Clift showed that the fossil mammals from the Australian caves were closely allied to the living marsupials of that continent; that Owen demonstrated the same pattern in South America (also manifest 'even to an uneducated eye', presumably including his own); and that he, Darwin, was so impressed by all this that he insisted in 1839 and

1845 on this 'law of the succession of types', on this 'wonderful relationship in the same continent between the dead and the living'. (The quotes are Darwin's words.[11])

It is customary to dismiss Australia's contribution to the gestating theory of organic evolution in keeping with Darwin's own famous last words: 'Farewell, Australia! ... I leave your shores without sorrow or regret.'[12] But a 1980 paper, 'Darwin and *Diprotodon*: The Wellington Caves fossils and the law of succession', argues that Darwin's initial insight came directly from the Australian discoveries.[13]

Inspect my schematic rendering of southern terrestrial biogeography including the fourth dimension of time (Figure 3.5).

Organisms survive and prosper by maintaining adaptation to their environments. In a natural theology of life's supernatural creation and the world's repopulation after the Deluge, it would be reasonable to predict, first, widespread affinities before the Deluge and a new set of widespread affinities afterwards. One would predict, second, affinities within the worldwide realms each of the tropical continents, the temperate continents and the oceanic-island habitats, respectively. Several biogeographers observed the patterns that falsified that latter prediction. Clift, Owen and Darwin compounded that falsification by adding in the Pleistocene megafaunas (as later so-called). All three grasped the law of succession of types. But only Darwin took the leap from regional geographic similarities to evolutionary centres of origin. As he explained in a letter to the biologist Ernst Haeckel in 1864:

> In South America three classes of fact were brought strongly before my mind. Firstly, the manner in which closely allied species replace species in going southward. Secondly, the close affinity of the species inhabiting the islands near South America to those proper to the continent. This struck me profoundly, especially the difference of the species in the adjoining islets in the Galapagos Archipelago. Thirdly, the relation of the living Edentata and Rodentia to the extinct species. I shall never forget my astonishment when I dug out a gigantic piece of armour like that of the living armidillo.[14]

---

11   Darwin (1964, p. 339).
12   Darwin, *The voyage of the* Beagle (1959, p. 434).
13   Dugan (1980).
14   Taken from Haeckel (1876).

Interlude, mid-century ferment: Six big British books and two others

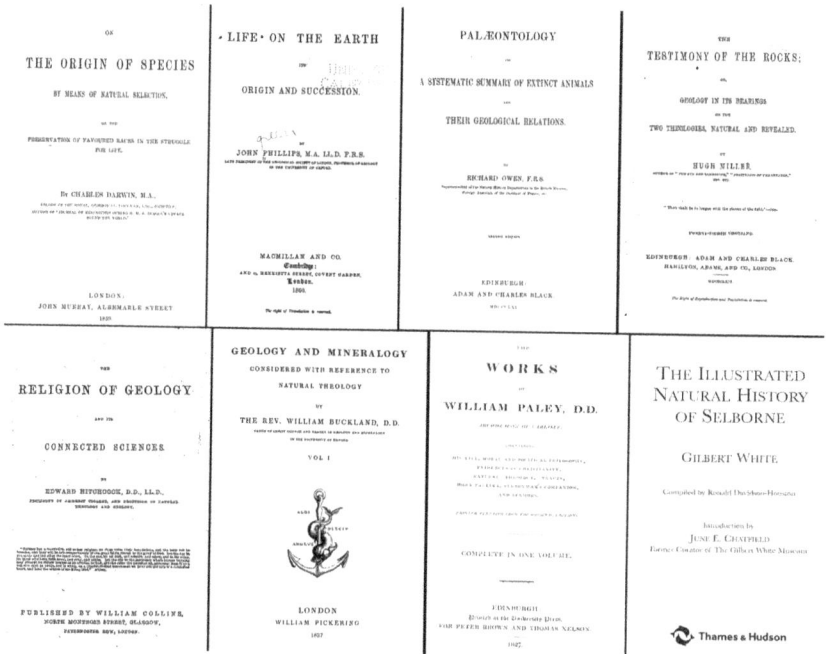

Figure 3.6. Discovering evolution: Eight books from the Anglosphere.
Source: All but one are from the nineteenth century; Gilbert White's late-eighteenth-century effort is shown here in a modern printing.[15]

---

15   In these times of anxiety about the decline of the West and the need for coherent curricula on Western Civilisation in the better universities, the Western canon of Great Books is mentioned frequently. Also very clear is the necessity to read, and read, and read. For that you need what Jewish grandmothers called the *Sitzfleisch*, you need the buttocks. Charles Darwin was a voracious and omnivorous reader. The teenager George Gaylord Simpson read the *Encyclopedia Britannica* in its entirety. Martin Glaessner was expected to read and absorb 200 pages a day as a postgraduate in Vienna. EO Wilson read both the Hebrew Bible and the Christian Bible, twice, while soaking up all the great evolutionists. To begin work on *The triumph of the Darwinian method*, Michael Ghiselin set about reading every published word by Charles Darwin. You didn't stock your mind with dot points and googled snatches. Alas, almost half a century ago I set a reading list for an honours student; counting the pages (225) she burst into tears at the enormity of the task; reactions to intense reading down the ensuing decades were unenthusiastic, if less dramatic. Here, however, is a book display of substance from yesteryear, capturing the discovery of geohistory and biohistory. You see DD degrees below and MA degrees and FRS bestowals above; and the list climaxes in a magisterial trio who could hardly be more divergent: Owen, Phillips and Darwin; and the Darwin is universally a 'great' book. Alas again: the list is only in the Anglosphere and the great Frenchmen Buffon, Lamarck and Cuvier are missing.

Very close in time, about 1860, close in space, in England, written in English by Englishmen, and exemplifying three very strong strands of our anastomosing narrative, are the three books on biostratigraphy, vertebrate palaeontology and organic evolution by John Phillips, Richard Owen and Charles Darwin, respectively (Figure 3.6, top far left through top centre-right).

Meanwhile a succession of books mark the tradition known as natural theology. The natural theologian sought evidence of the existence of a loving Christian God by studying His creation. Natural theology is as old as the religious contemplation of nature itself, much older than the Abrahamic monotheisms based in the Hebrew Bible, but this very English version of the endeavour arose in Christianity and most visibly among Protestants, keen and perceptive observers of nature, in the late seventeenth and eighteenth centuries and climaxing in the nineteenth.

Thanks to the imperishable *The natural history of Selborne* (Figure 3.6, bottom far right), the Reverend Gilbert White (1720–1793) is the best-loved and best-remembered of the parson-naturalists serving the agricultural societies of eighteenth-century England. Born in the Hampshire village of Selborne, educated and smallpox-infected at Oxford, he lived out his life as a curate, died and was buried in Selborne. His milieu was the natural history of the eighteenth-century Enlightenment, meaning the study of organic diversity, the natural kinds of organisms, their description, naming and classification; but White went further than anybody in several ways. He restricted himself intensively to a small but environmentally diverse patch of Hampshire (including woodland, sheep-grazed land on the chalk, farms, wetlands and streams); he spent more time out-of-doors experiencing all the seasons than did the other serious naturalists of the times; he went beyond classification and identification to pioneer activities labelled in due course as behavioural science, ecology and environmentalism; and he wrote it all up, reporting his results in letters to a zoologist and an amateur naturalist, letters which were assembled for publication as this book. White reports occasionally on fossils, such as exhumed ammonites; and by the time of his ordination in the mid-eighteenth-century it was entirely possible to believe in earthly time extending into the millions of years and still be a faithful or pious Anglican. But his natural environment was rural southern England.

We come to the foundation text of natural theology. The Reverend Dr William Paley (1743–1805) famously opened his *Natural theology* (contained within *The works*, Figure 3.6 bottom centre-right) by likening the universe to a watch. Contrasting our reactions to tripping over a stone on the heath to tripping over a watch in the same place, he argued that the watch's order, complexity and purpose, its design, must imply the work of an intelligent designer. (If Paley ever tripped over a fossil, he took little notice.) Likewise, the order, the complexity and the purpose in the universe must imply the work of a creator. Written in reaction to what he perceived as the atheism infesting the philosophy of the times, exemplified by the ominous David Hume, *Natural theology* is a long discourse on structural and functional biology, on the ever more exquisite solutions enabling each kind of organism to survive, flourish and reproduce in its environment. Paley's chosen speciality was human anatomy. Although he includes some comparative anatomy (discussion of which flirted briefly with and dismissed the sinister doctrine of transmutation, later called evolution), there is nothing in the book that might be called historical biology and no hint of a very great age of the earth.

In these respects Paley's *Natural theology* contrasts hugely with Buckland's natural theology, written after three decades of the new geohistory and biohistory (*Geology and mineralogy*, Figure 3.6, bottom centre-left). But natural theology was by no means done for—it adapted to the world's new dimension, deep time. The Earl of Bridgewater, a gentleman naturalist, commissioned eight Bridgewater Treatises to enquire during the 1830s into 'The Power, Wisdom, and Goodness of God, as manifested in the Creation'. Treatise VI, on *Geology and mineralogy*, was prepared by the Reverend Dr William Buckland (1784–1856) who, during the 1820s, became a geohistorian and biohistorian to be reckoned with, not least in his influence on Murchison. It has been argued that Buckland was the effective founder of a distinctive school of stratigraphy and geohistory and biohistory at Oxford that contrasted with the mathematical and physical bent at Cambridge.[16] Having at his disposal an entire biogeohistorical discipline of stratigraphy and palaeontology, a discipline having only just become worthy of the name in Paley's time, Buckland's central objective in Treatise VI remains the same as Paley's had been—to hold the line on the Grand Design by the Grand Designer—but now extended deeply into the organic remains of a former world.

---

16   Rupke (1983).

Buckland's exuberant Preface announces that natural theology has now arrived in deep time:

> The myriads of petrified Remains which are disclosed by the researches of Geology all tend to prove that our Planet has been occupied in times preceding the Creation of the Human Race, by extinct species of Animals and Vegetables, made up, like living Organic Bodies, of 'Clusters of Contrivances', which demonstrate the exercise of stupendous Intelligence and Power. They further show that these extinct forms of Organic Life were so closely allied, by Unity in the principles of their construction, to Classes, Orders, and Families, which make up the existing Animal and Vegetable Kingdoms, that they not only afford an argument of surpassing force, against the doctrines of the Atheist and Polytheist; but supply a chain of connected evidence, amounting to demonstration, of the continuous Being, and of many of the highest Attributes of the One Living and True God.

Nor does Buckland leave things there, for the sinister whiff of transmutation is still in the air in 1836:

> Theories which have been entertained respecting the Origin of the World; and the derivation of existing systems of organic Life, by an eternal succession, from preceding individuals of the same species; or by gradual transmutation of one species into another. I have endeavoured to show, that to all these Theories the phenomena of Geology are decidedly opposed.[17]

(Long gestating in Buckland's mind, however, were doubts about the widespread evidence in sediments claimed to record Noah's Flood, the Deluge, and by 1840 he was convinced of the reality of the Pleistocene Ice Age.) Mrs Buckland's Anglican anxieties over these career-threatening intellectual developments notwithstanding, Buckland DD FRS duly became dean of Westminster.

Forward another couple of decades, and we encounter two more titles in natural theology, by Edward Hitchcock (Figure 3.6, bottom far left) and Hugh Miller (Figure 3.6, top far right), one a distinguished academic and divine, the other a self-taught stonemason, both with something of palaeontological substance to contribute, that is, scientific credentials. We can say something about Buckland, Hitchcock and Miller as a group,

---

17   Buckland (1836, pp. vii–viii).

because their similarities beyond a suffusion of piety are stronger and more significant than their differences. Their field boots are well worn and their fingernails dirty, and they know their rocks and their fossils. They reject the young earth or any implication of an age of only a few thousand years. (Miller is especially stinging in a chapter on the geology of the anti-geologists. By the 1850s a three-event biblical history of Creation, Deluge and Armageddon has been well gone from science—it was dead science—for a century.) They subscribe to what we now call deep time and to the fossil-driven geological column. They believe in geohistory and biohistory at very long time scales on a very old earth. How long and how old are unknown, but many sedimentary accumulations are to be measured in kilometres' thickness, and time is accepted to be in the millions of years. However, they accept neither infinite time nor that notion of an eternal earth that goes back to Aristotle. They reject all theories of transmutation (subsequently called organic evolution). For them, species are very real historical entities with beginnings and endings in time, meaning extinctions, but no matter how similar an older shell, say, is to a younger shell, and whatever metaphysically was in the mind of the Creator and unknowable to us, the two shells have no actual, organic, genealogical connection. Biological organisation and adaptation existed on this earth long, long before the advent of 'Man'; even so, there is strong belief in progress—progress towards and preparation for the advent of 'Man'. And proof of design by an intelligent designer is everywhere and permeates everything.

But the fossil record added a twist to that faith in the reality of progress. Citing the evidence of the early fish and the early cephalopods, Miller and Hitchcock argue that the first organisms in their respective great group (or taxon) were perfectly designed and fully adapted; that the subsequent story was of decline, of retrograding from the complex to the simple, from the more perfect to the less perfect; that geology abounds in such stories and those stories are deemed to be utterly fatal to the hypothesis of development ('development' meaning transmutation and evolution). It was only in the longer narrative of each group being supplanted by another that overall progress is perceived. Extinction was long regarded as a blasphemous inference of the Creator's second-guessing, of His flying kites; but our natural theologians had safely negotiated that hazard en route to deep time.

Organic evolution did not happen: Species are immutable

There were roughly six decades from Cuvier's discovery of extinction in the 1790s, as we have seen, to Darwin's theory of the origin of species (1859). Those decades spanned the Palaeontological Synthesis, during which the lithological time scale of the mid-eighteenth century was supplanted progressively with the evidence of the fossil succession; disputes were resolved by fossils; new territories were mapped and prospected with fossils; and the hopes for an intercontinental and global geological time-scale and a broadly global history of life were justified and fulfilled by fossils. Martin Rudwick has called this happening, this magnificent efflorescence of biohistory and geohistory, the Cuviero-Lyellian Revolution. I wrote in Chapter 2 that the bollard from which all this hangs was biostratigraphy. John Phillips realised that a geological time scale applicable worldwide could only be based on the fossil succession, and he proposed in 1841 the three great eras, the Palaeozoic, Mesozoic and Cenozoic (see Figure 2.9).

But why did biostratigraphy work so well?

Palaeontologists since Cuvier found repeatedly from plotting fossil distribution in the stratigraphic record that animal and plant species appeared in the seas or on the lands of the time, existed for a while, and then disappeared. The palaeontologists did not need an act of faith, a proof of some grand design to accept this pattern. Their everyday work was challenging, confirming, correcting, improving and strengthening their tools of trade, namely the orderly pattern of fossil species distributed in space and time. And this biostratigraphic foundation of the Cuviero-Lyellian Revolution arose without any consensus as to why it really worked so well. That is, it was a robustly empirical foundation with no securely agreed theoretical support, no successful answers to the questions as to why did species come (the births or origins of species)? and why did they go (the deaths or extinctions of species)?

Since ancient times there have been four kinds of theory of the origin of species. One theory said that species are eternal, so it was known as Aristotelian eternalism; a second was that they were created or unfolded according to a divine blueprint; in a third, they arose spontaneously from unformed matter; or, fourth, they evolved. Aristotelian eternalism was no longer alive during the rise of biostratigraphy, but the other theories were indeed alive. We tend to think of spontaneous generation as maggots emerging spontaneously from rotten meat, the theory destroyed by Pasteur and the other pioneering microbiologists; but there was a different theory going by that name or as 'autogenesis'. Several well-known geologists and biologists, some labelled as 'forerunners of Darwin', have recently been identified as autogenists. In one clear example the natural theologian Edward Hitchcock constructed two trees of life, an animal tree and a vegetable tree (Figure 3.7).

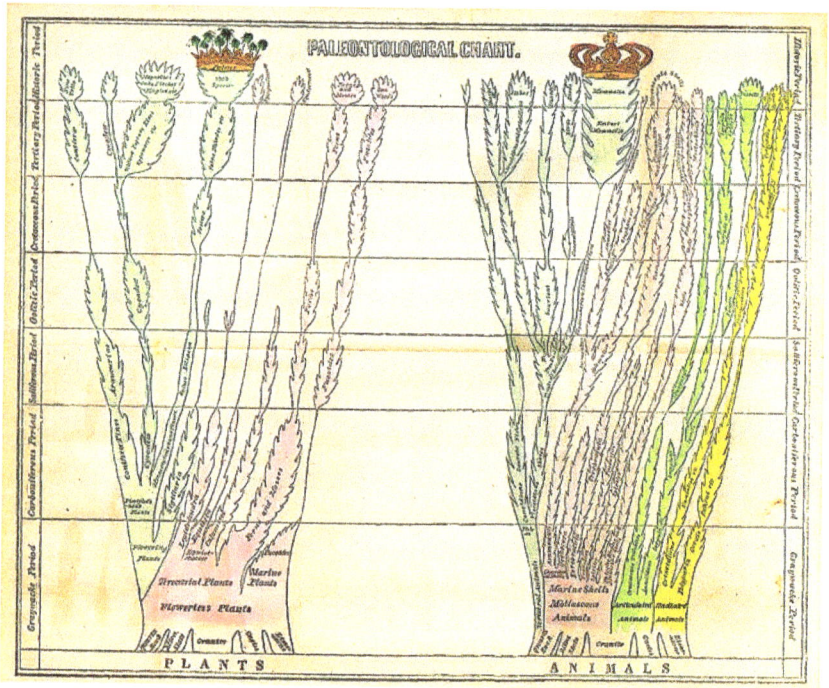

Figure 3.7. Hitchcock's family trees erupted from the rocks. This originally hand-coloured 'paleontological chart' was first published in 1840.

Alternatives to the notion of organic evolution in deep time during the six Cuvier-Darwin decades were serial creation, spontaneous generation and reverent silence; Aristotelian eternalism had disappeared. In 1840 Edward Hitchcock depicted Cuvier's four *embranchements*, his four great streams of animal kinds, against a geological time scale, and added the vegetables. No fossils were known from below the 'Graywacke Period' (the Cambrian-Silurian) and Hitchcock shows us why: the streams of life actually erupt out of the igneous rocks (granites) and out of the quartzites, gneisses, schists and limestones. In the Historic Period, man is crowned in the animals and the palms are crowned in the vegetables.

Source: From Archibald (2009, Fig. 5).

The former displays four great trunks, each trunk being one of Cuvier's *embranchements*, and each emerging during the 'Greywacke Period' (Cambrian–Silurian, Early Palaeozoic Era) from tree roots composed of metamorphic and igneous rocks, namely quartz rock, mica schist, granite, gneiss and limestone (not in the rocks). Likewise and at the same time a second trunk of flowerless terrestrial and marine plants emerges, soon adding the flowering plants. Not *in* the rocks but *of* the rocks; Hitchcock's notion expired.

So what were the palaeontologists and geologists believing in, as they toiled for six decades between early Cuvier and mature Darwin in this splendidly fruitful project of mapping the history and ancient geography of life on the Earth? (And I include the naturalists, anatomists, agriculturists and surgeons, the neontologists; but these people in their daily routines were hardly more perturbed than were, say, the chemists or the bishops, by the yawning abyss of deep time confronting daily the palaeontologists.) Probably the majority belief was in a serial, naive creationism (a Creator tinkering busily through the aeons); probably the minority belief was a reverent silence (geology is thriving, so be silent and get on with your biostratigraphy); a distant third would have been some version of organic evolution.

Lamarck discovered organic evolution.[18] He was the quintessential scientific naturalist, expert in organic diversity and a productive publisher in biosystematics, first in terrestrial plants, then in marine animals and in deep time. But his actual theory of change did not prosper. 'Lamarckian', pertaining to the simple inheritance of characters acquired in use and disuse by the parent (the blacksmith's muscles), is a grossly misleading use of a brilliant and tragic person's name; but Cuvier then Lyell did such

---

18   From Henry Fairfield Osborn's (1894) *From the Greeks to Darwin: An outline of the development of the evolution idea* to Rebecca Stott's (2012) *Darwin's ghosts: In search of the first evolutionists*, there have been numerous attempts to demonstrate that Darwin was less than original and less than candid about his antecedents. But until we get to Lamarck the overriding impression is of cultural stasis, with glimmerings of insight into change during deep time and wondering about the meaning of organic similarity and organic diversity. If Pierre Belon's remarkable demonstration of the bird–human similarity is comfortably ascribed to minor changes in the divine blueprint for animals with backbones, then need we enquire any further about cultural stasis through the millennia?

a demolition of Lamarck that the adjective has fastened as a pejorative.[19] One could say the same about 'transmutation' itself. Evolution was sustained through the years of the Cuviero-Lyellian Revolution by a thin red line (the first of two thin red lines, as we shall see). Darwin could recall but one evolutionist from his own formative years, Richard Grant in Edinburgh, an excellent marine biologist who was a strong and beneficial intellectual influence on the young Darwin.[20] Cuvier had two strong reasons for rejecting the transmutational belief of his time. The ecological reason, encapsulated in his slogan 'conditions for existence', was his belief in the highly integrated nature of an organism, every organism, each adapted to its environment, which implied that any change by transmutation unavoidably was a perturbation, a profound dislocation. Cuvier's biostratigraphic reason was that Lamarck thought of species as being ever-shifting and inherently unstable lineages, whereas Cuvier's establishing of biohistory required a discoverable succession of inherently stable and reliable species. Grant and others in Edinburgh excepted, the scientists of the generation after Lamarck and Cuvier mostly believed in some combination of Cuvierian objection and natural-theological rejection. Hitchcock and Darwin, one gleeful, the other rueful, published lists of the eminent and distinguished biohistorians and geohistorians who were quite at ease in accepting deep geological time while concomitantly rejecting transmutation.

But it was the very scientists on those lists who brought off the Cuviero-Lyellian Revolution initiated by Lamarck and Cuvier!

Which brings us back to Owen, the major figure who neatly fits neither this section in our narrative nor the next (Figures 3.8 and 3.9.).

---

19   Cuvier rejected any organic connection between Lamarck's fossil molluscs in the Paris district and modern species. Most of the second volume of Lyell's *Principles of geology* is devoted to biological arguments, rejecting Lamarck's evolutionism being high on the agenda. Irony upon irony: it was Lyell's accompanying Prévost in the field around Paris that encouraged the Frenchman to progress beyond Cuvier and Brongniart and produce that splendid reconstruction (recall Figure 2.6); and the same Lyell whose third volume of the *Principles* spelled out the changes in the molluscs, beginning with the vindication of Lamarck's work on his fossil shells as the basis for the Eocene Epoch (see Figure 2.7).

20   The metaphor 'the thin red line' does not refer to the modern warning in international diplomacy or politics, 'do not cross this red line'. It refers to an episode in the battle of Balaclava in the Crimean War in 1854. A Scottish regiment of red-coated soldiers held fast in a thin line against a much larger opposition. This is how I see the thin but tenacious lines from Lamarck to Darwin, when the 'Wernerian' Robert Chambers and the 'Lamarckian' Robert Grant sustained a thin red line of historical thinking and evolutionary belief in Edinburgh, the town of the radically ahistorical James Hutton, and from Darwin and Wallace to Matthew and Simpson during the dark days for natural selection. See Chapter 10.

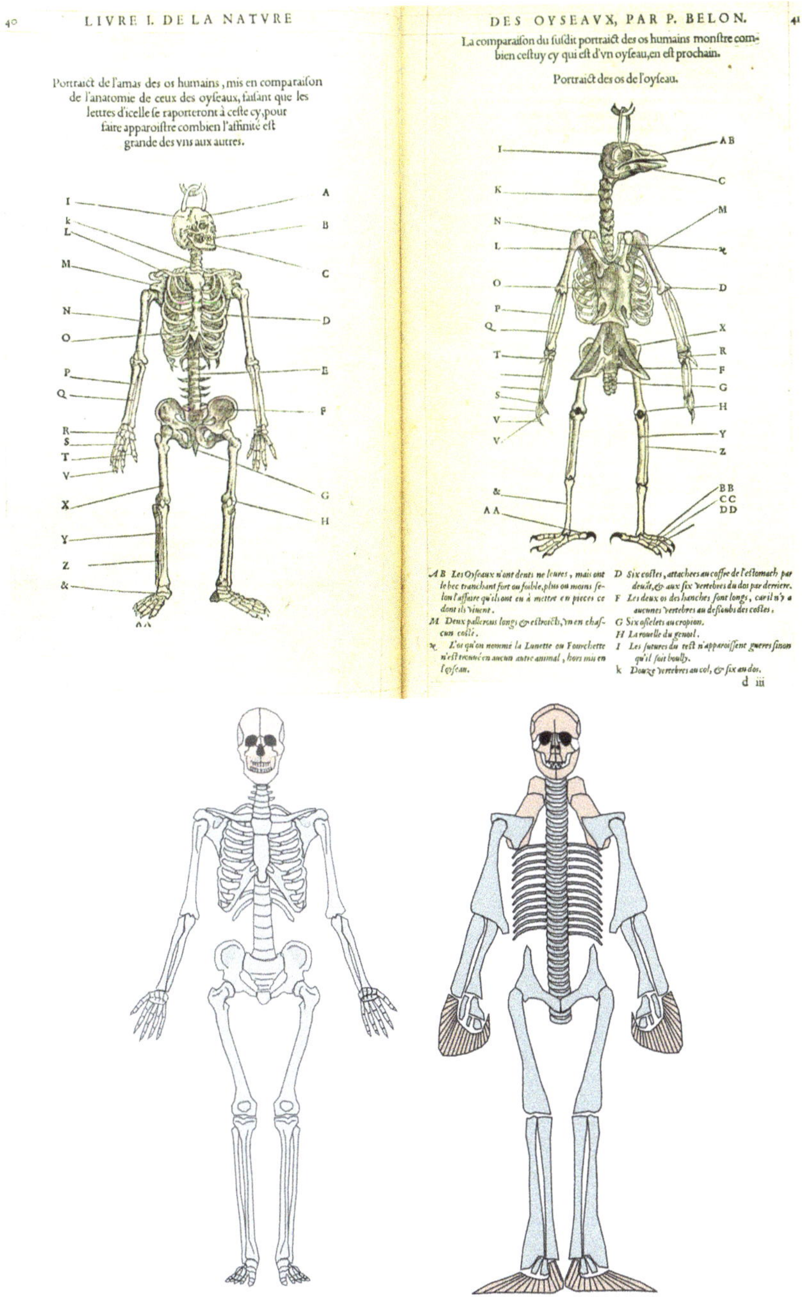

LIVRE I. DE LA NATVRE

Portraict de l'amas des os humains , mis en comparaison de l'anatomie de ceux des oyseaux, faisant que les lettres d'icelle se raporteront à ceste cy, pour faire apparoistre combien l'affinité est grande des vns aux autres.

DES OYSEAVX, PAR P. BELON. 41

La comparaison du susdit portraict des os humains monstre combien cestuy cy qui est d'vn oyseau, en est prochain.

Portraict des os de l'oyseau.

AB Les Oyseaux n'ont dents ne leures , mais ont le bec tranchant fort ou foible, plus ou moins selon l'affaire qu'ils ont en à mettre en pieces ce dont ils veunt .

M Deux petites langs, & estroictes, m'en chascun costé .

x L'os qu'on nomme la Lunette ou fourchette n'est trouvé en aucun autre animal , hors mis en l'oyseau.

D Six costes , attachees en caisse de l'estomach par deux, & aux six vertebres du dos par derriere.

F Les deux os des hanches sont longs , car il n'y a aucunes vertebres au desfoubs des costes ,

G Six osselets au cropion.

H La rouelle du genoul .

I Les sutures du test n'apparoissent gueres sinon qu'il soit bouilly.

k Douze vertebres au col , & six au dos.

d iii

Figure 3.8. Homology, from Belon 1555 to *Gogonasus* Man 2011.

The vertebrate skeleton has been known since antiquity for its similarities and differences. Its bone-by-bone matching emphasises the similarities implying an underlying body plan, known as homology, a *Bauplan*. This was not controversial, but the question was: does the unity of type demand actual relationship? The differences imply obvious and subtle adaptations: do they also imply evolutionary divergence? *Above*, in 1555, Pierre Belon compared the bird skeleton with the human, point by point. *Below*, John Long (2012) rescaled the bones of his 'fish' *Gogonasus* (~375 million years ago) as 'tetrapod' *Gogonasus* Man in a humbling comparison with *Homo sapiens*; overlook the demands of gravity and walk on dry land, and you have a plethora of similarity between swimmer and walker. Here is Darwin: '[Morphology] is the most interesting department of natural history, and may be said to be its very soul. What can be more curious than that the hand of a man, formed for grasping, that of a mole for digging, the leg of the horse, the paddle of the porpoise, and the wing of the bat, should all be constructed on the same pattern, and should include the same bones, in the same relative positions?' (*Origin*, 1964, p. 434).

Source: Belon (1555, pp. 40–41) via Wikimedia Commons. *Gogonasus* Man, Long (2012) and image courtesy of John Long.

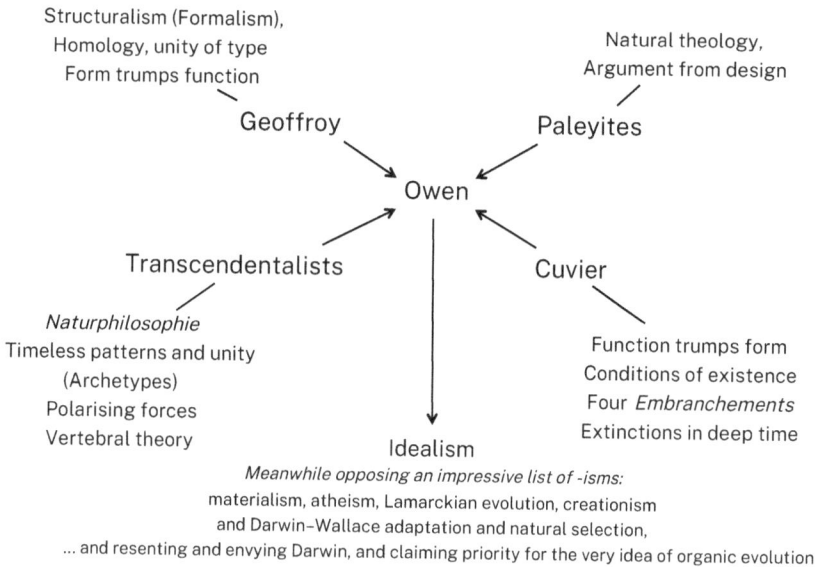

Structuralism (Formalism),
Homology, unity of type
Form trumps function

Natural theology,
Argument from design

Geoffroy

Paleyites

Owen

Transcendentalists

Cuvier

Naturphilosophie
Timeless patterns and unity
(Archetypes)
Polarising forces
Vertebral theory

Function trumps form
Conditions of existence
Four *Embranchements*
Extinctions in deep time

Idealism

Meanwhile opposing an impressive list of -isms:
materialism, atheism, Lamarckian evolution, creationism
and Darwin–Wallace adaptation and natural selection,
... and resenting and envying Darwin, and claiming priority for the very idea of organic evolution

Figure 3.9. The central position of Richard Owen.

It can help to visualise the currents and cross-cutting turbulences of intellectual ferment. This example concerns Richard Owen and the anatomy and history of the vertebrate animal in the 1820s–1840s. The influence on Owen closest to home was the natural theology of the English parson-naturalists, the 'Paleyites' pointing to abundant examples of adaptations, designed exquisitely by a designer. From Germany came the mystical, 'transcendental' ideas of *Naturphilosophie*. From France came the great debate between Geoffroy, who emphasised structure and unity of design (form trumps function), and Cuvier, who emphasised conditions of existence (function trumps form). All four streams include people who knew about animals, and here was the evidence for both evolution and anti-evolution widespread in those times, awaiting scholars determined to find one or the other. Meanwhile, Owen the complex and difficult personality amassed an impressive list of negatives while becoming a great anatomist. See Nicolaas Rupke, *Richard Owen: Biology without Darwin* (2009).

Source: Author's summary.

Cuvier drew together three disciplines of the vertebrate body, its structure (comparative anatomy), its development in each individual (embryology), and its history through deep time (vertebrate palaeontology). The outstanding figure in the next generation, Darwin's generation, was Richard Owen. Owen distilled a theory in which embryological development occurred under the influence of two forces, structural and adaptive. The structural force produced homologies, meaning the equivalent organ or structure in different animals, and what is inherited. The adaptive force produced individual distinctness, as one coped with the environment and survived. The forces together accounted for taxonomic diversity and variation in body parts. Homologies were the binding material in the concept of unity of type. As we can see here, such commonalities were well worked out long before biohistory, let alone evolution, was established. 'Archetype' is Owen's word for the notion of a basic architectural plan (German *Bauplan*) for a group, in this case, of vertebrate animals.

Look at the influences on Richard Owen.[21] We have the Parisian debate of Geoffroy's unity of type versus Cuvier's conditions of existence. We have the English 'Paleyites', the natural theologians. And we have the mystical 'German' *Naturphilosophie*, philosophy of nature, which sprang from the Romanticism so prominent in German thought and culture in the earlier years of the nineteenth century. Owen's work on homology is exemplified by his comparison of the limbs of the horse, *Hipparion*, and the Palaeothere, with the label, 'Derivation of Equinae' (Figure 3.10).

Such words as 'affinity', 'relationship', 'derivation' and 'equivalent' sprinkled in the palaeontological and biological literature of the decades preceding Darwin invite us to infer some belief in transmutation in some workers, in Owen most of all. Informed scholars are mostly agreed that Owen became a crypto-evolutionist behind the tactically pious utterances about the Creator that laced his writing in the best natural-theological tradition. But the natural theologians applied the pressure and Owen went silent on transmutation for a decade; scooped unknowingly by Darwin, he became seriously resentful. (Unaware of Owen's private evolutionary ambitions, Darwin thought he was in the creationist camp.[22])

---

21   Padian (2008); Rupke (2009).
22   In the Historical Sketch added under pressure to later editions of the *Origin*, Darwin confessed to having fallen into error, repeatedly, in his understanding of what Owen really thought about evolution, but he was consoled 'that others find Professor Owen's controversial writings as difficult to understand and to reconcile with each other, as I do'. See Rupke (2009).

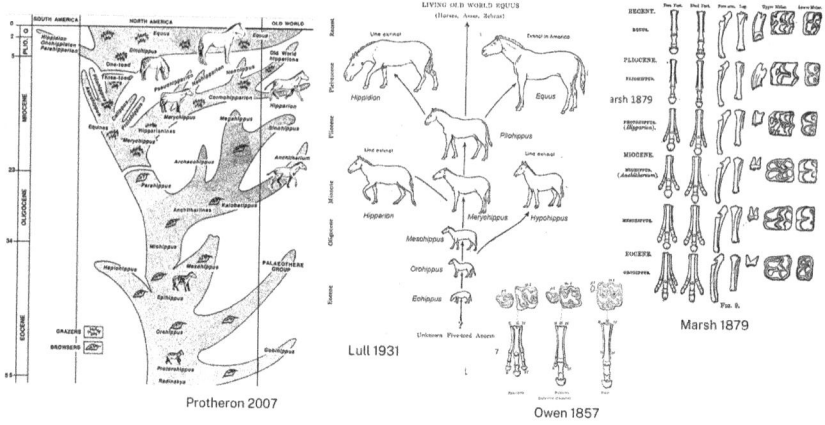

Protheron 2007

Lull 1931

Owen 1857

Marsh 1879

Figure 3.10. The evolution of the horse.

The story of the horses goes back to Owen in the 1850s. Think about it: in his diagram of the feet and teeth labelled 'Derivation of Equinae', Owen was showing changes in the feet and the teeth implying changes in environment and lifestyle — but he was not saying that derivation was evolution (while Darwin was about to write the *Origin*). Two decades and many tonnes of American bones after the *Origin*, it was all about evolution, when Marsh could make an informative sketch of evolution from just six sets of equine limbs and teeth spanning 40-odd million years (in twentieth-century numbers). This story reached its zenith in Matthew's (1926) fine synthesis of climatic change opening forests into grasslands, changing the diet from browsing to grazing and thus changing the teeth, changing the body size and feet as the arms race with predators escalated. But Marsh's and Matthew's elegant arrangements were interpreted into something more. Others (such as Lull [1931] but not Matthew) could visualise a straight line from *Eohippus* to *Equus* with side branches, the implication being a goal-directed evolution, something foreordained, a phylogeny planned for the noble *Equus* to arrive on the scene, big enough, strong enough and fast enough, and in good time to be mastered and harnessed by the crowned mammal, Man. Or so it seemed to the old natural theologians such as Edward Hitchcock. But modern reconstructions of the horse family tree are bushy and speciose (many species), with no proud central trunk thrusting purposefully upwards to modern times.

Source: See citations in figure.

Organic evolution did happen: Species are not immutable

Charles Darwin was a geologist. Darwin was also lucky. Born into a lineage and tradition of lively intellectual curiosity and worldly accomplishment, he grew up collecting beetles, riding horses and shooting partridges in conditions of rural comfort; he was mentored by outstanding scientists and scholars, especially Grant's marine biology in Edinburgh, Henslow's botany in Cambridge and Sedgwick's geology in a crash field course in

Wales (one-on-one with the Cambridge professor for two weeks), so that he was at ease around the globe in terrestrial and marine environments and biotas and in deep time (and socially at ease in diplomatic, naval and colonial environments); he could easily have missed out being on the boat but did not; he might have been in a strenuous social situation on Fitzroy's *Beagle* but was not (seasickness notwithstanding); his father financially covered a five-year global field trip and all its collecting and dispatching; his return was eagerly expected by his peers in London and he was ushered into the Geological Society and Royal Society; his personal life and domestic situation were strong; and never needing an honest paid job or to do the housework freed him for decades of preoccupation and obsessive concentration. And not to be forgotten are the four fierce supporters who were there when he needed them most—Charles Lyell, Joseph Dalton Hooker, Thomas Henry Huxley and Alfred Russel Wallace. Much has been written about Darwin's terrible bouts of ill health and his spiritual and emotional and mental turmoil. Much has been written about his responsibility for the socio-political perturbations in the hierarchies in Victorian England (one outcome was no knighthood); about relativism, the hollowing out of morals and standards and meaning as men turn out to be mere monkeys; about stinking privilege and scientific injustice (rich and connected Charles Darwin under-acknowledging his antecedents, the true discoverers of evolution and natural selection, then ripping off poor, isolated, marginalised Wallace); about the difference between a jumped-up field naturalist (i.e. Darwin, Wallace) and a real scientist (one, that is, who runs controlled experiments in a laboratory). This mindset is still with us; but still we scribble ...[23]

Darwin gestated a revolutionary and heretical notion for almost a quarter-century, until he was prodded into writing the book that brought off the most profoundly world-changing, evidence-based, successful theory ever proposed. In one mouthful the unified supertheory of evolution is tree of life produced by natural selection. I find it helpful for getting a grip on the actual science in 'Darwinism' to see the supertheory (one long argument, in Darwin's words) as an integrated cluster of five theories. The first two are

---

23   Were I asked to recommend some ways into the enormous corpus of Darwiniana, I might begin with four: Charles Darwin's *On the origin of species* (first published 1859), Janet Browne's *Darwin's Origin of species: A biography* (2007), Michael Ghiselin's *The triumph of the Darwinian method* (1969), and Ernst Mayr's *What evolution is* (2002). Ghiselin and Mayr are about theory and evidence, the actual geological and biological progress leading towards the present; that is, they are proudly Whiggish and, in the eyes of the sociologists and historians of Darwiniana, strongly biased and profoundly naïve. So, for that matter, is my own Whiggish effort (2013b). See also Chapter 10.

historical theories, meaning, what was the outcome in the course of geological time? The other three are causal theories, meaning, how has evolution happened, how is it happening? Nothing in earthly biology is untouched by evolution. It is clarifying for me to construct a visual representation of input and output, of fields of biological enquiry in Darwin's time and whence his evidence; perhaps the diagram will help the reader too (Figure 3.11).

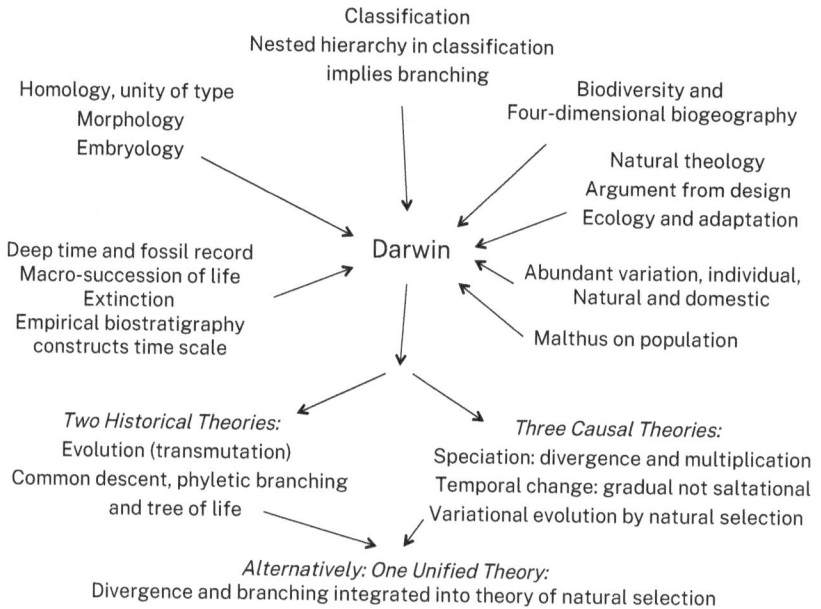

Classification
Nested hierarchy in classification
implies branching

Homology, unity of type
Morphology
Embryology

Biodiversity and
Four-dimensional biogeography

Natural theology
Argument from design
Ecology and adaptation

Deep time and fossil record
Macro-succession of life
Extinction
Empirical biostratigraphy
constructs time scale

Darwin

Abundant variation, individual,
Natural and domestic

Malthus on population

Two Historical Theories:
Evolution (transmutation)
Common descent, phyletic branching
and tree of life

Three Causal Theories:
Speciation: divergence and multiplication
Temporal change: gradual not saltational
Variational evolution by natural selection

Alternatively: One Unified Theory:
Divergence and branching integrated into theory of natural selection

Figure 3.11. The central position of Charles Darwin.

Available to Darwin since his Cambridge days was an array of data and argument in natural theology, the 'Paleyite' enterprise. Darwin had to read and absorb Paley. It is often overlooked in complaints about the gaps in the fossil record that biostratigraphy was becoming global and highly successful, and demanding a crowning explanation of its success. Likewise with systematics: why was the classifying of the multitudinous kinds of life on earth so successful? Likewise with adding the Pleistocene to construct four-dimensional biogeography, as we have seen. Likewise with bones and embryos, and the sheer success of Cuvier and Owen, whose very success demanded an overall explanation. All that grand science came to Darwin along with the pressure to produce a successful grand theory. The outcome can be deconstructed as the two historical theories (that were rapidly accepted) and the three causal theories (that languished until the next century); or as the one grand unified theory, which is how Darwin saw it and how we should, too. Perhaps most usually today, Darwinian evolution is seen as the two-part natural selection (the essential process) and tree of life (the outcome, the pattern).

Source: Author's summary.

The first theory, evolutionism itself, the evolution of life on an ever-changing planet, was not original to Darwin, but to Lamarck. Certainly original was Darwin's powerful and meticulous assembling of a wide-ranging case to prove that evolution happened. Only a small fraction of that huge dossier appeared in the *Origin*, which Darwin saw as an extended abstract.

Second, the tree of life was not original iconography[24] (Lamarck was there, too), but this tree, the branching transmutation of species, was the first to be compelling. The fossil record impressed Darwin, the geologist, with deep geological time and the great age of the earth, with the large-scale, macro-succession of life on earth, and with the enormous empirical success of biostratigraphy in exploiting that orderly succession to construct the global geological time scale. Darwin forced a brutal choice upon the people, but especially upon those of natural-theological bent: is it to be creation after creation after iterating creation? Or is it to be organic speciation and descent from ancestors and extinction?

Third: change through time is one thing but multiplication, the emergence of two or more species from a common parent is quite another; and Darwin came to realise that a theory of speciation (the word itself came later) was essential. The principle is ecological, coming out of economics via agronomy, and arguably it is the keystone of the *Origin*. Biogeography grades into ecology and adaptation, which is the terrain of the English natural theologians' arguments from design—and Darwin's prescribed reading list at Cambridge had included Paley. By the 1840s Darwin was armed with geographic speciation ideas from the *Beagle* voyage, and he knew quite a lot about a broad swathe of geology and, in due course, of biology. But the botanist Joseph Hooker challenged him to accrue his own biosystematic credentials, to get the real depth of insight possible only by doing the hard yards of assembling, sorting and classifying a difficult group of organisms. And so began an eight-year project on the systematics of barnacles. Darwin commenced the project as a tyro and emerged as a full-blown comparative morphologist and pioneering evolutionary or genealogy-based taxonomist. He discovered that branching evolution gives the natural nesting of groups within groups within groups, and in the 1850s he shifted his emphasis somewhat from geographic speciation to ecological diversification and speciation.

---

24 Archibald (2009).

The fourth theory was gradualism. Darwin, like Lyell before him, was opposed to the notion that change was essentially abrupt and even catastrophic. Advocating for gradualism was essential to calm down (in people's minds) Cuvier's and d'Orbigny's disruptive revolutions, and to head off the outright catastrophism that was still attractive to many (not to speak of biblical creation ex nihilo or extinction in the Noachian Deluge, both well dead in science); and he strove to establish the case for gradual not saltational change through geological time.[25]

The fifth theory, variational evolution by natural selection, has been the most-discussed component of this supertheory of evolution. There is huge if often cryptic variation within species; reproductive success varies markedly, and superior adaptive features are selected via reproductive success. Darwin talked widely with animal and plant breeders, people such as the pigeon-fanciers in the East End of London, people who knew that domesticated organic species had vast stores of variability, rapidly exposed by artificial (i.e. human) selection, and that species in the wild surely did too. Lacking the science of transmission genetics, he could not explain why beneficial characteristics would be preserved by selection, not lost in blending, like red and blue inks blending into a muddy brown; but he did know that the characteristics of male and female held stable and true through countless generations with little blending. Therefore, whatever mechanism was achieving sexual segregation, would not that mechanism be available also to conserve beneficial characteristics?

It took Darwin almost a quarter-century of obsessive hard labour to weld those five components into the supertheory of organic evolution by natural selection. In the decade or two after the *Origin*'s publication, the evolution of life as a fact was becoming a non-issue for virtually every competent and professional biologist and, more tardily, every geologist, and for most more-or-less-informed citizens in the Greco-Judean milieu.

---

25   But we have seen that the 'revolutions' in the Paris district were already being toned down or smoothed over in the 1820s, by Prévost (see Figure 2.6).

Near-death of Darwinism, another thin red line and the Darwinian Restoration

Fully respecting the entire corpus of anatomy and embryology and palaeontology from the eighteenth century to Owen, Darwin interpreted homology and affinity, meaning everything that might indicate a carpenter's blueprint or an intelligent designer's template adapted to different species, as due simply to common ancestry. Likewise, taxonomic descriptions and classifications, also compiled mostly by non- or anti-evolutionists and culminating in Linnean systematics, slotted into place as genealogies. And the third major category of labours by the anti- and non-evolutionists was the fossil record and its scale of deep time. Palaeontology presented Darwin with a massive compendium of the succession and history of life on earth and the insights of half a century's wrestling with the discipline of geohistory. Fossils revealed the rise and fall of major groups of organisms. (Admittedly, the opening chapters on early life were missing.) In return, palaeontology received from Darwin and Wallace a comprehensive explanation of that document, a great boosting of the discipline.

Organic evolution was simultaneously: (1) an all-pervading fact of life on earth, (2) a scientific supertheory and (3) a research program of limitless potential. But what was accepted after 1859 was evolution as tree of life alone—without natural selection! Owen was more influential in those seven decades of fossil discovery (1860s–1920s) than was Darwin. It took all those decades to complete the Darwinian Revolution of organic evolution by natural selection. Those times have been called the non-Darwinian decades, including even the death of Darwinism. What happened? It is not helpful to classify interested persons simplistically as pro-evolution or anti-evolution. It is too easy to invoke Max Planck's rule, whereby younger scientists accept new scientific ideas with greater alacrity than do their elders; or, in other words, that science progresses one funeral at a time, as the old reactionaries shuffle off the stage, still nursing their antiquated world views. What happened was that only transformational evolution and saltation could be 'seen' in the fossil record, in embryology and in comparative anatomy. All the new knowledge expanded and clarified the tree of life; nothing much seemed to demand natural selection, and the fossil record seemed to be too coarse and fragmented to clarify the evolutionary processes. Meanwhile the pioneering geneticists saw no implications in inheritance (transmission genetics) for natural selection and variational evolution.

The field naturalist Alfred Russel Wallace defended and promoted Darwinism, meaning speciation via variation and natural selection (he saw himself as more Darwinist than Darwin himself), into the twentieth century, at a time when there were very few others.[26] It really was a thin red line. In palaeontology there were very few Darwinians. In our story the most noteworthy was William Diller Matthew,[27] a shining Darwinian exception to the North American anti-Darwinians in his understanding of Darwin's theory of variation and selection, in his grasp of rock strata, ancient environments and deep time as strongly as his grasp of teeth and bones, and in his mentoring of George Gaylord Simpson, who was to usher in the Darwinian Restoration and become the most significant palaeontologist since Cuvier. Simpson's *Tempo and mode in evolution* (1944) was the palaeontological foundation document in the Restoration.[28]

---

26 Wallace (1989).

27 William Diller Matthew (1871–1930) achieved the trifecta, uniquely in vertebrate palaeontology before Simpson: (i) he was enormously successful and respected by his peers in 'palaeomammalogy', namely the anatomy, systematics and phylogeny of the ancient bones and teeth, biological palaeontology; (ii) he was robustly Darwinian in the face of Darwinism's dark days in his grasp of evolution by natural selection in (but not only in) vertebrate palaeontology; and (iii) he was strong and insightful in his biostratigraphy, namely building the North American Land Mammal Ages on the succession of fossil assemblages, geological palaeontology. When kidney disease took him before 60, his bibliography had attained 352 entries. See Matthew, '*Climate and evolution*' (1915 and 1939 with Colbert) and studies by Rainger (1981, 1985, 1986). The biography by Colbert (1993), his son-in-law, portrays satisfyingly a fine and high-achieving human being but says all too little about the actual rich science.

28 *Tempo and mode* (Simpson 1944) was one of four books published on evolution by Columbia University Press between 1937 and 1950. The others were *Genetics and the origin of species* (Dobzhansky, 1937), *Systematics and the origin of species* (Mayr, 1942), and *Variation and evolution in plants* (Stebbins, 1950). This clutch together with *Genetics, paleontology, and evolution* (Jepsen et al., 1949; a symposium volume) and *Evolution, the modern synthesis* (Huxley, 1942) are considered to be the foundation volumes of the Synthetic Theory of evolution which, following Ghiselin, I prefer to call the Darwinian Restoration. In hindsight this rapprochement between disciplines noted for their low levels of intercommunication was deemed worthy of a two-workshop conference in 1974, *The evolutionary synthesis: Perspectives on the unification of biology* (Mayr and Provine, 1980). But palaeontology's shifting Zeitgeist is captured best in Simpson's 1944 'Introduction' to *Tempo and mode* (p. xxvii):

> Not long ago paleontologists felt that a geneticist was a person who shut himself in a room, pulled down the shades, watched small flies disporting themselves in milk bottles, and thought that he was studying nature. A pursuit so removed from the realities of life, they said, had no significance for the true biologist. On the other hand, the geneticists said that paleontology had no further contributions to make to biology, that its only point had been the complete demonstration of the truth of evolution, and that it was a subject too purely descriptive to merit the name 'science'. The paleontologist, they believed, is like a man who undertakes to study the principles of the internal combustion engine by standing on a street corner and watching the motor cars whiz by. Now paleontologists and geneticists are learning tolerance for each other, if not understanding.

And more pungently (p. xxix):

> [The geneticists] may reveal what happens to a hundred rats in the course of ten years under fixed and simple conditions, but not what happened to a billion rats in the course of ten million years under the fluctuating conditions of earth history. Obviously, the latter problem is much more important.

Tempo proclaimed palaeontology as the only four-dimensional science. As to 'mode', Simpson perceived organic evolution at three levels, namely micro-, macro- and mega-evolution, or more clearly in three modes, namely speciation (origin of species), phyletic evolution (change in a lineage through time) and quantum evolution (jumping from one adaptive zone, or lifestyle, to another) (Figure 3.12).

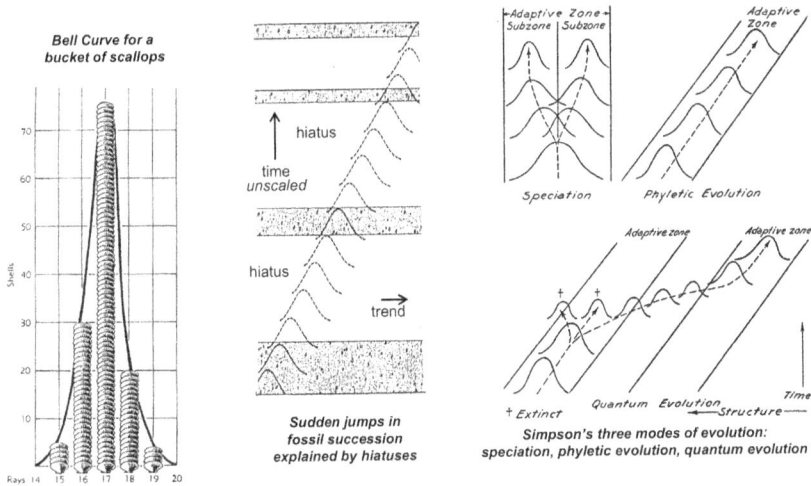

Figure 3.12. The bell curve in Simpson's modes of evolution.

The rays on the shells of scallops (*left*), a character that is easily preserved in fossils, is the example of normal distribution in a large sample, informally known as the bell curve. Thanks largely to Simpson's forceful arguments in 1937, the bell curve became popular as the symbol for variation in populations of a species in the 1940s when population thinking was being urged as the bridging concept between palaeontology, field biology and genetics. Sudden jumps in an upward succession of fossil samples (*centre*) might indicate not evolutionary jumps but gaps (i.e. hiatuses) between sampled strata. *Right*, Simpson's three modes of evolution. The adaptive zone is a species' place and space in the world. In speciation, populations are split, for a wide range of physical and ecological reasons, and henceforth go their separate ways. In phyletic evolution, the shift happens without splitting. Quantum evolution is driven by a major shift. These diagrams are conceptual and Simpson did not scale the time dimension or specify the structural dimension. See Chapter 10.

Source: *Left and centre*, diagrams by Norman D Newell. *Right*, Simpson's three modes (1944, Fig. 31), used with kind permission of Joan Simpson Burns.

Simpson's driving thesis (in his own words three decades later[29]) was:

> that the history of life, as indicated by the available fossil record, is consistent with the evolutionary processes of genetic mutation and variation, guided toward adaptation of populations by natural selection, and furthermore that this approach can substantially enhance evolutionary theory, especially in such matters as rates of evolution, modes of adaptation, and histories of taxa, particularly at superspecific levels.

And in Gould's words it was:

> as close to a 'one-man show' as any movement I know in the history of paleontology. It was idiosyncratic, unique, and surprising, and it both annoyed traditionalists (or left them utterly confused) and inspired a generation of younger workers.[30]

In the rise of transmission genetics and population genetics, and in emphasising variation in populations and species in the field and in the fossil record, the Darwinian Restoration successfully established variational evolution by natural selection. It would take several more decades for evolutionary developmental biology (evo-devo) to fill out the Restoration with a reinvigorated study of bodies, their development and their deep, deep interconnections of more than half a billion years ago (Chapter 10). But our story is going in a different direction—towards the microfossils, low in profile in the first four of the eight surges, then erupting to reinvigorate, even revolutionise, biostratigraphy and lately yielding insights into the environments staging and impelling the evolution of Cenozoic life.

---

29   Simpson (1976, p. 5).
30   Gould (1980, p. 157).

4

Microfossils: Ultimate archives of biogeohistory

The foraminifera: Micropalaeontology in southern Australia

Like life itself, fossils can be classified in various convenient ways. We have encountered the shells of animals without backbones, central but not unique to the development of the fossil-based time scale; we have seen the bones and teeth of vertebrates, central but not unique to the story of organic evolution. Shells and bones are familiar in a general sort of way to most of us. Not at all familiar, indeed, virtually unknown at the popular level, are a vast group lumped as the microfossils, meaning that they demand microscopic study. The microfossils are a very heterogenous group including spores and pollens from plants, parts of some animals especially teeth, and single-celled organisms.[1] This chapter is about the longest-known, single-celled organisms with shells making up a fossil record, the foraminifera (Figure 4.1a).

In the 1940s–1950s, Australian geology was invigorated by the need for comprehensive geological mapping. Mapping was needed urgently for minerals and fuels exploration but for many other reasons too; and both mapping and drilling into the rocks needed control—control in the form of answers to such questions as the rock relationships and the ages of the rocks—and micropalaeontology was called upon for many of the answers.

---

1   I like Maureen O'Malley's (2014) informal use of microbe for the microscopic single-celled organisms, the Bacteria and Archaea, plus the Protista, the single-celled eukaryotes, in contrast to the macrobes, all the multi-celled Eukarya.

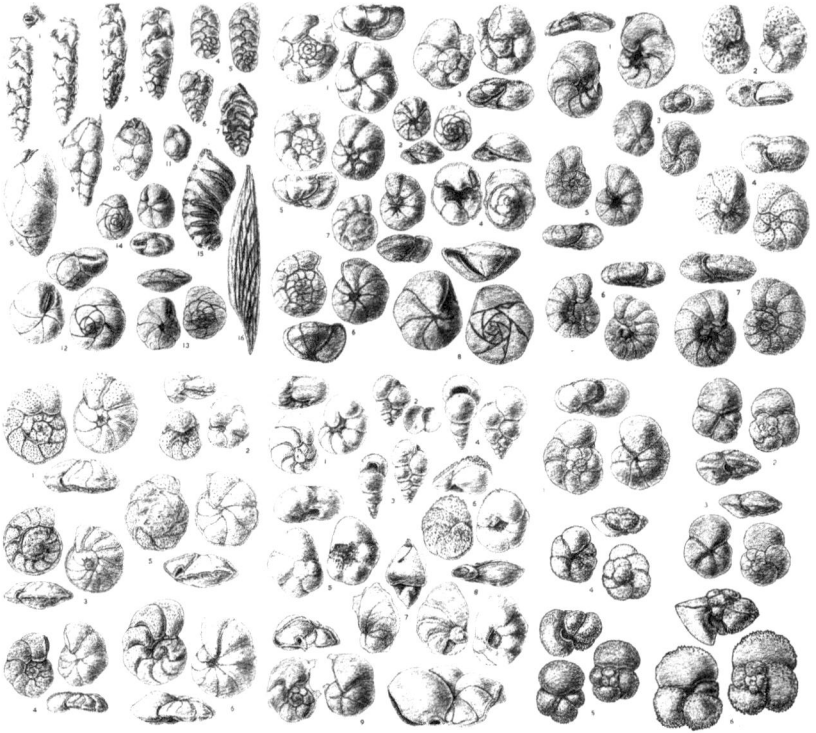

Figure 4.1a. Fossil foraminifer from the Otway coast in western Victoria.

Six plates of portraits of foraminiferal shells from strata across the Palaeocene/Eocene boundary, subsequently discovered to capture the extreme warming event introducing Hothouse Earth. About 120 species in all were found. The specimens are less than 1 millimetre in diameter, drawn under a stereobinocular microscope with camera lucida in one, two or three orientations. Such drawings were among the last to be published as the stereoscan electronmicroscope was about to become widely available.

Source: Author's drawings (McGowran, 1965).

The microfossil group leading the way at the time were the foraminifera, single-celled organisms with shells, more animal-like than plant-like in the binary division of biology into zoology and botany, but actually neither. I had bonded with my nautiloids and the response and encouragement from the international network of nautiloid workers to the publication was highly gratifying; but, so far as I could see, the nautiloids were not progressing the grand narrative of scientific research, quite unlike the foraminifera, which were also in industrial demand. So the foraminifera it was to be for my next project. Martin Glaessner produced a set of laboratory notes and a set of specimens assembled by the late Walter Parr (of whom more below) for a planned second paper—Parr's first having appeared in 1938. The materials came from Pebble Point, on the Victorian coast west of Cape Otway.

I discovered that I needed Parr's collection, used in his 1938 paper, from the Kings Park Shale below Perth in Western Australia. And the third necessity was a field collecting trip to Exmouth Gulf in Western Australia. Common to the three sets was a more or less similar geological age, in the Palaeocene and Eocene epochs in the Palaeogene Period.

I needed to find and extract and assemble the specimens from the sedimentary rock, to manipulate them under a stereo binocular microscope at magnifications of 50 times to 200 times, and to illustrate them. Illustrating primarily meant pencil drawings done with a camera lucida in which the microscopic field of view was superimposed on a white card; photography too often was unsatisfactory because it was not possible to capture a three-dimensional object entirely in focus and unblurred. Shells with trochospiral growth (meaning, growing like a snail) are shown most thoroughly in three views, conventionally above, below and sideways, although two views often will do (and words like top and bottom lack the meaning that they have in orienting animals).

The breakout species in my early work was a foraminifer found below Perth by Parr in 1938, discovered to be new and described as *Globorotalia chapmani*, and dated tentatively as Late Eocene (see Figures 4.1 and 4.2.). (It is now known as *Globanomalina chapmani* [Parr].) The individual foram grew its shell from an initial chamber, in this case adding about a dozen more to attain an adult diameter of 0.5–0.6 millimetres. Two decades later, I possessed a context, not available to Parr, in which to re-examine his species, the context being similar or identical specimens located in space and time elsewhere on the planet. Three things about *G. chapmani* became clear. The species was lurking elsewhere in the world under incorrect names; it was one of the select groups of forams that were planktonic in their habitat; and its presence indicated a Palaeocene age for strata worldwide that contained the species. This was the first robust determination of a Palaeocene age throughout southern Australia (as distinct from a few tentative, weakly supported suggestions). But what goes around, comes around. Just as I could correct Parr's age determination from Late Eocene to Palaeocene, so did I go wrong when comparing Pebble Point with King's Park and needed correcting in my turn. Many of the fossils were common to both, but *G. chapmani* did not look quite right—close and related, to be sure, but not the same. I plumped for dating Pebble Point as a bit older than Kings Park—but that conclusion was upended when our Pebble Point species turned out to be *Globorotalia australiformis*, a species newly described in New Zealand and younger than *G. chapmani*, making Pebble Point a bit younger than Kings Park.

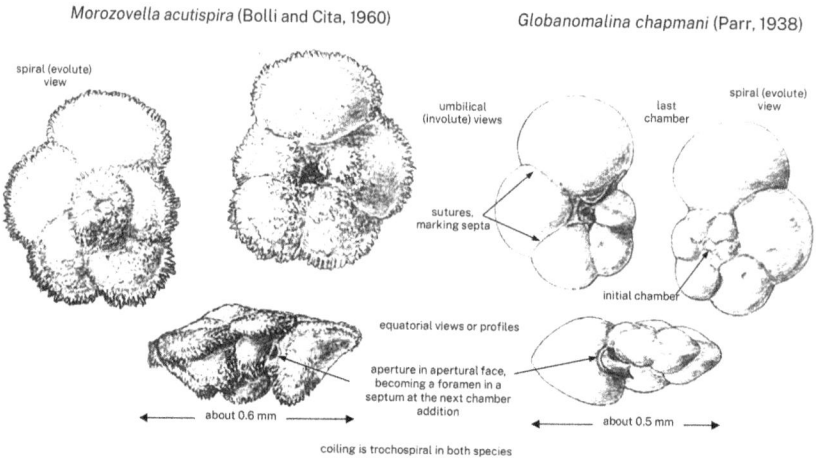

Morozovella acutispira (Bolli and Cita, 1960)

Globanomalina chapmani (Parr, 1938)

spiral (evolute) view

umbilical (involute) views

last chamber

spiral (evolute) view

sutures, marking septa

initial chamber

equatorial views or profiles

aperture in apertural face, becoming a foramen in a septum at the next chamber addition

about 0.6 mm

about 0.5 mm

coiling is trochospiral in both species

Figure 4.1b. The trochospiral shell of many foraminifera, in three standard views.

Two specimens of planktonic foraminifera, Palaeocene in age, from the Boongerooda Greensand in Western Australia. Trochospiral shells ('like a snail') are seen most completely in three views: along the axis of coiling from both the evolute (spiral side) and involute (umbilical side) directions, and sideways, or parallel to the axis, or the equatorial view. (One or two views are sufficient for some purposes.) The shell is grown chamber by chamber.

Source: Author's drawings.

That study was a modest contribution to the grand global project of correlation and age determination—of getting the ages right, upon which all of geohistory and biohistory rests. But what about the actual tools of this trade, the discrimination of the fossil species themselves in space and time? My first impression of microfossils was of a bewildering array of forms, of variety or diversity, and of the sheer numbers available. Where the available specimens of the nautiloids were very few, the foraminifera came in their hundreds to thousands. The micropalaeontologist learns the trade by sorting the small specimens washed from a sample of mud or mud-rock into groups, which look like distinct species, and mounting the sorted groups with water-soluble adhesive on an assemblage slide. My second impression then was of the variation to be seen within each species.

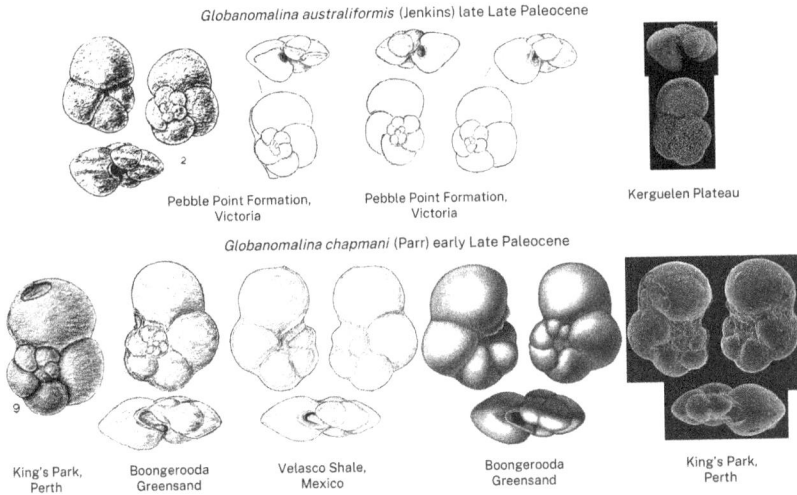

Globanomalina australiformis (Jenkins) late Late Paleocene

Pebble Point Formation, Victoria   Pebble Point Formation, Victoria   Kerguelen Plateau

Globanomalina chapmani (Parr) early Late Paleocene

King's Park, Perth   Boongerooda Greensand   Velasco Shale, Mexico   Boongerooda Greensand   King's Park, Perth

Figure 4.2. Two species of *Globanomalina* in the Late Palaeocene.

Illustrating two Palaeocene species. *Right*, two stereoscan figures, which made camera lucida drawings (all the others) largely superfluous. The drawings are the author's, except for one specimen from the Boongerooda Greensand, easily distinguished by its professionalism, which was by Lawrence Isham at the US National Museum.

Source: *G. australiformis*: drawings are from McGowran (1965), and stereoscan (*far right*) from Olsson et al. (1999, Pl. 33, Figs 10, 11). *G. chapmani*: holotype (*far left*, McGowran, 1964); Boongerooda and Velasco specimens by the author (unpublished); second Boongerooda, from Berggren et al. (1967); and the Kings Park stereoscan from Haig et al. (1993, Pl. 2, Figs 15–17).

The two faces of a species, variety and variation; simultaneously the same yet different; I was becoming aware of some misfit here, some cultural dissonance. On the one hand, the study of fossil foraminifera was being driven and funded by the practical needs of geology and the economic geology of minerals and fossil fuels exploration and exploitation. It seemed intuitively reasonable that finely delineated species could allow finely delineated correlations and ages of strata—precision in species determination would lead to precision in geology and to precision in communication between palaeontologists. Hence arose the practice of 'splitting' in biosystematics, of carving up the variation to narrowly define and name the 'species' in the service of geology. On the other hand, my head was full of the 'new systematics' which was part of the mid-twentieth-century resurgence of Darwinian biology. The new systematics was less concerned in defining species than in discovering them in nature, variation and all—for the variation within a population was intrinsic to evolution by natural selection. So I had to make sense of this clutter of specimens from the Palaeocene of Australia the only way I knew how—by drawing and assembling clusters of similar specimens (Figure 4.3.).

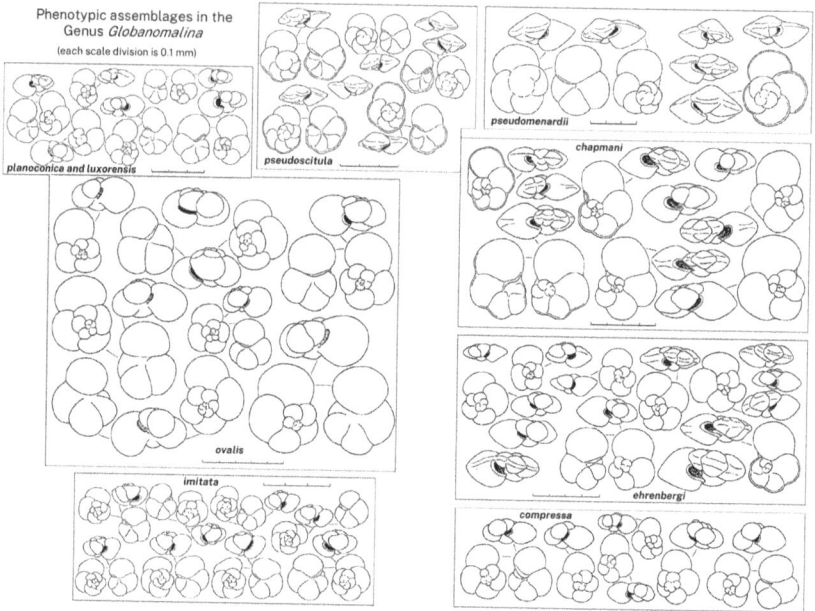

Figure 4.3. Assemblages of species in the Genus *Globanomalina* (scale division 0.1 mm).

These are collections of specimens, assemblages, each assemblage from a single sample of sediment. They are microfossils: note the scales. The lines link views of the same specimen. The outline drawings in about 1960 are intended to show (i) variation within the assemblages and (ii) differences between the assemblages. There is overlap between the assemblages, making it problematic to identify a single isolated specimen as of one or another species. The strongest example of overlap is between *Globanomalina pseudomenardii* and *G. pseudoscitula*. Which raises the urgent question: what are the ages of these assemblages? How do they fit into a time frame? See Figure 4.4.

Source: Author's drawings; McGowran (2005a).

The clusters could be arranged in chronological order to construct a likely genealogy, or theory of evolutionary relationships in what was to be renamed the genus *Globanomalina* in due course. Now, compare the simple outline sketches of *ehrenbergi* with *australiformis* (Figure 4.4.). Very similar though they are, they are not directly related, at least according to this genealogical theory. Likewise, *pseudoscitula* looks very much like *pseudomenardii*. The theory claims that the *planoconica–australiformis–pseudoscitula* cluster is related to the preceding *ehrenbergi–chapmani–pseudomenardii* cluster only by way of the diverging of *planocompressa* from *archaeocompressa*, millions of years before.

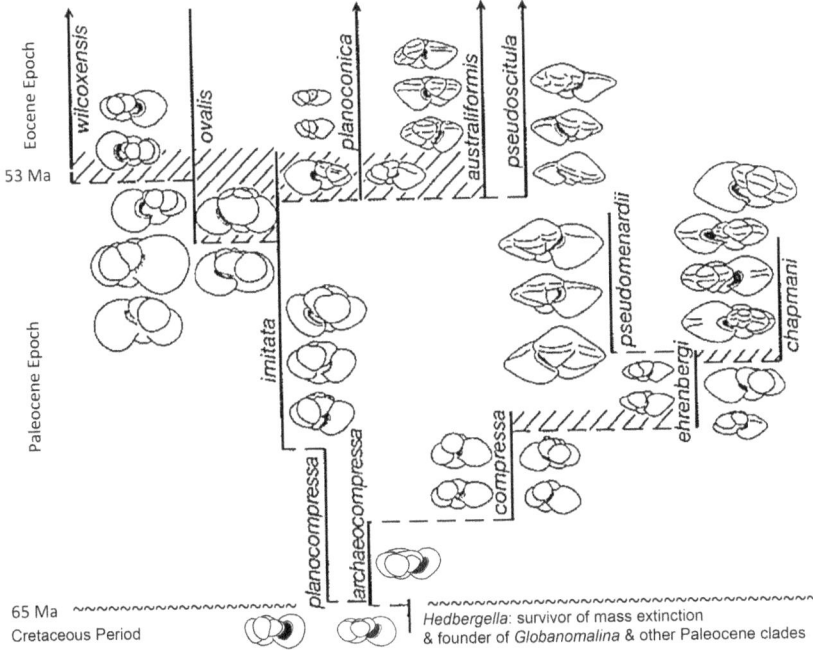

Figure 4.4. Species of *Globanomalina* interpreted as cladogenesis, or family tree.

The assemblages' given species names are plotted against a time scale. The founder species of *Globanomalina* arose from *Hedbergella*, a survivor of the catastrophe and mass extinction at the end of the Cretaceous; it speciated during the Palaeocene; it looks as if there was a cluster of speciations at or near the end of the Palaeocene. This could explain why *G. pseudomenardii* and *pseudoscitula* are so similar: the shells are simple, the range of observable characteristics is limited, selection pressures surely will recur, so repetition of bodyplans through time is likely. This allochronic evolutionary convergence is indeed a common occurrence among the planktonic foraminifera. And the timing, onset of Hothouse Earth at the end of the Palaeocene, is suggestive (Chapter 7).

Source: McGowran (2005a).

Among various inferences that might be made here, I mention three. One is the repetition through time in the similarities of the shells, a theme that will recur in this narrative. The second point is a localised and personalised example—I got the Pebble Point age wrong by confusing *australiformis* with *ehrenbergi*.

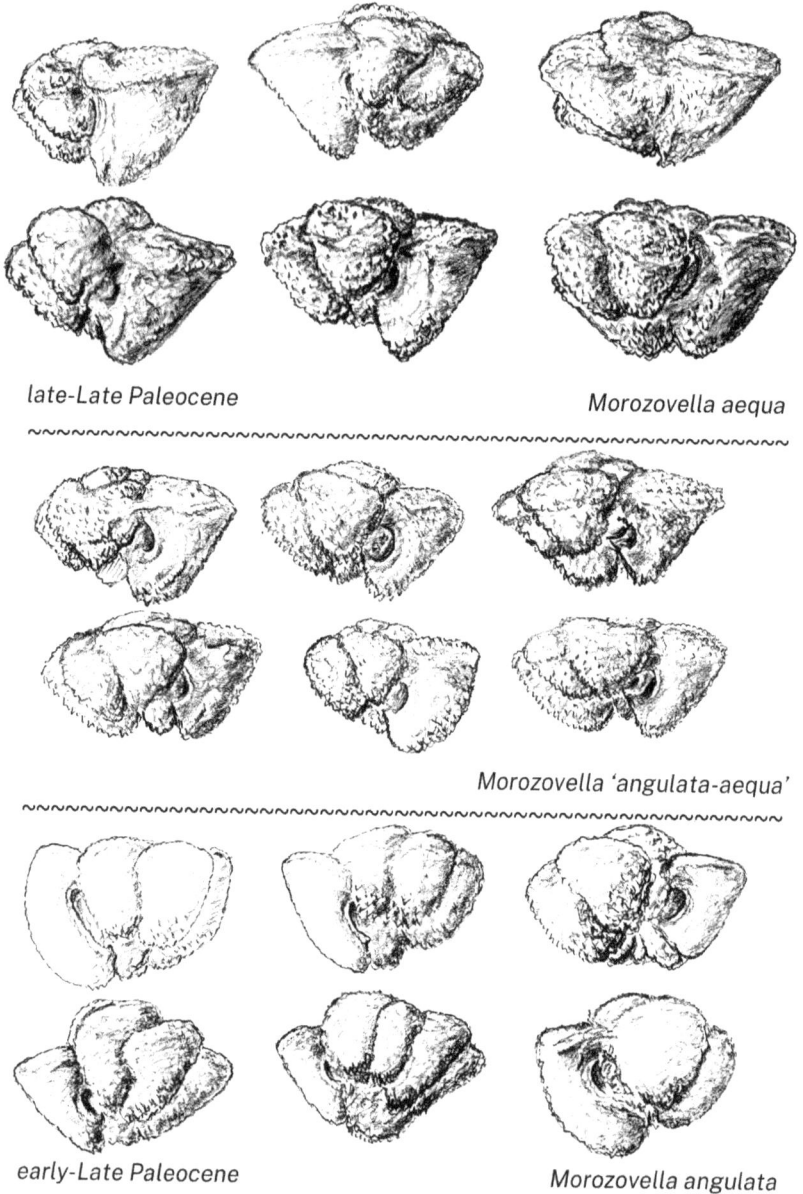

late-Late Paleocene   *Morozovella aequa*

Morozovella 'angulata-aequa'

early-Late Paleocene   *Morozovella angulata*

Figure 4.5. *Morozovella angulata-aequa* interpreted as anagenesis.

Biostratigraphers have intuitively believed that the narrower the species concept, the sharper the tool for getting the age of the strata. That has meant accepting that the 'morphospecies' will be an artificial slice of the 'real' or 'biospecies'. Instead, I tried sorting specimens in samples of slightly different ages, and found that the range of variation shifted through time, as indicated in these equatorial views in 18 camera lucida sketches.

Source: Author's depiction.

The third point about the reconstructed genealogy of *Globanomalina* is that it is a theory of cladogenesis, or branching evolution. The diagram implies that a species arises, proceeds to exist without changing the form of the shell very much, and then goes extinct. This pattern in time is the ultimate basis for the theory of biostratigraphy, namely, that strata can be dated by their contained fossils—a stratum can be no older and no younger than the life span of the contained fossil species. This theory predated Darwin's *Origin* by half a century. But what about another well-established evolutionary principle, namely, that species vary and the range of variation shifts through time? Competing with the between-species theory of cladogenesis is the theory of anagenesis, meaning the within-species shifts through time. Illustrating this is another flurry of pencil drawings, of specimens variously identified as the species *angulata* and *aequa* (as shown) and as at least three others (not shown). In this example I tried and failed to show that the morphospecies as named in the literature fell into consistent separate clumps—they intergraded too much! But the range in form of the shell shifting though time could, it seemed, be given two legitimate or correct names with an informal or unofficial intermediate (Figure 4.5).

The slate is never blank

When Richard Feynmann's first seminar as a postgraduate in Princeton was looming, he heard that Wolfgang Pauli would be there, and John von Neumann and Albert Einstein. He got through it, he did well; Pauli tried to put him down, but Einstein politely put Pauli down. In a more modest academic setting but probably as nervous as Feynmann, I gave my first postgraduate paper at a Geological Society meeting in Adelaide. Still to do the field work south of the Exmouth Gulf, I had already found, in a couple of samples supplied by Mac Dickens of the Bureau of Mineral Resources in Canberra, clear evidence of a break, a disconformity representing a large hiatus in time (several million years of time) across the boundary between the Late Cretaceous Miria Marl (with many Cretaceous ammonites) and the Palaeocene Boongerooda Greensand (with *Globanomalina pseudomenardii*); and Martin Glaessner suggested that the discovery was well worth airing. Almost none the audience were familiar with planktonic foraminifera, which encouraged me to hold forth and proselytise quite freely. But Eric Rudd, the professor of economic geology, arose and said that he had studied and mapped those rocks in the early 1930s; that clearly there was no disconformity, no break in the succession; that he had been there and I had not and I was wrong. So Glaessner stood up and politely put him down with a brief and pointed discourse on the biostratigraphic recognition of hiatus and disconformity. In the front row was Mac Dickens, who had supplied the samples; by then I had my tail up and I invited Mac to comment. He was a palaeontologist; he had been to the outcrops and he politely put Rudd down too. Shortly after that meeting I went into the field determined to find that gap, the disconformity that the forams indicated must be there. The field evidence couldn't have been better—numerous ammonites filled with lithified (hardened) Miria Marl and Cretaceous forams had been picked up and redeposited, to become embedded in the Boongerooda Greensand with Late Palaeocene forams, along with lithified slabs of marl. Although the two thin formations run consistently in parallel for tens of kilometres around two low and large anticlines, there was indeed a break, representing several million years. My sharp lesson that evening (beyond the adrenaline rush of an academic stoush) was that we don't begin with solemn observational procedures in scientific method in field and laboratory. We never begin anything with a mind empty of ideas; in the well-known metaphor of child-rearing and educating, the philosophers' 'blank slate' to be written upon never existed.[2] We begin with a head full of, well, hypothesis or prejudice or nonsense; and our heads need a good rinsing in reliable evidence, one way or the other. The pain comes when your beautiful theory is slain by an ugly fact. But I felt no pain, not that time.

So these are the foraminifera!

Chambered coiled shells frozen in rocks were known in classical antiquity, such as in the limestones in the Nile valley where they were known as petrified lentils (from the workmen's lunch) or Pharoah's beans; elsewhere they were seen as petrified coins. But most of the shells discovered since the invention of the microscope were less than a millimetre in diameter. The coiled shells resembling nautiloids and ammonites were identified as miniaturised cephalopods (Figure 4.6). They differed in not having the connecting tube, the siphuncle, seen in *Nautilus*; instead, there was a small opening between the successive chambers, a 'foramen', hence the 'Foraminifera'. When it was noticed in the 1830s that living individuals displayed nothing like an animal's organs, the foraminifera were reassigned to the single-celled Protozoa—still believed to be animals, but animals of 'very low' organisation, or 'primitive'.

---

2   Steven Pinker, *The blank slate* (2002).

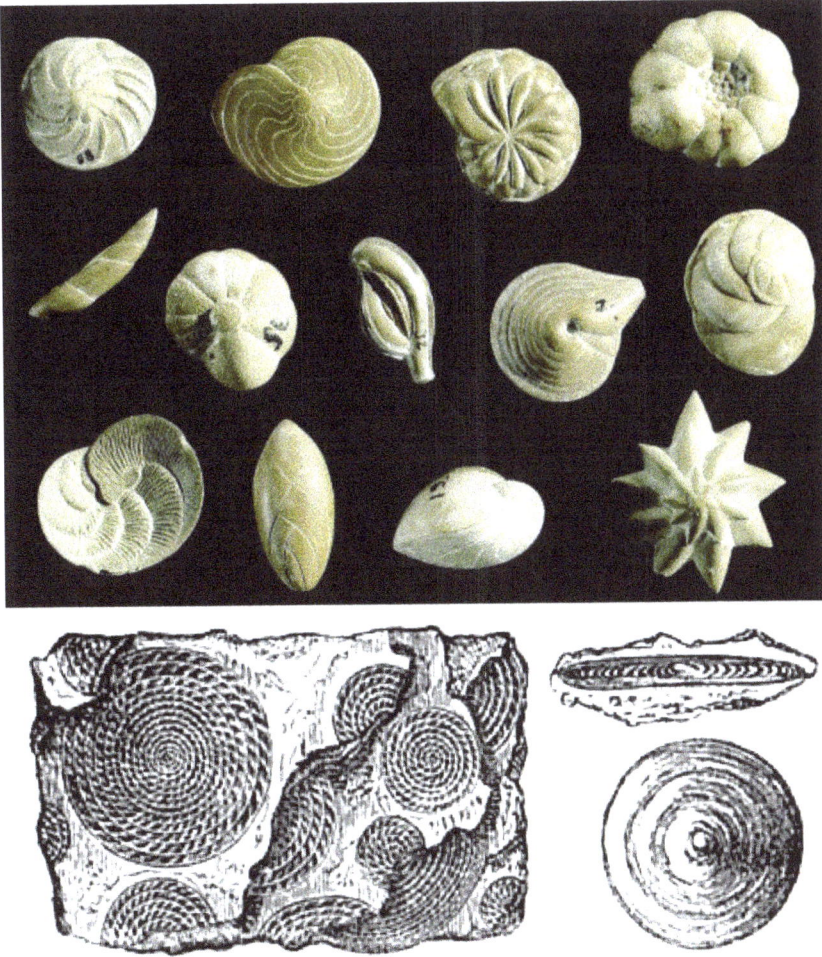

NUMMULITES LÆVIGATA.
(Pharaoh's Beans.)

Figure 4.6. Foraminifera models (*above*) and Pharaoh's beans (*below*).

Above, d'Orbigny's models of foraminifera, helpful for getting the feel of these shells and for people lacking microscopes. *Below*, Pharaoh's beans? Petrified lentils from a stone-worker's lunch? Miniature *Nautilus*-like cephalopods? None of these, but, as realised in the next century, foraminifera, employing the strategy of photosymbiosis. These shells, measured in millimetres to centimetres, were packed densely into Eocene limestones. The shells were packed during life with photosynthesising microbes, possibly cyanobacteria or naked diatoms of dinoflagellates (Chapter 7).

Source: Models, Marie-Thérèse Vénec-Peyré (2004), with permission; beans, Hugh Miller's *Testimony of the rocks*.

Being highly observable, rocks, shells, bones and leaves can be assimilated into popular culture and general knowledge. Needing a microscope, the study of microfossils has been below the popular radar for two centuries and more. Living, recently dead and fossil, foraminifera are available in their thousands and millions, and one might think that their potential as bearers of information on the environment, the age of strata, and the deep history of life on earth was correspondingly apparent as those disciplines were expanding in the nineteenth century. But the potential was to remain unfulfilled for much of the century. To be sure, there was much to be done by way of description, of distribution geographically and ecologically, and of classification, but there was more to the lack of progress than that.

Fit to rank proudly with the French scientists already mentioned—Lamarck, Cuvier, Prévost, Deshayes—was Alcide d'Orbigny (1802–1857), in due course to be dubbed the father of micropalaeontology and founder of biostratigraphy. His father, naval surgeon and naturalist, introduced the boy to microscopy and to the wonders of the multitudinous foraminifera in the modern muds and, living in a port town in France, young Alcide had the nous to ask mariners and naturalists to bring him muds from ports in far distant lands. As his interests broadened and his collections extended back into the fossil record, he became a prodigious publisher of taxonomic palaeontology. But his collecting, describing and naming of so many new species was driven by a vision: a vision of a geological time scale built of a succession of stages in the development of life on earth.

He perceived the stages as natural assemblages of fossils, distinguished one from the other by rather sharp, not to say drastic, changes (or transitions, or turnovers). Where Darwin (after Lyell) regarded breaks in the fossil succession as failures of the record removing critical evidence of descent and relationship, d'Orbigny sought the breaks out as intrinsic to the succession, as essential components of the grand temporal tapestry. Less like Lamarck and more like Cuvier, he was more interested in the discontinuities in life and sudden extinctions than in the origin of species and their gradual replacement in a continuum. To fulfil his vision he assembled a huge collection of fossils from France and abroad, from his own efforts and more than 200 donors. Between 1824 and 1860 he published three massive and seminal works, *Paléontologie Française, Prodrome de Paléontologie Stratigraphique* and *Cours Élementaire de Paléontologie et Géologie Stratigraphiques*. As for the foraminifera—a minor part of his output—he illustrated them; he modelled them (to make them more accessible to those lacking microscopes); he described them (claiming in 1826 to have increased the count of known species from less than 100 to 600–

700; hundreds more were still to come, modern and fossil); and he established their first comprehensive classification. And then there was his field work, pithily acknowledged by Darwin thus, in his *Beagle* journal of 1839:

> When at the Rio Negro, we heard much of the indefatigable labours of this naturalist. M. Alcide d'Orbigny, during the years 1825 to 1833, traversed several large portions of South America, and has made a collection, and is now publishing the results on a scale of magnificence, which at once places himself in the list of American travellers second only to Humboldt.[3]

Two wives and this stupendous output from his 'indefatigable labours' later, d'Orbigny died at 54. But reaction before and after his death to his accomplishments was hostile and multi-pronged. French colleagues, especially Deshayes, were abusive about his foraminiferal taxonomy. So were the British. More broadly, colleagues in zoology and botany thought that palaeontology was not a science, comprising merely decayed fragments of zoology and botany. His modern champion, Marie-Thérèse Vénec-Peyré, has pointed out (in very modern terms) that d'Orbigny was a victim of his avant-garde belief that palaeontology must be both multi-disciplinary and conscious of its place in both the global geosphere and the biosphere.[4] By frequently using the ambiguous word 'creation', d'Orbigny acquired the various reputations of being a creationist in the biblical sense, and a catastrophist in the anti-uniformitarian sense, the sense that Lyell had set out to overcome in *Principles of geology*. It did not enhance French self-esteem that by 1859 two British beetle-collectors had trumped the might of French and German science in establishing the fact and theory of organic evolution. In Cuvier and d'Orbigny they (the French) had two convenient scapegoats depicted as getting organic evolution disastrously wrong with their catastrophist philosophies, thereby damaging the national intellectual and cultural brand. And still we are not done with this gloomy narrative. Digging for fossils was revealing the existence of dinosaurs, the rise of the land plants and in due course the flowering plants, the genealogy of the horses, the links predicted between reptile and bird and man and monkey, and much more in the way of progress in the development of life on this earth. Meanwhile, the multitudinous forams seemed to be going nowhere, as invisible intellectually as physically. D'Orbigny accepted the discovery that they were single-celled protozoans; but Darwin's Glossary[5] tagged the

3   Darwin (1959, p. 88).
4   Vénec-Peyré (2004); Lipps (2002); Laurent (2002).
5   In the Sixth London Edition of the *Origin* (Library of Liberal Classics, New York, 1898).

Protozoa as the lowest division of the animal kingdom, and within them the foraminifera as a class of very low organisation; and Carpenter capped the tale by insisting that the foraminifera have made no progress at all since Precambrian times. Carpenter was an outspoken critic of d'Orbigny's work and he had Darwin's ear. No matter, said Darwin: the foraminifera were ideally fitted for the simple conditions of life. There was no need for them to go anywhere, evolutionarily speaking, no need to make any progress at all. As well as severely distorting d'Orbigny's legacy, all of this cast a severe nineteenth-century chill on the potential of this richly diverse and abundant record of the fossil foraminifera for biostratigraphy in economic geology and geohistory, and for organic evolution and biohistory.

On a piece of chalk — and a new realm of earthly science

In 1868 Thomas Henry Huxley, attending the meeting of the British Association for the Advancement of Science, gave a lecture *On a piece of chalk* to 'the working men of Norwich'. In his plain and compelling prose Huxley took his audience downscale to the multitudinous microscopic skeletons of planktonic organisms making up the chalk. He took the audience upscale to the chalk mass exposed in 'that long line of white cliffs to which England owes her name of Albion', and signifying the sea flooding across much of Europe and lands beyond during the age of the dinosaurs and giant marine reptiles. He went further—he drew a comparison of the ancient chalks with the recently discovered *Globigerina* ooze carpeting extensive tracts of the modern deep Atlantic Ocean. Huxley's agenda was that the world was much older than the years allotted biblically; that there was a genuine continuity in the stream of environments and life from geological epochs past to the present day, uninterrupted by vast catastrophes; that life on earth has evolved (this was less than a decade after *On the origin of species*); and—most importantly!—that the excitements of evolutionary theory and deep-ocean exploration were quite accessible for the interested population at large, such as the working men of Norwich. Published, Huxley's lecture became famous.[6]

---

6   And rightly so. It was Huxley at the top of his form as lecturer, populariser, orator, advocate. *On a piece of chalk* was published as an essay in 1868, included in various collections, and issued as a book of that title a century later with an Introduction by Loren Eiseley and a lavish assemblage of halftone illustrations by Rudolf Freund (see Eiseley, 1967). These are excellent, with the unfortunate exception of *Globigerina bulloides*, which gave its name to the *Globigerina* ooze of the deep sea but here looks more like what Huxley called 'a badly grown raspberry'.

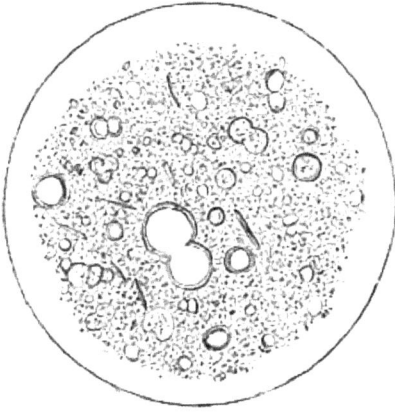

FIG. 88.—Microscopic section of chalk from Sussex. Magnified about 220 diameters.

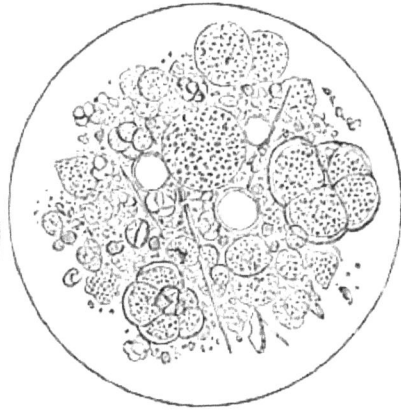

FIG. 89.—Atlantic ooze from a depth of 2,250 fathoms. Magnified about 220 diameters.

Figure 4.7. Huxley's comparison of Cretaceous chalk and Atlantic ooze.

TH Huxley compared the newly appreciated *Globigerina* ooze from the depths of the North Atlantic Ocean (at 2,250 fathoms) with the chalk of Cretaceous age in southern England. The sediment actually is mostly coccoliths and better called calcareous ooze, although the various planktonic foraminifera are more obvious at low magnifications.

Source: Huxley (1878, Figures 88 and 89).

The intellectual adventures deconstructing earth and life history hitherto were based on the shells accumulating in the shallow seas (the neritic realm) and the bones and plants in strata of lakes and wetlands (the terrestrial realm). The peoples of maritime tribes and nations were of necessity well aware of the global pelagic realm and its moods, but the deep ocean was dark and mysterious (even today, people compare the deep ocean knowledge-wise with the dark side of the moon). Huxley was excited by the recent reports of a fine-grained, sticky sediment from the depths referred to as 'ooze', which appeared to be dominated by *Globigerina*, a genus of foraminifera named by d'Orbigny. He was also excited by the strong microscopic similarities of the *Globigerina* ooze with the Cretaceous chalk on Europe and other continents (Figure 4.7).

Did the similarities imply similar environments? Was the chalk actually not of the neritic realm at all, but of the pelagic realm? If so, had the Cretaceous seafloor been hoisted upwards or the modern seafloor depressed? And was modern *Globigerina* truly a planktonic organism in the surface waters or actually a benthic organism living down in the muds?

With too little evidence available at the time, Huxley guessed wrong on some of these questions. No matter! Much more important, he was getting people's attention, and he and several other scientists of influence succeeded in having Britain correct that ignorance. Britannia had been ruling the waves since Nelson at Trafalgar but others, Germany and the United States, were becoming interested in ocean science. The British responded hastily and the Royal Navy vessel HMS *Challenger* circumnavigated the world between December 1872 and May 1876 in the first systematic, grand-scale, scientific investigation of the global ocean. The *Challenger* was a wooden, steam-assisted sailing ship and most of its gun bays became scientific laboratories, workrooms and storage space. It had five scientists, an artist, 20 officers and about 200 sailors. Covering 68,890 nautical miles from 1872 to 1876, the ship visited 362 stations. The expedition gathered an enormous body of data which was worked up and published in 50 volumes between 1880 and 1895. In greatly expanding our knowledge and raising more questions than it answered, this feat established the modern science of oceanography. Included in this accomplishment was detailing the fauna of the deep sea, the largest biotope on the face of the Earth, and our interest here focuses on Henry Bowman Brady's 1884 *Report on the Foraminifera dredged by HMS Challenger*.[7] Inspect the numbers: descriptions, figures and distribution data on 915 species of foraminifera (15 per cent of the total number of extant species) in 368 genera (44 per cent of extant genera) in 814 pages and 116 colour plates.

In a quantum leap Brady's endeavours opened up two areas of knowledge of the modern Earth, in turn enabling a deeper understanding of Cenozoic history. Broadly, the areas are oceanic ecology and environment and oceanic biogeography. 'Bathymetry' refers to depths, and 'palaeobathymetry' to depths in times past. The ocean spills across the margins of the continents as shallow seas on continental 'shelves' (the neritic), to an extent varying through time. More comprehensively than any other group of organisms, the shelly foraminifera enabled a division into inner, middle and outer shelf. Out to sea beyond and below the shelf is the continental slope, which could be divided into upper, middle and lower bathyal zones. And distinct again were the faunas of the ocean floor proper, the vast abyssal plain. Thus we had,

---

7   Brady's massive 1884 achievement on the *Challenger* foraminifera has had much attention in the ensuing decades, unsurprisingly so, as their value to the study of modern and ancient oceans has grown, and our understanding of their systematics—identifying and classifying all those species—has progressed accordingly. Robert Wynn Jones republished the plates with comprehensive revision of the species and an informative introduction and background (Jones and Brady, *The* Challenger *Foraminifera* [1994]).

thanks to Brady and the *Challenger*, seven bathymetric zones spanning the neritic and pelagic marine realms. Brady (1835–1891) himself had already interpreted a fossil foraminiferal fauna from Fiji as having lived at depths of perhaps 200 fathoms (~350 m), and another fossil fauna from Barbados as having lived on the sea-bottom of a depth perhaps of 500–1,000 fathoms (~1,000–2,000 m). And I, having once collected samples from the top of the spine of New Ireland (Papua New Guinea), found that their contained foraminifera were of the lower bathyal zone, whence they were hoisted veritable kilometres up into the air in the very recent past—mountain building virtually before wide human eyes, indeed. Palaeobathymetry is part of the broader question of palaeoenvironments and their changes in deep time—changes in geography, in advances of the sea across the continental margins, in climatic shifts and in critical ecological factors such as salinity, oxygen and food supply. Ideas on all of these things had arisen in enquiring minds, and all received a huge boost from the expansion of foraminiferal studies in modern and ancient oceans.

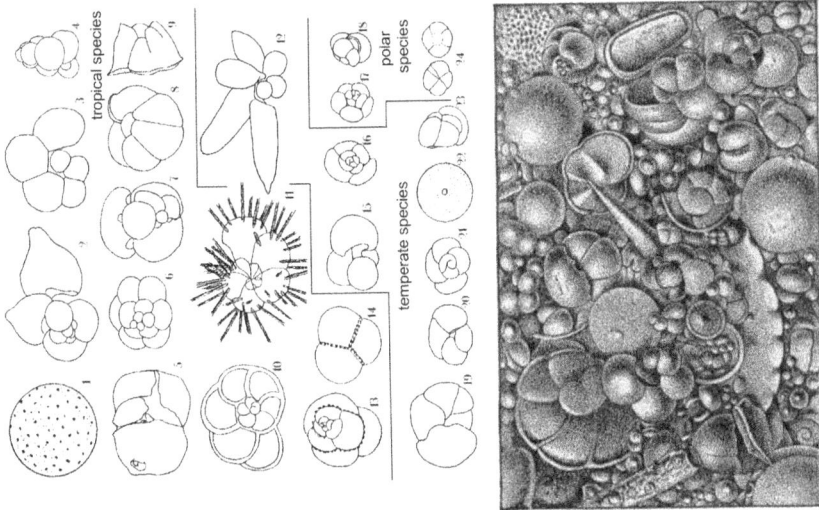

Figure 4.8. Modern planktonic foraminifer, and John Murray's biogeographic identifications.

Left, this (modified) figure from (Sir) John Murray's *The ocean* (1912), his popular account of the *Challenger* expedition (he was one of the three naturalists on board), demonstrates the potential of planktonic foraminiferal biogeography. One could predict that the tropical belt with its more diverse assemblage would expand and contract in response to global warming and cooling. Confirming that prediction indeed became a fertile enquiry in sorting out the ice ages of the Pleistocene Epoch, and in due course in the Cenozoic Era. *Right*, winnowed *Globigerina* ooze, an illustration for an article in *Popular Science Monthly*, 1893–94. by an unknown author citing Murray and Renard (1891).

Source: Murray (1912) and Wikimedia Commons.

Benthic organisms live on the mud or in the top few centimetres, or attached to a stem or a shell. Most species of foraminifera are or were benthic in their habitat. The planktonic species live in or just below the mixed upper waters of the open ocean, also in or just below the photic zone, that is, in areas reached by sunlight. There are many fewer planktonic species than benthic, but they delimit a pattern in their circumglobal biogeographic distribution. John Murray's illustration shows in simplified form a threefold pattern of tropical, temperate and polar species (Figure 4.8.).

Imagine a map with no polar species at all, only temperate and tropical species advancing polewards. That would suggest a time distinctly warmer than the present, prompting in its turn a search for parallel patterns in fossils from the neritic and terrestrial realms. Such was the promise of progress in geohistory and biohistory, thanks to the 1870s *Challenger* voyage giving us glimpses into the vast and mysterious realm of the global ocean.[8]

Mapping and drilling: The twentieth-century demand for foraminiferal services

The twentieth century saw the vast expansion of geological mapping and prospecting for minerals and fossil fuels and water. A major expense in exploration and development was drilling, and it was economically desirable to extract and analyse all possible information as to the ages of the strata drilled, their environments of deposition, and their subsequent geological history, such as faulting and folding and metamorphosing (stressing and cooking). Microfossils, beginning with the foraminifera, were in demand!

The global-scale habitats of certain foraminifera imply a biostratigraphic versatility not possessed by the founding fossils of the discipline of dating and correlation, such as the molluscs in the neritic realm, where we left things in Chapter 3. And that versatility was increased as the planktonic foraminifera, floating way above the mud, drifted from the oceanic ocean into the neritic realm. Many died and dropped to the bottom; many others were ingested by swimming predators of all sizes and the ensuing rain of turds emplaced the shells in a wide range of marine environments. And any number of entombed specimens were available to the micropalaeontologist simply by washing samples from outcrop, dredge or drillhole.

---

8   Richard Corfield, *The silent landscape: In the wake of HMS Challenger 1872–1876* (2004).

M. F. GLAESSNER

I II III IV V VI VII VIII IX X XI XII XIII XIV XV

Globigerinella aspera
Globigerinella micra
Globigerina pseudobull.
Globigerina compressa
Globigerina cretacea
Globigerina bulloides
Globigerina bulloid. var.
Globigerina inflata
Globigerinoidea conglob.

membranacea
angulata
aragonensis
aragonensis var.
crassata
crassata var.
pseudoacitula
pseudoacitula var.
aff. crassula
(Globorotalia)

aff. appenninica
marginata
linnei
formicata
contusa
arca
rosetta
stuarti
(Globotruncana)

Schackoina cenomana
Schackoina multispinata
Hantkenina liebusi
Hantkenina alabamensis

Gümbelitria sp.
Gümbelina globulosa
Gümbelina tessera
Gümbelina crinita
Gümbelina budensis
Pseudotextul. elegans
Pseudotextularia eggeri
P. elegans var. acervul.
P. elegans var. varians

Fig. 6. Die Verbreitung der Planktonforaminiferen in der Oberkreide und im Paläogen des Kaukasus.
I Cenoman, II Turon, III Turon? IV Emscher, V Santon, VI Untercampan, VII Ober-campan, VIII Maastricht, IX Dänische Stufe, X Paleozän, XI Untereozän, XII Mitteleozän, XIII und XIV Obereozän, XV Unteroligozän. Die Altersangaben für die Zonen II, IVb, VII, VIII, XIII, XIV und XV beruhen auf Funden von Makrofossilien, in allen finden sich reiche bezeichnende Foraminiferenfaunen.
- - - - - Vorkommen unsicher, — · — · — Vorkommen in anderen Gebieten.

Этюды по Микропалеонт. I, 1   1937   Studies in Micropaleont. I, 1

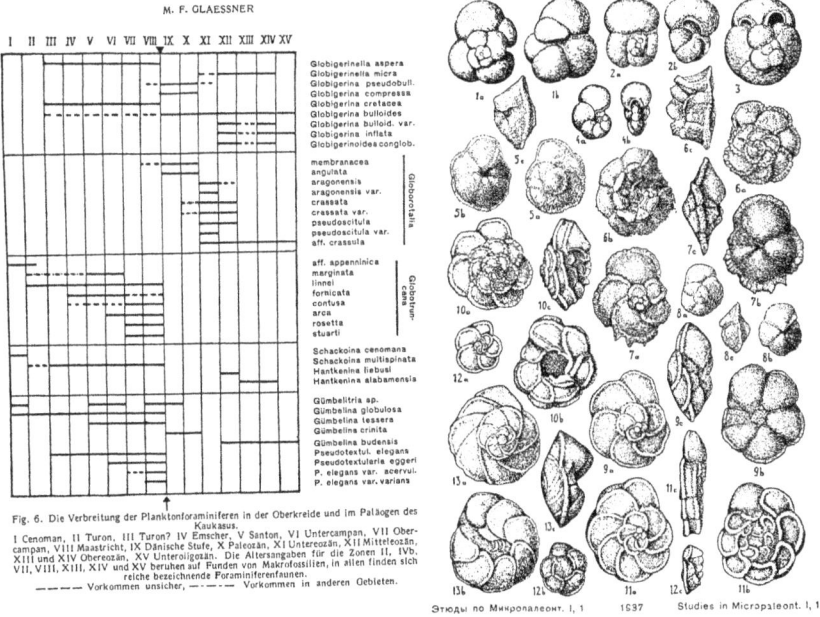

Figure 4.9. Martin Glaessner's biostratigraphy, Cretaceous to Palaeogene, Caucasus.

Micropalaeontological data were accumulating in the earlier decades of the twentieth century, in company and governmental files wherever there was petroleum exploration and development, but Martin Glaessner's range chart of planktonic foraminifera was the major advance of the 1930s. It was compiled from intensive exploration in the Caucasus, but thanks to World War II, the Iron Curtain and the Cold War, it never attracted the recognition it deserved. The incomings and outgoings of species, assembled from many outcropping and drilled sections of strata, were used to divide the succession into biozones I to XV (oldest to youngest). I added the arrow at the zone VIII/IX boundary which is the Cretaceous/Palaeogene boundary. Most of the species living in the upper waters of the oceans at that time disappeared suddenly. But this spectacular oceanic event was ignored for decades until the extinctions of the (terrestrial) dinosaurs and the (especially neritic) ammonites, glimpsed as far back as the eighteenth century (Buffon and Cuvier), became scientifically popular. For a modern view of this oceanic event, see Chapter 10.

Source: Glaessner (1937a) and McGowran (2005a, 2013c).

Scrutinise two figures from a study by Martin Glaessner,[9] who was invited to Moscow in the early 1930s to establish research laboratories supporting petroleum exploration in the Caucasus and Crimea. The first figure (Figure 4.9) summarises the biostratigraphic succession divided into zones. Looking cool and objective, this range chart was the outcome of field work then prolonged searching, sorting and identifying microfossils from the samples under the microscope, and compiling, testing and checking the ranges of the species through the strata.

---

9   Glaessner (1937a).

The 15 fossil-based divisions (zones I–XV) younging to the right comprise eight Cretaceous, two Palaeocene, four Eocene and one Oligocene in age. This study in Glaessner's laboratory in Moscow in the 1930s was a particularly impressive piece of work, anticipating the Caribbean studies which were synthesised after the war. Note the great change where I have inserted the arrow. Here Glaessner also anticipates by half a century the scientific furore over the mass extinctions involving the ammonites at sea and the dinosaurs on land, and the bolide hypothesis. The second figure (Figure 4.10) displays Glaessner's interpretation of the evolution of the genus *Globotruncana*, the first such serious contribution from the foraminifera to such efforts in phylogeny, so many decades after the reactionary setbacks of the mid-nineteenth century.

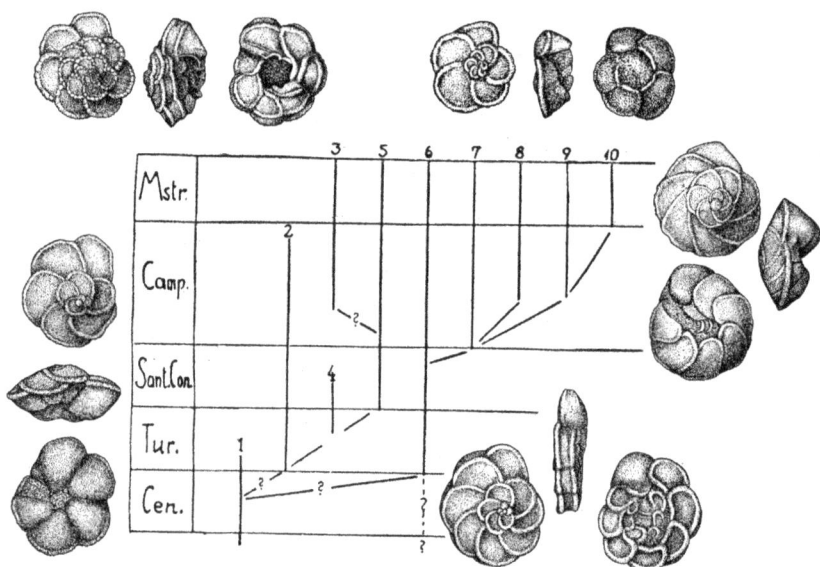

Die hypothetischen Verwandtschaf tsbeziehungen der Globotruncanen. 1—Gruppe der *G. appenninica*, 2—*G. marginata*, 3—*G. contusa*, 4—*G. cf. contusa*, 5—*G. fornicata*, 6—*G. linnei*, 7—*G. arca*, 8—*G. rosetta*, 9—*G. stuarti*, 10—*G. conica*.

Figure 4.10. Glaessner's suggested phylogeny in *Globotruncana*.

Glaessner also drew a range chart for the genus *Globotruncana*, displaying the extension though time of the contained species (the divisions Cenomanian to Maastrichtian are stages of the Late Cretaceous Period). He turned the range chart into an inferred tree-of-life chart for the evolution and extinction of *Globotruncana*, and this too was a first, after a century of underappreciating the evolutionary richness of the foraminifera. Clockwise from lower left, the illustrations are of the species named and keyed as 1, 7, 8, 9 and 6.

Source: Glaessner (1937a) and McGowran (2013c).

This was one of six papers published by Glaessner in the 'in-house' journals of the Paleontological Laboratory in the Moscow State University.

The Cenozoic epochs were established in Europe on Deshayes's synthesis of the neritic molluscan faunas, as we have seen. But the biostratigraphic correlations broke down where the environments were different, particularly in the Alpine mountain belts region where the Eocene and Palaeocene rocks were thick, dark and lacking the necessary neritic macrofaunas. That problem was solved by the micropalaeontologists in the 1930s, especially Glaessner in Crimea and the Caucasus and Otto Renz in the Apennines. These planktonic foraminifera provided the evidence of the pelagic realm in deep time, of ancient deep oceans now preserved in the cores of Alpine mountain belts—modern oceanography was to become underpinned by palaeoceanography. This is my surge #IV. World War II and the Cold War broke down communication between scientists and there was confusion as to the correct names of species being described in Russia, North America and Europe.[10] In the postwar surge in micropalaeontology[11] and our new appreciation of the pelagic realm, Glaessner's published research never received the recognition it deserved.

Return to Australia

In 1881 the Reverend Walter Howchin migrated to Adelaide, preparing to die of tuberculosis. His expiry took 56 years, during which time he made well-known the two great glaciations with a geological record in South Australia, the Sturtian and the Gondwanaland glaciations, of Neoproterozoic and Permian age, respectively, and he established the stratigraphic succession in the Neoproterozoic–Cambrian Adelaide Geosyncline, the rocks of the Flinders – Mt Lofty ranges. Before that, though, and more significantly than anyone has realised, Howchin wrote the first really insightful paper on southern Australian micropalaeontology.[12]

---

10   Also in some limbo was Otto Renz's study (1936) of these microfossils as exposed in the Apennines in Italy.

11   Berggren (1960); McGowran (2013a).

12   McGowran and Hill (2015).

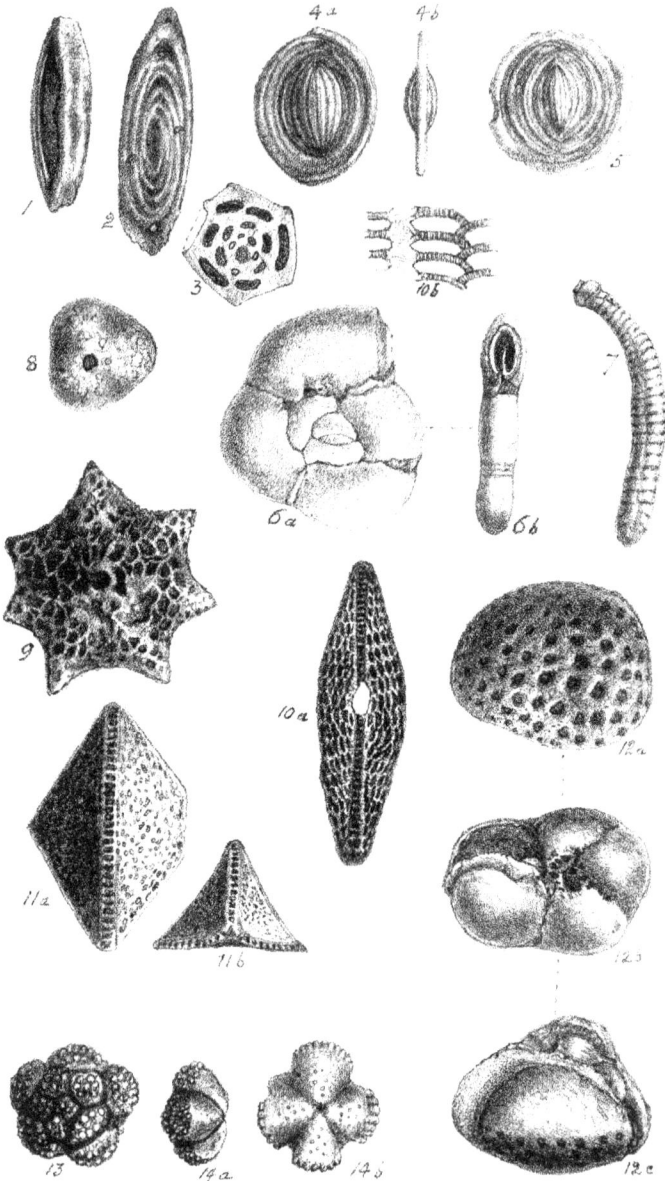

Figure 4.11. Howchin's plate of foraminifera from Muddy Creek.

Walter Howchin came to Australia to die, presumably of consumption, but early in that process (it took half a century) in 1889 he wrote up the foraminifera of Muddy Creek in western Victoria. Howchin spotted their palaeoclimatic significance, especially the species now known as *Lepidocyclina howchini* (9, 11a, 11b; and 10a and 10b as seen in a thin section of the shell); see Chapter 9.

Source: Howchin (1888).

In 1888 he described the foraminifera from Muddy Creek (Figure 4.11) (whence my kangaroo bone) and concluded that the district was warmer at that time. He deserves a generous quote (p. 18):

> With regard to climatic conditions, the majority of the Muddy Creek Foraminifera point to a higher temperature prevailing in the locality of their deposition than is proper to such latitudes in the present day. A very large proportion of species are characteristically tropical, and a decided majority in each case have their geographical range, in the present, restricted to the tropical and warmer temperate zones.

And also (p. 19):

> The general resemblance, which the Muddy Creek Foraminifera bear to the recent species now inhabiting the northern and north-eastern shores of tropical Australia, is very striking, and would appear to indicate that in early Tertiary times either the tropical currents of the ocean bore more directly on the southern shores of the continent, or that the zone of tropical heat reached nearer to the Pole in the Southern Hemisphere, as it appears to have done in the Northern, at the beginning of the Tertiary period.

As we have seen (Chapter 2) Frederick McCoy and Julian Tenison Woods had wrestled unsatisfactorily with the eternal problem of age versus environment, and Howchin's 'early Tertiary times' would turn out to be Miocene instead. But I deem Howchin's paper to be the most perceptive statement about Cenozoic environments to come out of nineteenth-century southern Australian biogeohistory. The key to his insight and confident palaeoclimatic inference was his previous association with HB Brady and his grasp of Brady's *Challenger* foraminifera and their biogeographic implications for the fossil record. But Howchin did not build upon this substantial and ultimately decisive advantage over his molluscan colleagues in solving our age problems in southern Australia. He moved instead into other geology and became broader and broader in his scope, rather than deeper and deeper. So, our problems with the ages of our southern limestones sat in stasis and disrepair, where I left them at the end of Chapter 3, thanks to the lack of macro-fossil species in common with the faunas of the Tertiary epochs on the other side of the planet. This all changed with our knowledge of the foraminifera through the work of Walter Parr in Melbourne and the advent of Martin Glaessner.[13] (Figure 4.12.)

---

13 For concise information on these people, see Quilty (2013); also McGowran (2012, 2013c).

Walter Parr
1894–1949

Irene Crespin
1896–1980

Martin Glaessner
1906–1989

Nell Ludbrook
1907–1995

Alan Carter
1926–1989

Mary Wade
1928–2005

Murray Lindsay
1929–2004

David Taylor
1930–2011

Brian McGowran
1936–

Charles Abele
1937–

John Cann
1937–

Pat Quilty
1939–2018

Qianyu Li
1956–

Graham Moss
1957–

Amanda Circosta
1958–

Stephen Gallagher
1966–

Figure 4.12. Some foraminiferologists in southern Australia.

Some forerunners, mentors, colleagues in southern Australian foraminiferal micropalaeontology.

Source: Parr, unknown; Crespin (unknown photographer: Bartlett, 2006); Glaessner, Mary Wade; Ludbrook, Geological Survey of South Australia; Carter, courtesy of Jean Carter; Wade, courtesy of Sue Turner; Lindsay, courtesy of Margaret and Bruce Lindsay; Taylor, Mary Wade; McGowran, author's copy; Abele, pers. comm.; Cann, pers. comm.; Quilty, Australian Antarctic Division; Li, pers. comm.; Moss, pers. comm.; Circosta, pers. comm.; Gallagher, pers. comm.

Stalin's regime in Moscow demanded that Glaessner either become naturalised or leave. The Glaessners left, returning to Vienna in time for the Anschluss in 1938 and its persecution of the Jews, during which Martin cleaned the barracks windows for the recently arrived *Wehrmacht*. Thanks to a friendship sustained since university days in Vienna in the 1920s, he was hired as a petroleum micropalaeontologist to establish a laboratory in Port Moresby (and then, thanks to the Pacific War, to move to Melbourne) and to finish writing his textbook on micropalaeontology. Being moved on successively by the despots Stalin, Hitler and Tojo within half a decade, the Glaessners thus scored a spectacular if unsought hat-trick.

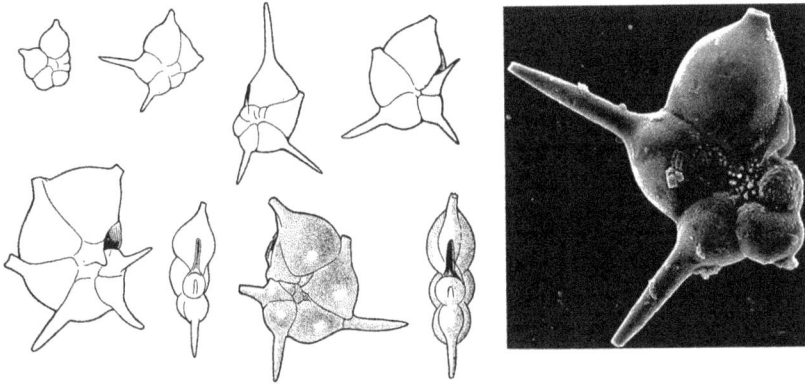

Figure 4.13. *Hantkenina*, Parr's key to unlocking the recalcitrantly provincial Tertiary.

Walter Parr's (1947) discovery of *Hantkenina alabamensis compressa* in western Victoria and south of Adelaide was the most significant foraminiferal paper in southern Australia between Howchin (1888) and Glaessner (1951). With all respect due to our conchologists, Parr demonstrated the Late Eocene age after a century's biostratigraphic struggling in Indo-Pacific provincialism. The stereoscan image (*right*) is from Lindsay (1981) of a specimen from beneath Adelaide.

Source: Drawings from Parr (1947); scanning electron microscope image from Lindsay (1981).

In Melbourne Martin met Walter Parr, who became a friend and highly esteemed colleague for a sadly brief time, terminated by Parr's death in 1949. A successful bureaucrat in Melbourne, Parr was a gifted and productive amateur student of the foraminifera, publishing excellent work from 1926 until his splendid, posthumous monograph on Antarctic foraminifera, based on samples from Douglas Mawson's 1929–1931 expedition.[14] Parr had assembled a large collection of beautifully prepared assemblages of foraminifera and he and Glaessner had made ambitious plans for taxonomic and biostratigraphic research. Parr's key discovery, breaking through the century-long doubts and uncertainties over how our strata fitted together in the Cenozoic time scale, was the genus *Hantkenina* in south-western Victoria (Figure 4.13), and then at Maslin Bay south of Adelaide.

14   Parr (1950).

Glaessner's 1937 range chart shows the species *Hantkenina alabamensis* ranging through the Upper Eocene strata of the Caucasus, and it was also known to be of Late Eocene age in North America and elsewhere. Now we had a peg, a bollard to stabilise the age of the limestones and associated marine sediments in southern Australia. *Hantkenina* was an Eocene index fossil. Under Glaessner's supervision, Alan Carter exploited Parr's superb samples to establish a biostratigraphic succession from the Late Eocene to the Miocene epochs, which not only sorted out our jumble of fossils and strata, but also established links with the wider world. Sure, the benthic foraminifera in this part of the world had many provincial characteristics but numerous species were geographically wide-ranging. The planktonic species led by *Hantkenina* went one better, floating far and wide and establishing links among environments from the deep oceans to the marginal seas, and from the tropics to the higher latitudes. Mary Wade, Nell Ludbrook and Murray Lindsay extended the data base and strengthened those links[15].

Mary also contributed significantly to the evolution of the planktonic foraminifera. Recall that the succession of fossils in strata could build the geological time scale in about six decades without a coherent theory as to why that succession had actually happened in deep time. Recall too that although Darwin supplied the fact and theory of evolution, it was another seven decades before Glaessner produced a persuasive family tree for a group of (Cretaceous) planktonic foraminifera. Beginning in the 1940s attention focused on the evolution of one of d'Orbigny's most prominent species living in the global ocean, *Orbulina universa*, and the potential importance of that speciational event as a worldwide biomarker within the Miocene. By 1956 we could celebrate the ancestral–descendant relationship in what became known as the '*Orbulina* bioseries'. (Figure 4.14.)

For it was one thing to establish the empirical fact that one fossil species consistently succeeds another fossil species in the strata, this empirical configuration giving us a very solid building block in the geological time scale; it was quite another thing to be able to define that block as beginning with a speciation, all laid out in strata distributed from the North Atlantic region to the far south of New Zealand. But that defining rapidly happened, beginning with the reconstruction by Walter Blow. Our two diagrams contrast Blow's fine splitting of the fossil populations into 'morphospecies'

---

15   Wade (1964); Ludbrook and Lindsay (1969).

with Wade's more relaxed view of the intergradation from one identified morphospecies to the next. The same evidence from nature; two perceptions by highly competent micropalaeontologists.[16]

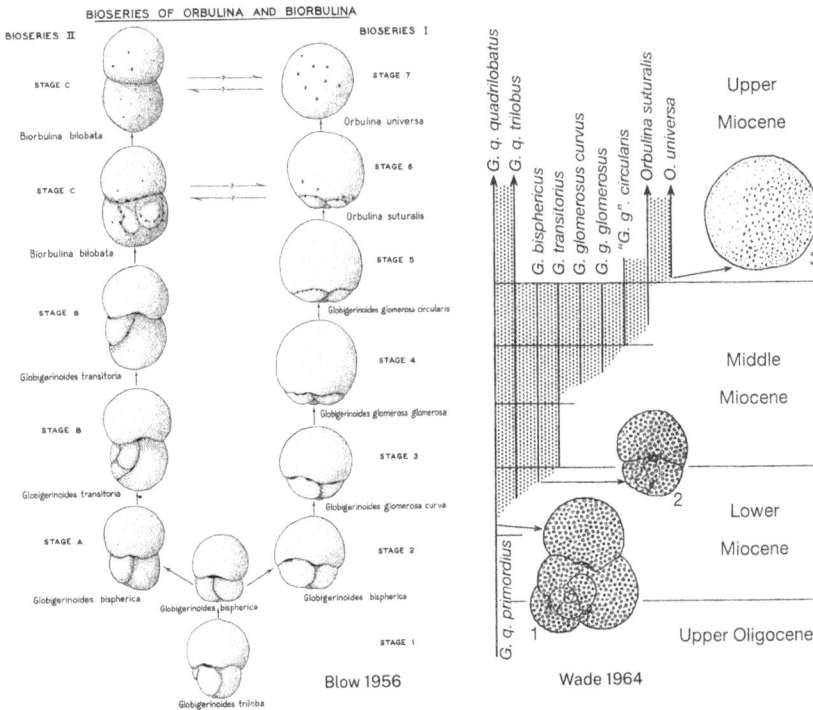

Figure 4.14. Reconstructing the *Orbulina* lineage in two cultures: Two ways of presenting a Miocene lineage of planktonic foraminifera.

In the 1940s two micropalaeontologists in New Zealand and the 'East Indies' recognised the evolutionary emergence of *Orbulina* as a particularly strong, 'worldwide' biostratigraphic event (subsequently to be known as a 'datum'). In Venezuela and the Caribbean region, Walter Blow reconstructed in some detail two fossil successions (bioseries) leading to *Orbulina*. In southern Australia, Mary Wade reconstructed the same bioseries differently. She saw the bioseries as a broadening array of intergrading forms (morphotypes) until, suddenly as it would seem, the array was hollowed out between the two survivors, *Globigerinoides* at one end and the brand new *Orbulina* at the other. Marinaded in evolutionary biology as I was, I saw the Wade approach as more in tune with nature while conceding nothing in biostratigraphic power.

Source: *Left*, Blow (1956); *right*, Mary Wade (pers. comm.), also in McGowran (2005a).

---

16   Micropalaeontology has always been highly 'applied' or 'practical', owing to its power in addressing the perennial deep-time problems of the age of rocks and the analysis of ancient environments. It long seemed reasonable to define our 'species' as finely as possible, rather than treating samples as populations displaying bell curves (Chapter 3). This diagram by Blow is an extreme example. Wade's more biological approach loses nothing in the 'practical' matter of correlation and age determination, as I have argued (McGowran, 2005a).

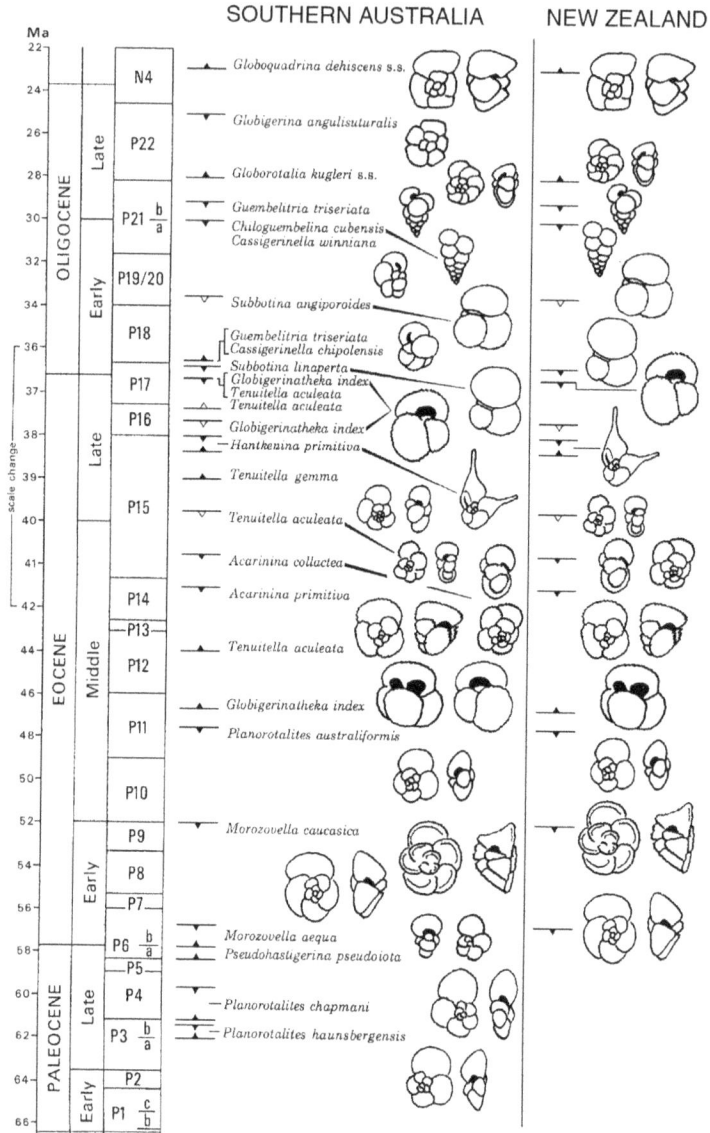

Figure 4.15. Trans-Tasman comparison of bioevents.

When I began looking at foraminifera in 1958 we knew virtually nothing about the Cenozoic in southern Australia below the Late Eocene: that is, 25 million years of a 65-million-year era were almost a blank. This chart shows the situation in planktonic foraminifera in the Palaeogene a quarter-century later. The succession of bioevents shows several parallels with New Zealand, thereby strengthening the framework of geological time and correlation. Teeth up, first appearance; teeth down, last appearance; hollow symbol, somewhat mysterious incoming or outgoing.

Source: McGowran and Beecroft (1985).

On modern estimates, these charts of *Orbulina* evolution span less than 1.5 million years. This is my first reference to the quantification of earth history, to numerical dating. Rather poetically, the postwar acceleration in Cenozoic geohistory received another boost almost simultaneously with Parr's *Hantkenina* discovery: this was Arthur Holmes's 1947 estimate that the Eocene Epoch spanned the dates 58 to 38 million years ago (the base of the Eocene was the top of the Cretaceous). By 1960 the beginnings of the Oligocene, Eocene and Palaeocene epochs were dated at the well-rounded 40, 60 and 70 million years, respectively. The efforts of Parr, Glaessner, Carter and Wade by about 1960 had fleshed out much of the Cenozoic record from the Late Eocene onward: let's say, from 42 million years ago to the present day. Thus most of the Eocene and all of the Palaeocene in southern Australia, 40 per cent of Cenozoic time, was lacking strong biostratigraphic or any other dating constraint. This was motivation enough for a tyro palaeontologist inheriting Parr's unfinished Eocene and Palaeocene business. By 1970 (Figure 4.15) the fossil foraminifera in southern Australia had yielded a succession of first and last events, mostly but not all seemingly speciations or extinctions, each set against a scale in millions of years.[17]

Developing modern Cenozoic time scales is a highly interdisciplinary affair, as is demonstrated by this sample (Figure 4.16).

---

17   Until the 1960s, our stratigraphic charts had time running up the left-hand side, but unscaled. Biozones were usually given equal duration on our charts, or the duration might be merely guessed at. By the late 1960s meaningful radiometric age determinations were becoming robust enough in their geological significance for WA Berggren to begin producing charts with realistic numerical scales in Ma in the left column (Ma = mega-annum = one million years before the present day). (i) Cenozoic series, epochs and ages were matched with (ii) fossil zones, first the foraminifera, then other microfossil groups with preservable skeletons (reliable knowledge of all being hugely boosted by the drilling in the deep oceans). (iii) On the continents zone fossils came from bones and teeth and pollens and spores. (iv) Reversals in the polarity in earth's magnetic field through deep time gave us a geomagnetic time scale, and this became a kind of backbone to the various successions in the continental, neritic and pelagic realms. Most recently has arisen cyclostratigraphy, based on the rhythms in the earth–moon–sun system. Many people have contributed their offerings large and small and often contentious, but Berggren has been the pre-eminent Time Lord down the decades, because everything, from rates of organic evolution to growth rates of oceans, has relied upon those time scales. Our integrated Cenozoic scale for southern Australian sporomorphs and planktonic foraminifera was among the first regional scales to use the Berggren scale (McGowran et al., 1971). However, Berggren had difficulty with funding for the construction of modern time scales; certain geophysicists, so help me, believed that this endeavour 'was not true science'.

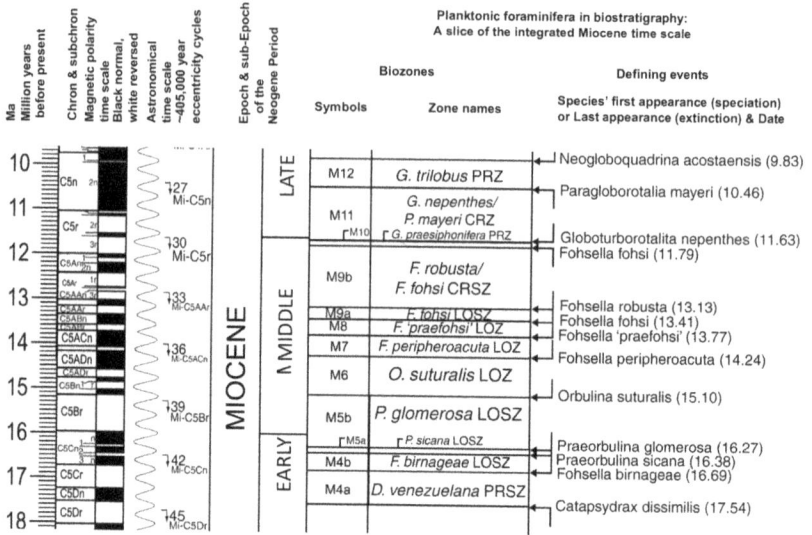

Figure 4.16. Multiple attacks on Miocene chronology.

This slice of Miocene chronology, from 18 to 10 million years ago, exemplifies the modern condition of the geological time scale for the Cenozoic Era. The magic word is integration — the disciplines of geochemistry, geophysics, palaeontology and geology mutually reinforce, cross-check, keep each other honest. Compiling speciations and extinctions in low-latitude planktonic foraminifera began in the 1930s. Other microfossils have followed, from deep-ocean protists and algae to micromammal teeth. The backbone of the scale, as it were, is now the reconstructed succession of Earth's magnetic polarities, the chrons, potentially preserved in terrestrial and marine rocks everywhere. The listed dates of the events down to 10^4 years' resolution come ultimately and circuitously from radiogenic minerals, such as in volcanic ashes sandwiched in fossiliferous oceanic sediments. The slice of chronology is taken from the Cenozoic synthesis by Wade et al. (2011).

Source: From Wade et al. (2011).

Four packets of strata and four unconformities

We can make generalisations about the distribution of strata in space and time—sweeping generalisations to be sure, but perhaps meaningful and even illuminating (Figures 4.17 and 4.18). The broadest pattern is that the Cenozoic strata fall into four chunks of Cenozoic time. The oldest are the sands, silts and clays of Palaeocene to Early Eocene age. Next are the strata of Middle Eocene to Early Oligocene age, displayed most spectacularly as the lower and older of the two limestones in the cliffs of the

Bight. Third are the strata of Late Oligocene to Miocene age, including the upper limestones underlying the Nullarbor Plain and exposed in river and coastal cliffs from Adelaide and the lower Murray River to eastern Gippsland. The fourth and youngest group of strata extend age-wise from the Late Miocene to the present day.

We have seen that Lyell and Darwin 'had agendas', in the modern argot, when they perused the known record of fossils and strata. For their respective reasons it suited them to regard the fossil record as highly incomplete. We can now state that the four-part record of Cenozoic history in southern Australia is (i) distinctly real, that is, not merely a haphazard artefact of destructive erosion and decay, being instead (ii) part of a bigger four-part panorama.

Figure 4.17. The four Cenozoic sequences with boundary unconformities.

Four outcrops in southern Australia showing the local manifestations of the four regional unconformities named in Figure 4.18, and the ages of the strata on either side: 1, Point Margaret on the Otway coast, western Victoria (Frieling et al., 2018); 2, Chapel Hill Road, McLaren Vale, South Australia; 3, Coast south of Sellicks Beach, South Australia; 4, Sellicks Beach, South Australia.

For abbreviations, see Figure 4.18.

Sources: 1, Steven Bohaty; 2, 3, 4, author's images.

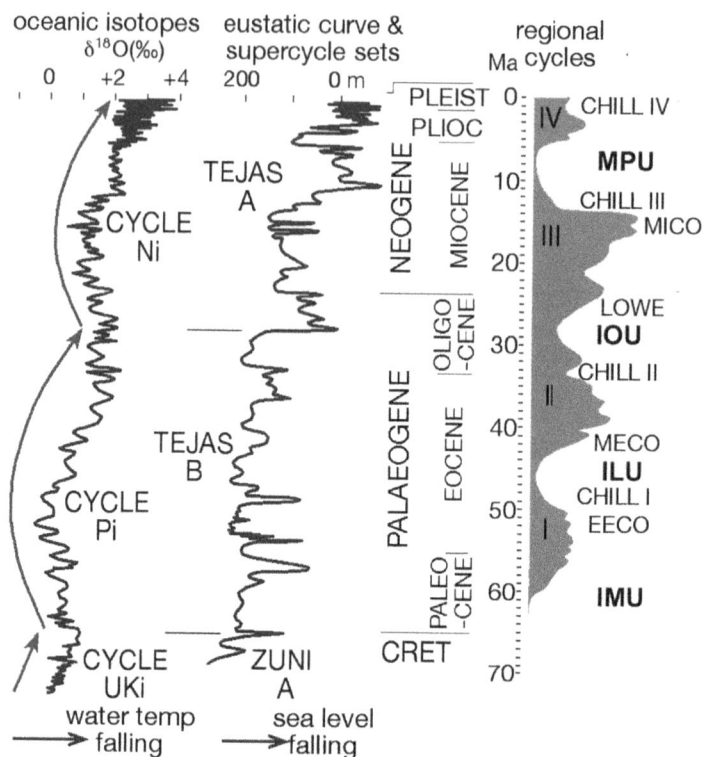

Figure 4.18. The fourfold Cenozoic sequences in global context.

In the 1970s we came to realise that the strata on the southern and western continental margins of Australia were in four 'packages' in terms of their age — two Palaeogene packages and two of Neogene age. Strata are distinguished and named as groups, formations and members in the essential housekeeping procedure known as lithostratigraphy, but it is more interesting to stand back and contemplate the real pattern through deep time. Packages of strata were found to be bounded by major breaks or unconformities, which on a numerical time scale show up as gaps or hiatuses. The grey sketch is an impressionist rendering by Qianyu Li primarily of sea level. The four regional unconformities in Figure 4.17 are shown as hiatuses here. Utilised in seismic mapping, they were named thus, in descending order: MPU, Late Miocene–Pliocene unconformity; IOU, intra-Oligocene unconformity; ILU, intra-Lutetian unconformity; IMU, intra-Maastrichtian unconformity. A global context (see Chapter 6) is given by 1980s compilations of oxygen–isotopic cycles Pi and Ni and eustatic supercycles Tejas B and Tejas A, saying something about temperatures and sea level, respectively.

Sources: Figure 4.18 is modified from McGowran et al. (2004). Oxygen–isotopic cycles, Abreu et al. (1998). Eustatic supercycle sets, Tejas A, B, Haq et al. (1988). Acronyms for the four regional unconformities and hiatuses, Holford et al. (2014).

The four packages of strata can be seen as representing four slices of deep time, namely Early Palaeogene, Late Palaeogene, Early Neogene and Late Neogene. Surprise, surprise, we have seen this fourfold pattern already, in Figure 0.3.

5

Drilled ocean and drifting continent

The Indian Ocean incident

Awakened around 4 am one morning in early 1972, I was required to inspect a strange core, just arrived on deck at our drilled Site 217 on the Ninetyeast Ridge. Pale grey in colour, the core was shot through with white, fibrous, asbestos-looking veins of some mineral. I knew identical rocks in north-western Australia as the Korojon Calcarenite, and the same swarms of disaggregated needles dominated samples under the microscope (Figure 5.1).

The veins actually were calcite, sections cut by the drill through heavily built clam shells of the genus *Inoceramus*, the species I knew attaining a metre and more in length. But those clams (unrelated to the modern species) lived in Late Cretaceous times on the continental shelf, in the neritic zone where water depths are measured or estimated in metres or tens of metres. In fact some of the sediment being drilled was of very shallow marine origin indeed, resembling the muds in the Coorong lagoons on the southern Australian coast and even displaying mud cracks indicating a briefly desiccating seafloor. So what was neritic *Inoceramus* doing under 3 kilometres of Indian Ocean water plus another half a kilometre of sediment of Late Cretaceous and Cenozoic age—sediment that had accumulated on the bottom of a deep ocean?

Figure 5.1. Two views of *Inoceramus*, a pseudo-clam from the Late Cretaceous limestones in western Australia.

This fragment, 16 centimetres in length, came from a shell more than a metre long, colonised by oysters in a shallow sea. The same species or a close relative was exhumed at Site 217. Some species of *Inoceramus* were the largest of all ancient clams; poster figures for the Cretaceous seas, none survived the Cretaceous Period. Note the needle-like crystals of calcite in the lower view; such needles in microfossil samples signify *Inoceramus*.

Source: Photo by Vicki Kramer.

This was not our first surprise on the submarine topographic feature known as the Ninetyeast Ridge. It was our third, but the three surprises formed a clear pattern glimpsing a bigger picture (Figure 5.2). (Reconstructing our profile looked like sheer effrontery at the time, for the geographic span of our three drilled sections was comparable to the distance from Adelaide to Darwin; and what a mind-boggling notion to produce a geological section across Australia from three points—a lesson in the contrasts between continental and oceanic geology.)

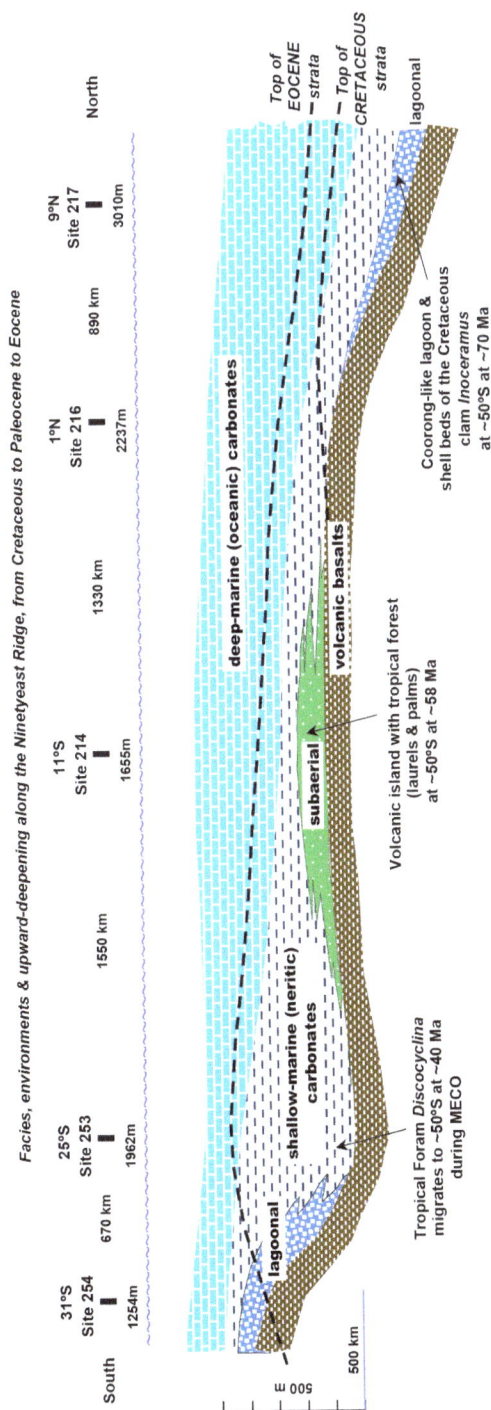

Facies, environments & upward-deepening along the Ninetyeast Ridge, from Cretaceous to Paleocene to Eocene

Figure 5.2. First geological section along the Ninetyeast Ridge.

This geological section south–north along the Ninetyeast Ridge was based on five drilled sections spanning 40 degrees latitude (locations on Figure 5.4). Contemplate the preposterous notion of a meaningful, north–south, geological profile of Australia from Tasmania to New Guinea based on five points. The notion is somewhat less absurd in oceanic than in continental geology. Vertical exaggeration is huge and the distances between sections are collapsed. Underlying each section is a neritic or nonmarine facies, meaning a suite of sediments reflecting the environment — a richly vegetated, tropical volcanic island, a shallow sea, a lagoon. Each locality is relentlessly sinking, with deep-oceanic facies of calcareous ooze and chalk taking over. And just as the ridge crest youngs to the south, so too does the belt of shallow facies. The upward-deepening is diachronous, being crosscut by boundaries in geological time, of which two only are shown. The long, linear and prominent Ninetyeast Ridge was built of volcanoes as the crust moved across a hotspot, a plume of magma far to the south (about 50°S), cooling and therefore subsiding as it moved north. This stratigraphic reconstruction and the inferred motion by the ridge corroborated the spreading inferred from the geomagnetic patterns on either side.

Source: Redrawn and extended from Pimm, McGowran and Gartner (1974) by including southern sites (Davies et al., 1974).

131

The Ridge at Site 216 (its crest now more than 2 km deep) included a volcanic island with subaerial lavas and tuffs; and as it sank beneath the waves while moving northwards the island was blanketed first by shallow-marine sediments and then, close to the end of the Cretaceous, by deeper, that is, oceanic, sediments. And at Site 214 there was not only a volcanic island, but it had been clothed in lush tropical vegetation[1] at a high palaeolatitude sufficient to produce the immature coal known as lignite. That island grew then drowned during the Palaeocene Epoch and the present depth of the sediment surface is less again.

Thus, from north to south we had this: (i) three islands grew volcanically in the same place at successively younger times; (ii) three islands drowned at successively younger times; and (iii) three islands were blanketed by hundreds of metres of oceanic sediment composed overwhelmingly of skeletons of calcite, secreted by microorganisms, planktonic foraminifera and coccoliths, living in the photic zone in the open ocean. Changes in global sea level were not relevant here. This was sustained subsidence of the earth's oceanic crust, kilometres of subsidence; and it is time to broaden our canvas.[2]

The 1950–1960s were the climactic times of oceanic exploration in marine geology and geophysics, yielding very dynamic-looking physiographic maps of abyssal plains, deep-ocean trenches, volcanic mountains, extensive submarine plateaus and circumglobal ridges broken by huge faults. I was particularly impressed by Augusto Gansser's *The Indian Ocean and the Himalayas: A geological interpretation* (1966). Gansser (1910–2012), doyen of Himalayan geology, made the connection between ocean and continental physiography (Figure 5.3) in this terse Abstract of 1966:

> The recently discovered lineaments of the Indian Ocean indicate a shift northwards shift of the Indian Shield which is directly responsible for the Himalayan Orogeny. The western and eastern N-S-directed major lineaments in the northern Indian Ocean continue on land into the Quetta Line (W) and Arakan Yuma Line (E) which are related to the western and eastern syntaxial bends of the Himalayas. The tectonic analysis of the Himalayas suggests a crustal shortening of at least 500 km, which corresponds to the minimum amount of northwards drift of the Indian shield that is indicated by the physiographic pattern of the Indian Ocean.[3]

---

1   Carpenter et al. (2010).
2   The first results of this voyage were published as von der Borch et al. (1974). Sites 253 and 254 were drilled on Leg 26 of the Deep Sea Drilling Project (Davies et al., 1974).
3   Gansser (1966, p. 831).

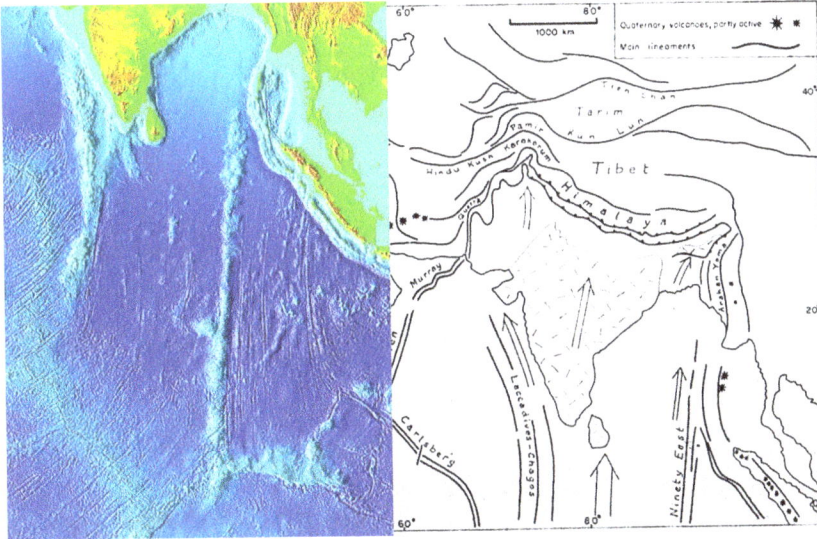

Figure 5.3. *Left*, **modern physiographic illustration of India's migration.**
Right, **Gansser's traces of India's flight northwards.**

Left, a modern physiographic illustration of the Ninetyeast and Chagos–Laccadive ridges. The twofold pattern of lineaments, the grain of the ocean floor, the sense of a north-west–south-east grain encroaching on a north–south grain was quite apparent in the 1960s compilations of physiography. *Right*, in this conceptual sketch of 1966 Augusto Gansser emphasised the unity of northwards shift in continent with ocean. Five arrows point northwards, on land and at sea. The roughly parallel lines with teeth on the upthrow side, <100 kilometres apart and extending >5,000 kilometres in the Himalayas, are the Main Boundary Fault and the Central Thrust. The big wrinkles at each end are the Quetta and Arakan-Yuma 'syntaxes'. It was crucial to Gansser's thesis that the continental wrinkles align with the oceanic ridges. In his metaphor the Chagos–Laccadive and Ninetyeast ridges were the rails, as it were, on which India moved—until it crashed into Asia. The stars are volcanoes; Gansser also observed that the earthquake-prone Himalayas have no volcanoes, unlike the island arcs to the south-east and the mountain ranges to the north-west. Gansser's splendid paper marks that early-1960s time when the oceanic physiography was becoming clearer, the theories of oceanfloor spreading, plate tectonics and the geomagnetic time scale were about to erupt, and drilling the deep oceans was about to commence—and our perceptions of our planet were about to change forever.

Source: *Left*, National Oceanic and Atmospheric Administration via Wikimedia Commons (commons.wikimedia.org/wiki/File:NinetyEastRidge.jpg); *right*, Gansser (1966).

Gansser's 'crustal shortening' referred to the compressive forces now manifested mostly in a belt less than 100 kilometres wide between two great thrust faults, the Main Boundary Fault and the Central Thrust. In the perceived shortening of at least 500 kilometres, the general sense in Himalayan geology of what happened was that ancient, crystalline peninsular India either was overthrust from the north or was underthrust from the south. Surveying the newly glimpsed submarine panorama, Gansser was convinced of the latter. And 500 kilometres lateral movement seemed hardly a major hurdle.

The drilling rigs of petroleum exploration and development were pushing out to sea and into the neritic environment—the shelves of the world's continents. In the 1960s science and technology came together to force the widespread acceptance of continental drift as a fact—acceptance, that is, by virtually all informed people; and plate tectonics arose out of continental drift as the newly ruling paradigm of the earth sciences.

The *Glomar Challenger* entered the Indian Ocean for the first time on our Leg (#22) of the Deep Sea Drilling Project; by then the ship had been drilling into the floors of the other deep oceans for three years. Among the cascading falsifications of theories, confirmations of other theories, and discoveries of new problems and new solutions in marine and global geology, one particularly interesting scientific underpinning was strengthening by the year. It was the interaction between three ways of determining the ages of sedimentary strata. Abundant microfossils produced biostratigraphic age determinations, isotopic ratios locked into fossils and rocks gave radiometric age determinations, and frozen reversals of the earth's magnetic field permitted geomagnetic age determinations. Biostratigraphy built the time scale with its Palaeozoic, Mesozoic and Cenozoic eras; and within biostratigraphy the quite different planktonic microfossils were mutually cross-checking (when things were going well, anyway). Radiometric age determination was putting ever-better numbers on the geological time timescale and on events in geohistory and biohistory. Discovering that the earth's magnetic field reversed itself from time to time, and that those reversals, frozen in rocks, could be dated, opened the possibilities of a geomagnetic time scale. In an ideal world one could dream of a pristine sample of igneous and volcanic rock from the top of the oceanic crust, with a geomagnetic signature, a radiometric signature and overlain by the first layer of deep-ocean sediment with one or more biostratigraphic signatures; and the three disparate data points would keep each other honest. In the geologically turbulent real world this ideal is difficult to attain, but even so our integrated time scales get better and better on this simple principle of mutual correction and reinforcement, of consilience, all nourished from the pipelines of new evidence.

The geomagnetic time scale of the late 1960s presented in both the temporal dimension in columns and the spatial dimension in maps. As oceanic crust formed at the oceanic ridges from subcrustal magma and cooled, it preserved the planet's magnetic signature of the moment; as the new crust moved away from the ridge, becoming denser with cooling and subsiding to form the abyssal plain, that plain preserved a plan of the geomagnetic anomalies, to be revealed by geomagnetic surveying.

Our conceptual model prior to drilling, then, was of a twofold succession in the generation of the floor of the eastern Indian Ocean (Figure 5.4).

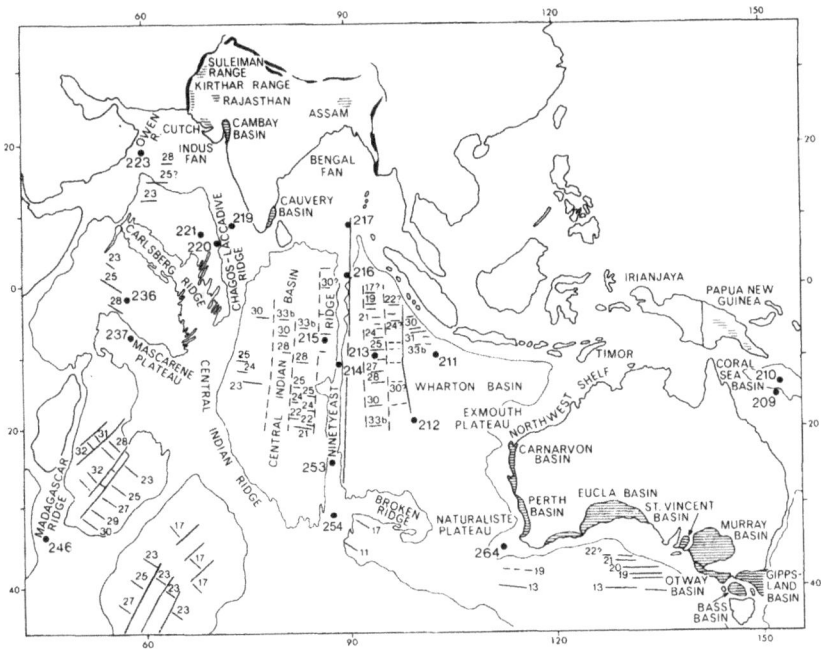

Figure 5.4. Locality map for the Indian Ocean, drawn in 1976.

Just two contours sufficed for the geography needed here, the modern coastline and the 4-kilometre isobath, the latter outlining the submarine basins, ridges and plateaus. Notice particularly the main contrast between old and new. The new is the north-west–south-east grain centred on the Carlsberg–Central Indian ridges tending to the south of Australia. This is Neogene oceanic crust. The old is the south–north grain of the Ninetyeast Ridge and the faults in the Central Indian Basin and Wharton Basin, with the numbered geomagnetic anomalies striking east–west. This is Cretaceous and Palaeogene oceanic crust.

Source: McGowran (1978).

In the early marine surveys, magnetic anomalies striking east–west in the oceanic crust were found to be of Mesozoic and earlier Cenozoic (Palaeogene) ages. To the 'Indian' west, they were successively older to the north, implying motion northwards. To the 'Australian' east, the pattern and motion were in the opposite direction. They were younging to the south. The oceanic fracture zones were striking south–north. In the later Cenozoic (the Neogene Period) the modern configuration of the Indian Ocean Ridge system was in place wherein strikes were strongly north-west–south-east and the system was passing between Australia and Antarctica.

A major objective of our expedition was the nature and origin of the Ninetyeast Ridge and how it fitted into the pattern of oceanfloor spreading. It is a huge topographic entity, 5,600 kilometres long and extending from 34°S to 17°N. Our reconstructed early history of the Ninetyeast Ridge corroborated the south–north spreading inferred from the geomagnetic pattern. That it was a ridge at all, constructed from the same igneous materials as the oceanic floor but extruded subaerially, happened because the spreading seafloor passed above a hotspot, a volcanic centre fed by a plume of magma from below the crust. The extruded lava cooled, increased in density and subsided: hence the deepening to the north; hence too the sinking of terrestrial and shallow marine strata with their fossils into the oceanic depths.

The rapid northward motion of the ridge during the Palaeogene implied that it was on the western plate together with India, opposing the eastern plate, more or less stationary and carrying Australia. But in the modern seafloor spreading system, India and Australia were believed to be on the same plate. When and why did the sticking together, perhaps the welding of the Indian and Australian plates, take place? Our biostratigraphy discovered a hiatus, a pause for several million years from the Late Palaeocene to the Early Eocene, implying a prominent interruption of the steady subsidence. Disrupting the steady subsidence would disrupt the steady accumulation of sediment at Site 214, perhaps by shifting the balance between sedimentation and erosion in favour of the latter. We suggested that this disruption marked a fundamental reorganisation of oceanfloor spreading in the region of the Indian Ocean.

While reviewing the drilling of ridges and plateaus in the Indian Ocean, I discovered that this hiatus was part of a wider pattern, being found on the Ninetyeast Ridge, Chagos–Laccadive Ridge, Mascarene Plateau and Naturaliste Plateau. As the philosophers say, correlation does not mean causation. True; but also true is that a widespread temporal pattern demands a comprehensive historical explanation. That an ocean-scale, oceanfloor spreading system might grind to a halt, become reorganised and start up again in a new configuration would seem to meet that criterion. Which raises the question: why? What would force such a reorganisation within the earth's crust? Since long before the rise of plate tectonics it was recognised that continental crust is less dense than oceanic crust, this being the most cogent argument against the notion of continents 'drifting' across oceans, or of granitic crust 'ploughing' through oceanic crust. In populist plate-tectonic terms, you can't stuff a granitic continent down a basaltic oceanic subduction zone. It is too buoyant. An array of geological evidence and argument seemed to converge on the Eocene Epoch as the critical time of Indian–Asian collision, the event being twofold, including first an island arc and then the continent. Well might one imagine a shuddering response at sea, such as a coeval interruption on various ridges and plateau, to this insult to the rigid crustal plates.

Meanwhile, marine geophysical surveys were beginning to glimpse a timetable for Australia's separating from Antarctica, especially the identification of the geomagnetic anomalies 19–22. More, the Coral Sea and Tasman Sea had ceased expanding by that time.

So! A decade or so after the establishing of plate tectonics theory and half a decade after the *Glomar Challenger* began probing the geographically complex and mysterious Indian Ocean, we had a more general geohistorical theory to enlighten this narrative of southern Australia. It ran like this: the Indian Ocean was forming and growing in later Cretaceous times as oceanfloor spreading freighted India northwards. Australia languished far to the south-east, separated from Antarctica by (as it was christened, years later) the narrow Australo-Antarctic Gulf. The spreading pattern in the Indian Ocean was north–south and south–north, punctuated by huge transform faults parallel to the Ninetyeast Ridge, itself growing volcanically in the south and subsiding northwards (and thereby splendidly validating spreading-with-cooling, hence subsidence). This system stalled during the Palaeogene Period, in the Palaeocene–Eocene transition, when India collided with Asia. Spreading resumed in the region in due course in a reoriented configuration, this time passing not to the north of Australia but to its south. Most significantly for our narrative, Australia's separation from Antarctica accelerated.

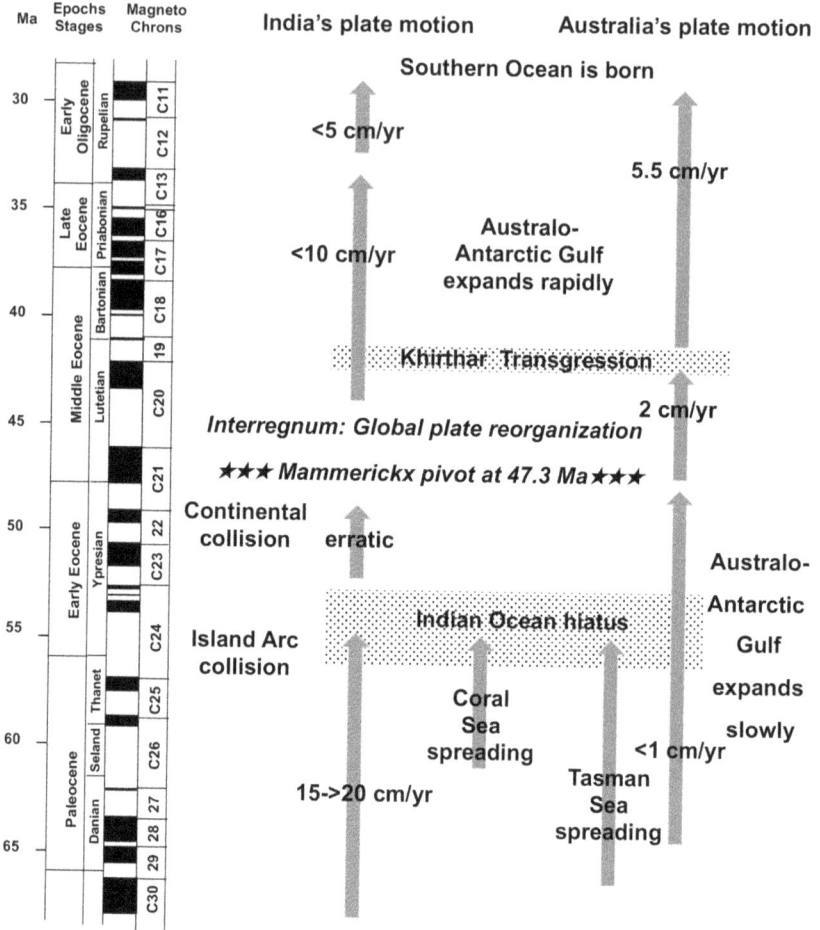

Figure 5.5. Chronological portrait of India–Australia reactions.

By the late 1970s – early 1980s this chronological scenario was in place (McGowran, 1989). The geological time scale now included the geomagnetic component (black is normal polarity, meaning like today's; white is Earth's polarity reversed). India's northward flight (which reached speeds of 22 centimetres/year in the Late Palaeocene) slowed markedly and wobbled (the 'erratic' motion during Anomaly 22–21 time was inferred from a 'chaotic' pattern on the ocean floor, a situation explained by the discovery of the Mammerickx Microplate a quarter-century subsequently). Australia and Zealandia had ceased separating. Then Australia's plate motion increased and the Australo-Antarctic Gulf widened. Capturing the 'Interregnum' of global reorganisation and the onset of the most profound changes in the history of southern Australia (see Chapter 8), 'fifty million years ago' was once a niftily appropriate title (McGowran, 1990) but the geological timetable improves relentlessly and '47.3 million years ago' is somewhat less euphonious.

Source: Modified after McGowran (1990).

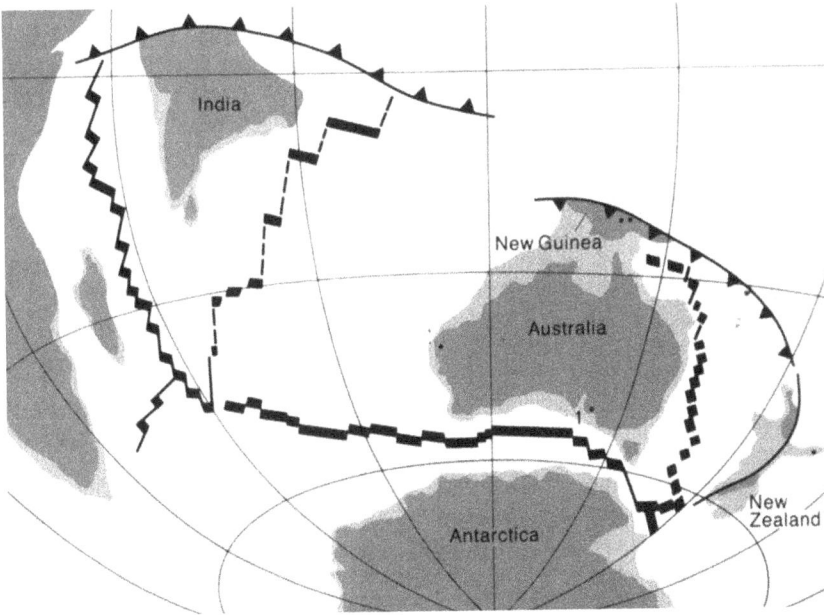

Figure 5.6. Geographic portrait of India–Australia reactions.

This is a geographic portrait of the tectonically turbulent Eocene times. India was moving northwards at speeds up to 22 centimetres/year in the Palaeocene and Early Eocene. But then, ocean floor spreading south-east of India about an east–west axis (thick lines), frequently fragmented by transform faults (dashed lines), was slowed, wobbled and halted when India collided with Asia in a zone of crustal convergence (teeth on the upthrust side). A parallel drama was playing out east and north of Australia, in the Tasman and Coral seas. 'New Zealand' is actually 'Zealandia' (Mortimer et al. 2017). Two extended and fragmented lengths of spreading axis extinguished. In their place the axis between Australia and Antarctica, born in the Cretaceous but barely moving in all that time, was jolted into action — and that crustal concatenation forced the birth of the Southern Ocean. Tethys died — so that the Southern Ocean and Icehouse Earth may live.

Source: McGowran (1990).

The essentials of this story were in place by the later 1970s (Figures 5.5 and 5.6). Subsequent developments have improved the narrative much more than damaged it. There has been recurring angst in the literature about the actual timing of India's collision (or better, collisions) with Asia and how the timing illuminates the rise of the world's highest mountain chain, the Himalayas. (It would be astonishing were this uncertainty not so, because these collisions extinguished the Tethyan Ocean, with cascading implications for Cenozoic history.) Especially significant has been the discovery, in the middle of the Indian Ocean, just west of the Ninetyeast Ridge and now at about 20°S, of a tiny piece of oceanic crust christened the Mammerickx Microplate. The microplate formed in the stresses brought on by the collisions far to the north. Its discoverers date this crunch at chron C21 at about 47.3 million years ago—and they claim this as the most precise and unambiguous date for the onset of the great collision and the enforced reorganisation of seafloor spreading.[4] The timing of the briefly active Mammerickx Microplate sharpens and corroborates the scenario outlined here (Figure 5.5), justifying the appellation Mammerickx pivot.

In due course it became clear that before the Palaeogene jolt, the Australo-Antarctic Gulf had a 50-million-year history as a narrow waterway, floored by slowly spreading oceanic crust. But we need to backtrack now in the Cenozoic story of southern Australia (indeed a quarter-billion years' backtrack).

Back story in Gondwanaland

At Hallett Cove near Adelaide the ancient rocks of the Mt Lofty Ranges display a polished and striated (scratched and gouged) surface, a surface being exhumed by modern erosion from beneath soft sands and clays. From time to time those sediments disgorge erratics, meaning large stones that don't belong there, stones that came from elsewhere, miles away—such as granites. This was the second discovery in the nineteenth century of a site on the Fleurieu Peninsula recognised as of glacial origin. The grinding, polishing and scratching of striations is the work of the sands, pebbles and boulders in the base of the slowly moving glacier, the hardest of the rocks themselves being faceted as only this particular geological process can achieve. In being soft and easily eroded and erratic-laden, the sediments strongly reminded our pioneering geologists of the recently recognised Pleistocene glaciations spread widely across the Northern Hemisphere. Although no fossils had been found and there was no other apparent means of age determination, it was soon clear (i) that these strata were below and older than the local strata containing Cenozoic fossils, and (ii) that similar strata were associated with marine and terrestrial fossils of Permo–Carboniferous age on the eastern and western margins of the continent. We expand the story with point (iii), that glaciations of similar age were being found at presently widely separated localities on India, South Africa and South America.

---

4   Matthews et al. (2015). The Mammerickx Microplate is just to the west of the Ninetyeast Ridge at about 20°S. Initiated 47 million years ago, it sits in its timing beautifully within the interregnum between the two slices of plate-tectonic history: the earlier slice, Cretaceous–Palaeogene recorded in all the north–south lineaments and the later slice in the Neogene north-west–south-east trends. Also, as the authors point out, it is far removed geographically from the collision and quite independent of all the geological evidence for the collisions, the death of Tethys and the prehistory of the Himalayas.

Land now known as southern Australia was littered with poorly consolidated materials of glacial origin, namely moraines, left by retreating ice, and 'fluvioglacials', aqueous as well as glacial. And one can still feel sympathy, and awe, when pondering that the discoverers of this ice age were out in their estimates of its age by a quarter of a billion years too young.

So far, so good; but there were problems with the big picture. If you take compass bearings on the direction of the striations at Hallett Cove, you are pointing south-east in the direction of Victor Harbor, which not only has glacial pavements and glacial landforms but also the granitic intrusions that supplied erratics found elsewhere. Most of the striations on the Fleurieu point north by west. And what extends way beyond the horizon to our south, where the glaciers seem to have arisen? The Southern Ocean, clear to Antarctica.

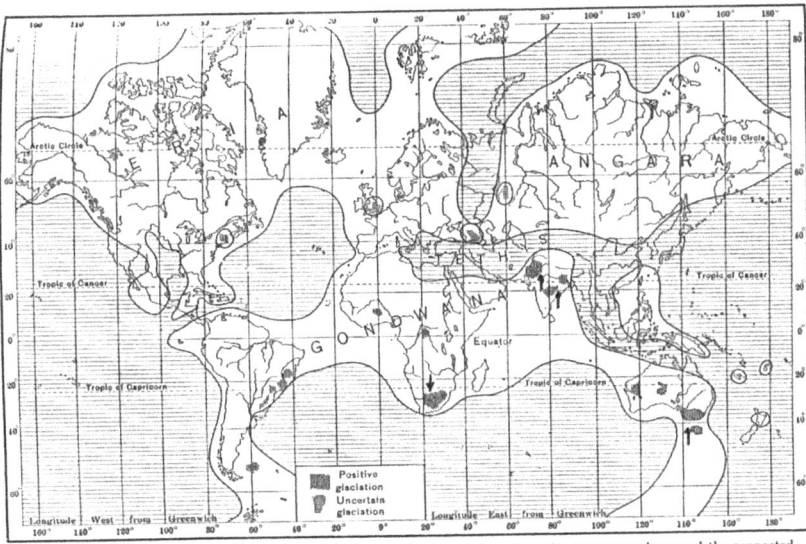

Paleogeography and glaciated areas of early Permian time. Note the transverse shape and the connected condition of the continents. Arrows indicate the direction of glacier flow. Carnegie Institution of Washington.

Figure 5.7. Palaeogeography in Permian times, as of a century ago: Schuchert's Permian land bridges.

Charles Schuchert, the most eminent historical geologist of his time, grappled with the theory of continental drift for inconclusive decades. Transoceanic land bridges were the main competing theory of ancient lands, climates and animals. This map from Schuchert's famous 1915 textbook displays glaciated Gondwanaland (lacking Antarctica!) in the Permian. The southern continent straddled the equator, if we read this map literally, and the ice on South Africa and India seems to have flowed away from, not towards the equator. Arrows indicate the direction of glacier flow. 'Uncertain glaciation' north of the Tethys seaway and in eastern North America is discounted.

Source: Schuchert (1915).

In the absence of a cogent mechanism for moving less dense granitic continents through more dense, basaltic oceanic rock, continental drift remained a minority explanation for explaining away the seeming source of continental-scale glaciations in the open ocean. More popular were the land bridges across the oceans (Figure 5.7), postulated to provide a source for glacial ice. Land bridges had the added attraction of solving certain problems in the distribution of plants and animals, to be mentioned below, but postulating sunken bridges raises the same intractable problem of how to submerge less dense continental rock within more dense oceanic rock.

It took almost a century for biostratigraphy to contribute to the story of the ice age in central-southern Australia, other than by referring to the fossils and strata far away on the eastern and western continental margins. Nell Ludbrook discovered an assemblage of foraminifera, but they were the agglutinated forms of the test (shell), those in which suitable grains selected from the environment were 'organically' cemented, not the secreted-calcareous forms; nor was there any sign of other calcareous shelly fossils known from neritic seas ever since Cambrian times. Soon added to our inventory of the microfossil assemblages were the pollens and spores and marine dinocysts which provided the biostratigraphic control on fossil fuel exploration and development in the 1950s and 1960s. We could see an ecological gradient here, from fully marine (calcareous shells) to marginal marine or brackish (agglutinated foraminifera, then dinocysts only), to freshwater (pollens and spores from land plants). And this exploration was delineating fault-controlled sedimentary basins in southern Australia—challenging the view that the landscapes of the times were predominantly carved by rock-charged ice operating like a titan's gouge and sandpaper.

In the early days of rethinking the geohistory and biohistory of southern Australia in the paradigm of plate tectonics and continental drift, I constructed a two-part theory (Figure 5.8).

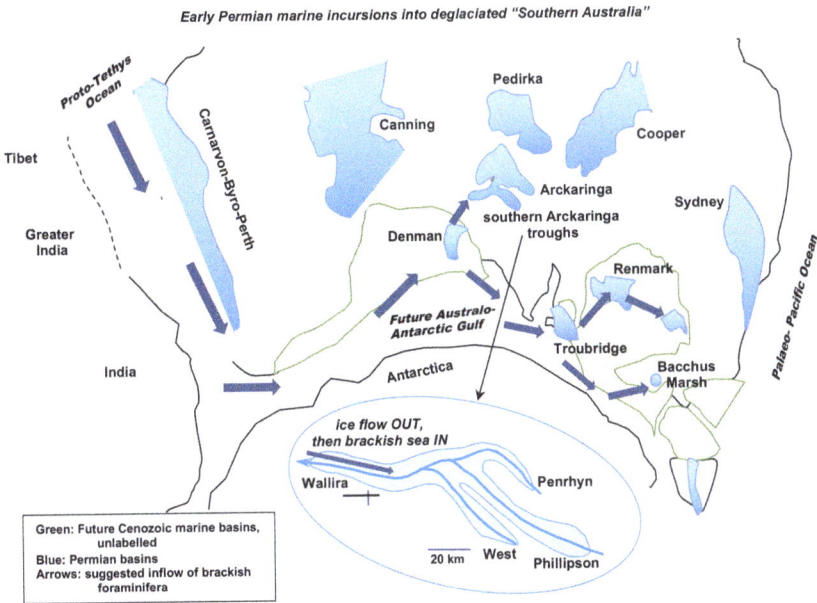

Figure 5.8. Marine invasion of fracturing eastern Gondwanaland.

Littered across Carboniferous–Permian Australia when it was still well embedded in Gondwanaland were sedimentary basins and ice centres. As the main icecaps waned about 290 million years ago, the rising seas penetrated the supercontinent rather deeply, reaching what is now central-southern Australia (and peninsular India and southern Africa). Fossil assemblages of foraminifera and marine dinocysts are evidence for brackish and oxygen-starved inland seas and estuaries — hostile conditions barring the molluscs, brachiopods and bryozoans (i.e. a normal marine 'shelly' fauna) inhabiting the seas bordering the Proto-Tethys and Palaeo-Pacific. The seas might have advanced from the north-west or the south-east, with little evidence either way, but I suggested instead (dark arrows) that they were channelled by fracturing patterns in Gondwanaland, 100 million years before the orthodox scenario of breakup during the Mesozoic Era. However, the inset depicts the troughs in the southern Arckaringa Basin as scoured by glacials, not formed by faulting (note the scale of these huge valleys!). The 2008 paper by Menpes, Korsch and Carr presents both the glacial and the tectonic scenarios.

Source: Transformed from original idea in McGowran (1973).

First, the extensive melting of polar and near-polar ice caused a glacioeustatic rise in sea level and shallow seas were able to penetrate deeply into Gondwanaland (in the latest Carboniferous or earliest Permian, about 300 million years ago). Thousands of kilometres from the open ocean, these seas were too brackish and too starved of oxygen to support the shelly biotas—the molluscs, the brachiopods, the bryozoans—adapted to normal (if very cold) marine conditions in Permian seas. Second, I could see a pattern in the distribution of the marine fossils which looked like it might be foreshadowing the generally accepted breakup of Gondwanaland

143

during the Mesozoic Era. That is, that the lineations of the future Australia, its future sedimentary basins, and the future Australo-Antarctic Gulf were emerging as long ago as late in the Palaeozoic Era. I envisaged a sea not sprawling shapelessly across a continent, instead following channels, hollows, depressions, sometimes glacial, sometimes tectonic.[5]

Figure 5.8 illustrates a version of the eternal metaphor in biogeohistory— did the land rise or did the sea fall? The present case is glacial carving versus tectonic shaping. Both happened, but can we disentangle them?[6]

Australo-Antarctic Gulf: Birth and death in Gondwanaland crackup

In due course sea became ocean (Figures 5.9, 5.10, 5.11). That is, the slabs of the disintegrating supercontinent, continental crust, were pushed apart by new oceanic crust, forming the Australo-Antarctic Gulf. For 40 million years this process happened extremely slowly while the real action in oceanfloor spreading was elsewhere, such as in the fast-forming Tasman Sea. Very little of the present width of the Southern Ocean occurred during that time. For the next 50 million years accelerated crustal growth in the Gulf was part of the newly reorganised global spreading system. The Australo-Antarctic Gulf widened and inevitably was subsumed in the newly emerging Southern Ocean.

---

5   Based on McGowran (1973). The discovery of Permian foraminifera in southern Australia is in Ludbrook (1957) and Harris and McGowran (1973). The arrows indicate the basins where foraminifera are known. Haig (2003) described from the western margin of Western Australia these foraminifera, which are agglutinated, meaning that the shell is made up of grains selected from the environment and stuck together organically, that is, not calcite crystal secreted by the organism. Haig further observed the quite remarkable morphological similarity of these species to modern species characterising modern seas prone to lowered salinity and lowered oxygen, pointing to their 'great conservatism in evolutionary and ecological development'—300 million years' conservatism, indeed.

6   McGowran and Alley (2008) published an image of a 70-metre deep glacial valley in Backstairs Passage (between Kangaroo Island and Fleurieu Peninsula). The seismically delineated troughs in the Arckaringa Basin have been presented in both scenarios by Menpes et al. (2010) as due to faulting and as huge glacial valleys. A book on Permian deglaciation was sedimentological and insufficiently stratigraphic (López-Gamundí and Buatois, 2010).

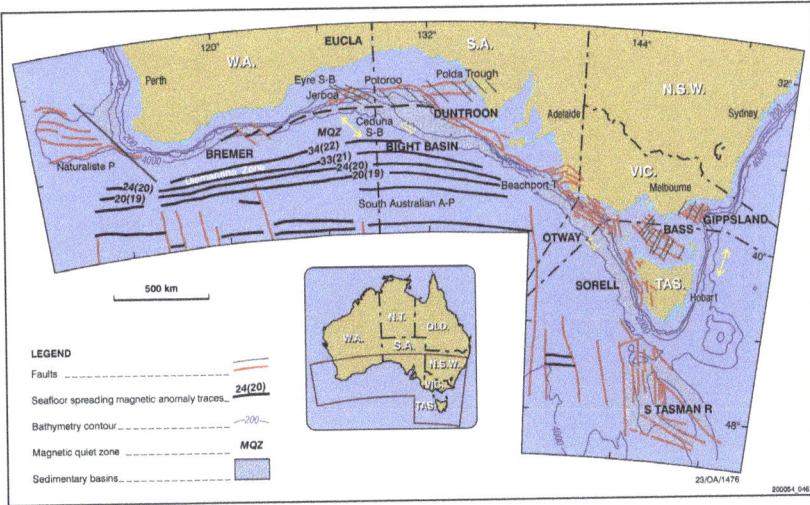

Figure 5.9. The southern margin of Australia, formerly the north flank of the Australo-Antarctic Gulf.

The swarms of faults are known as the Southern Rift System. The four sets of yellow arrows show the sense in which the crust has been stretched during continental breakup and oceanfloor spreading. Double numbers on the seafloor spreading magnetic anomalies are 1972 interpretations (bracketed) and 1982 revised interpretations (unbracketed) (the 1972 identifications are seen also on Figure 5.4). In recent years the Bight Basin and its Eyre and Ceduna subbasins have been of scientific, economic, conservational and political interest. Note the Jerboa and Potoroo drill-sites, all-too-rare penetrations of ground truth.

Source: From Totterdell et al. (2014, Fig. 4.2) in the *Petroleum geology of South Australia*.

Continental margins formed by breakup and separation were known as passive margins or trailing edges, in contrast to the active margins or leading edges. It was a neat dichotomy between the undramatically placid and the dramatically turbulent. 'Passive margins' invoked processes of pull-apart, tension and normal faulting, and non-volcanism. Uplift and the development of hilly topography sat uneasily in the scenario. 'Active continental margins' were the domain of collision and compression, thrust faulting, arcs of volcanoes and earthquakes and trenches, and of metamorphism and mountain building. Breakup in eastern Gondwanaland began some time before 100 Ma in the Mesozoic Era with rifting, producing an uplifted rim at the nascent midocean ridge and an adjacent rim basin, under what is now the Ceduna Plateau in the Bight (Figures 5.12 and 5.13).

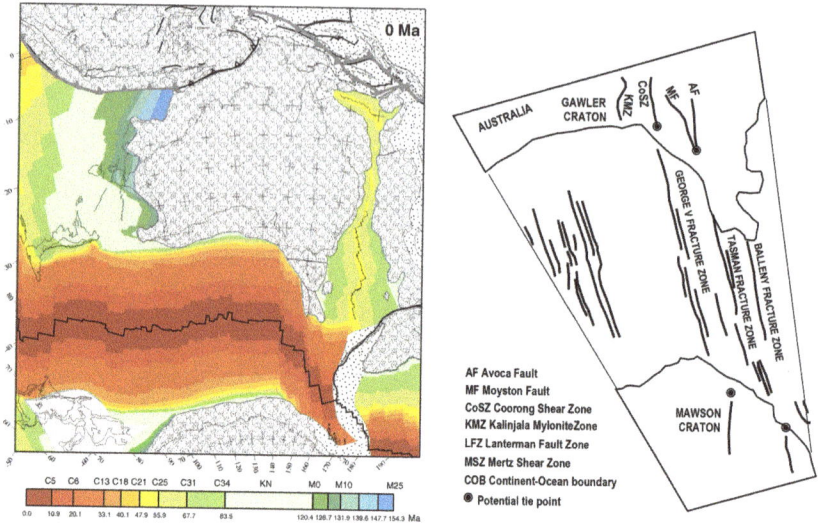

Figure 5.10. Seafloor spreading between Antarctica, Australia and Zealandia in *time* and *place*.

The scale gives dates of ocean floor in millions of years before the present and the geomagnetic chrons. *Time*: Spreading ceased in the Tasman and Coral Seas at the time shown in yellow. As shown by the orange strips, this is the time of reinvigorated spreading in the Australo-Antarctic Gulf. It is the change from the old to the new spreading regime in the region. It is also the time of the Mammerickx pivot (47.3 Ma). *Place*: It seems not so absurd to suggest a link between the Mammerickx pivot in the Indian Ocean and the breakout of a logjam caused by the Coorong Shear Zone. The accelerated spreading is accompanied by three leaps to the south-east across three fracture zones, as seen in the sketch. The potential tie-points not only suggest a plausible reconstruction of east Gondwanaland but also imply that ancient fractures in old continental crust influence and shape the destiny of young fractures in new oceanic crust.

Source: Map kindly supplied by Dietmar Müller, with the accompanying sketch extracted from a review by George Gibson et al. (2013).

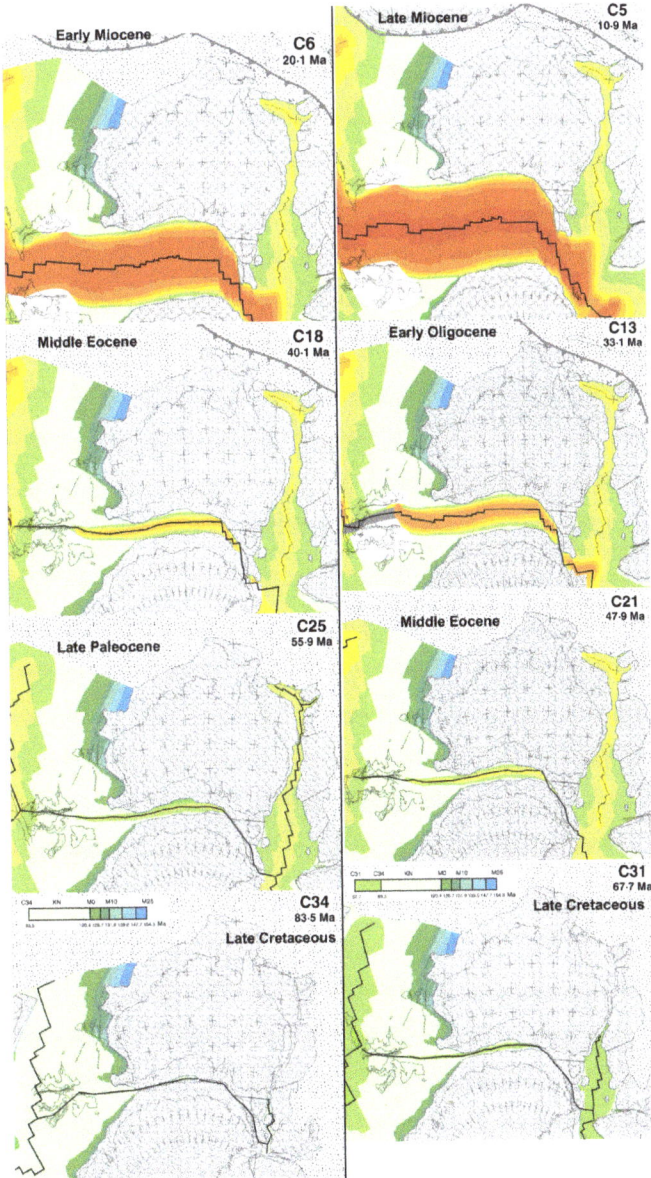

Figure 5.11. The Australo-Antarctic Gulf from birth to death.

This zigzagged series of eight snapshots of seafloor spreading was assembled in the synthesis by Dietmar Müller and associates in 2000. The date in Ma is accompanied by the geomagnetic chron. For the scale, see Figure 5.10. Notice how slowly the Australo-Antarctic Gulf was growing during the first 40-odd million years (when the Tasman Sea was opening) in contrast to the next 40 million.

Source: An assembly of eight palaeogeographic figures by Müller et al. (2000), but displaying coloured versions supplied most kindly by Dietmar Müller. See also Müller et al. (2016).

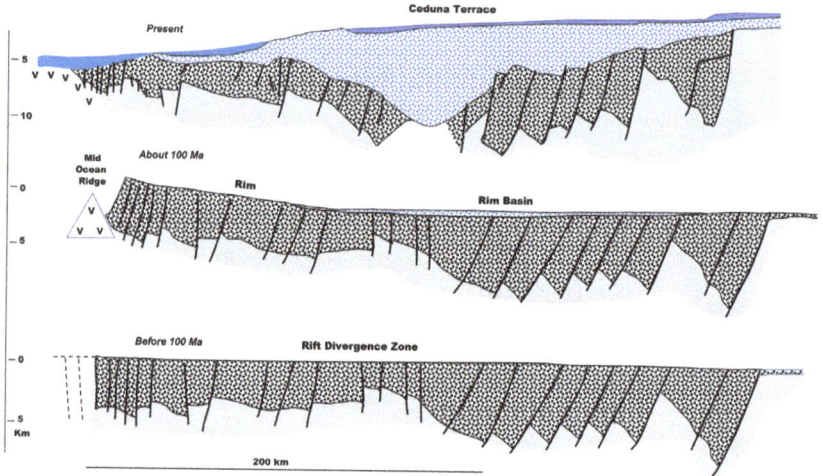

Figure 5.12. Stretching and tension: Veevers's southern Australia as pull-apart continental margin.

Southern Australia was regarded as a classical example of a pull-apart, rifted, 'passive' continental margin, a rift valley between two diverging continents when a supercontinent breaks. This time-series located near the WA/SA border reconstructs stages in the rift divergence zone (there is no 'rift valley' in this series by JJ Veevers). Grey is 'basement'. Black stipple with swarming faults is sands and muds accumulated before breakup. Blue stipple is the Bight Basin (see also Figures 5.9 and 5.13) of recent interest for its possible fossil fuels and their perceived environmental menace. The Bight Basin accumulated thicknesses of 15 kilometres and more of estuarine and marine muds and sands when great rivers drained much of the wet and richly vegetated Australian continent during Late Cretaceous and Early Palaeogene times. Omitted here, its swarm of faults is seen in Figure 5.13. The thinner and younger strata include the limestones forming the modern cliffs, the only outcrop. Vertical scale is in kilometres.

Source: Redrawn from Veevers (2000).

Oceanic crust formed beneath the Australo-Antarctic Gulf from mid-Cretaceous, about 83 million years ago. As the Ceduna Delta grew, supplied from draining most of wet Australia, the growing pile sagged under its own weight, explaining the swarms of normal faults which grew and propagated. Gravity tectonics were adding to plate tectonics, reinforcing the perception of tension, of pull-apart.

But it is not quite so simple as pull = tension plus normal faulting whereas push = compression plus reverse faulting or thrusting. Consider the reconstructed section through strata in the Otway Basin in south-west Victoria (Figure 5.14) ranging from almost flat to gently folded.

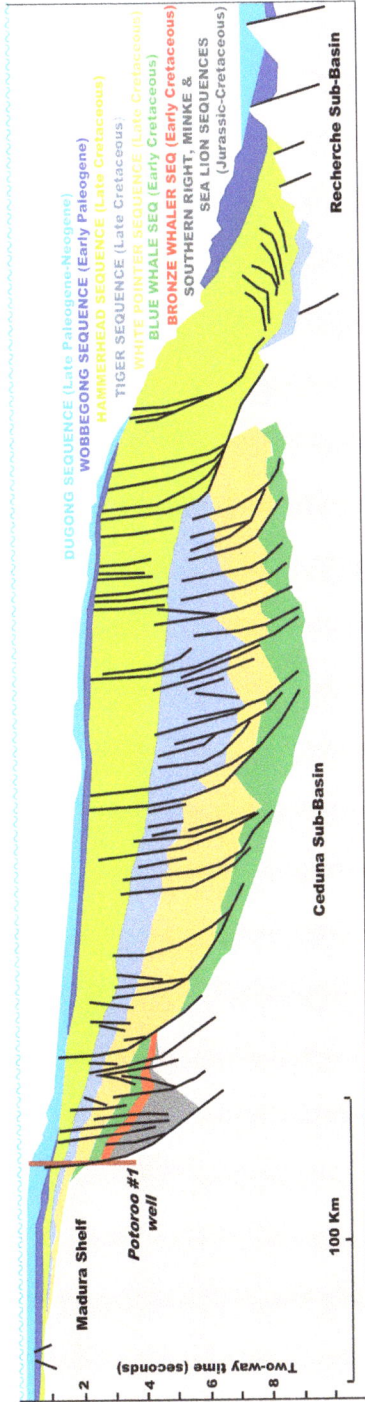

Figure 5.13. Madura–Ceduna seismic sequences in cross-section.

The vast pile of strata comprising the Cretaceous Ceduna deltas is organised into chronostratigraphic packages called sequences and named after marine creatures of the southern seas. Sagging under its own weight, the pile is faulted and the normal faults are initiated then grow as the pile grows. Thus these gravity–tectonic growth faults are not directly related to the underlying faults displayed in Figure 5.12. Vertical scale can only be seismic–geophysical: two-way time in seconds, from which depth can be calculated. The relatively thin Dugong and Wobbegong sequences are of most interest in the present narrative. The section includes the Poteroo exploration well with its all-important information on the ages of these strata. Referring to the map in Figure 5.9, this section of about 500 kilometres strikes south-east through Potoroo and across the Ceduna Sub-basin.

Source: Redrawn after Totterdell et al. (2014, Fig. 4.20) © Commonwealth of Australia (Geoscience Australia) 2019.

149

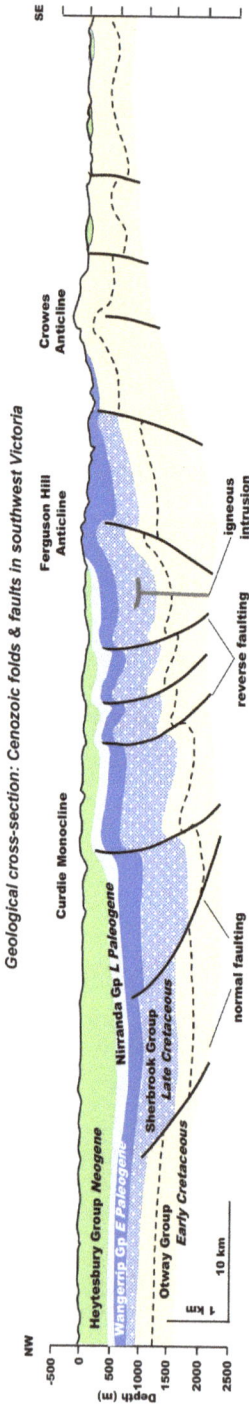

Figure 5.14. Holford's compression and squeezing in central southern Australia.

Faults and folds in the Otway Basin in south-west Victoria from Port Campbell to the Otway Ranges display evidence concurrently of tension and compression and imply that such forces vary markedly in space and in time.

Source: Redrawn from Holford et al. (2014).

In what surely are a family of closely related structures, some faults seem to be normal and some are reverse. This pattern suggests that compression and tension might intermingle in space and in time. The same fault might reverse its sense of movement. Although the units making up the global crust are called 'plates', they are not simply rigid structures moving through space and time, either intact or broken, sometimes carrying a thick but less dense slab of granitic continent, sometimes not. Thus such appellations as 'quiet', 'passive' or 'nonvolcanic' can be misleading.

The Flinders and Mt Lofty ranges are surrounded by large flat basins once hosting neritic seas to the east and west, the lowest part of the continent with salt lakes to the north, and an ocean to the south. The ranges are made up of the rocks of the Adelaide Geosyncline. Its sediments are of Neoproterozoic age, notable for its Sturtian ice ages and Ediacaran first animals, and Cambrian age, notable for its Archaeocyath limestones and fossils of the 'Cambrian explosion' in animal evolution. The Geosyncline was wrapped up by the Delamerian Orogeny of faulting, folding, metamorphism, granites and volcanics, and uplift. The geosyncline became the orogen, which is now the Sprigg tectonic domain. The Delamerian Orogeny was the last major event in forming and consolidating the supercontinent Gondwanaland (whose breakup is prominent in the present tale). There was a mountain range. The 'Mount Lofty Ranges Mark I' of Palaeozoic age are sparsely documented, but we know that the mountains did exist because granites, emplaced at depths of double-digit kilometres in the earth's crust, had to have been uplifted by that much, to be exhumed and picked up and transported by the Permian ice (enormous amounts of ice from mountainous sources, if the Arckaringa troughs are enormous glacial valleys).

More to the point here is that the biggest and most influential Delamerian faults were reverse or thrust faults. What of the modern 'Mount Lofty Ranges Mark II'? It has long been believed that the Cenozoic basins flanking the uplands occupied a rift valley, implying tension, in a nineteenth-century simile (from Gothic architecture) of a fallen keystone in a perturbed arch. But it was also believed that the Cenozoic faults defining the Cenozoic basins were reactivated Delamerian faults—but can laterally compressive structures become laterally tensional structures, just like that? Yet another idea was that the Delamerian faults had little to do with the Cenozoic faults a half-billion years later, beyond their both occupying a periodically mobile zone in the earth's crust persisting for a billion years or more. Recently and belatedly we have acknowledged that Mount Lofty Ranges Mark II are uplifting in Anthropocene time under compression— squeezing—with old rocks pushing over very young sediment to the west in the west and to the east in the east. And doing it on ancient faults. Most recently, we have traced compression back from the present, down through the Neogene and into the Palaeogene, within striking distance of the earth-changing events of the Middle Eocene. It is reactivated compression, all the way down in the Cenozoic Era, of faults an order of magnitude older than that.[7]

But why this tectonic domain and this mountain range? (Figure 5.15)

---

7   These topics are discussed in the geological introduction to the *Natural history of Gulf St Vincent* (McGowran and Alley, 2008), in a discussion of the tectonic ideas of the polymathic Reg Sprigg (McGowran, 2013a), and in a somewhat expansive field guide (McGowran et al., 2016), but the magisterial account of active faulting recurring through half a billion years is in Preiss, *The tectonic history of Adelaide's scarp-forming faults* (2019).

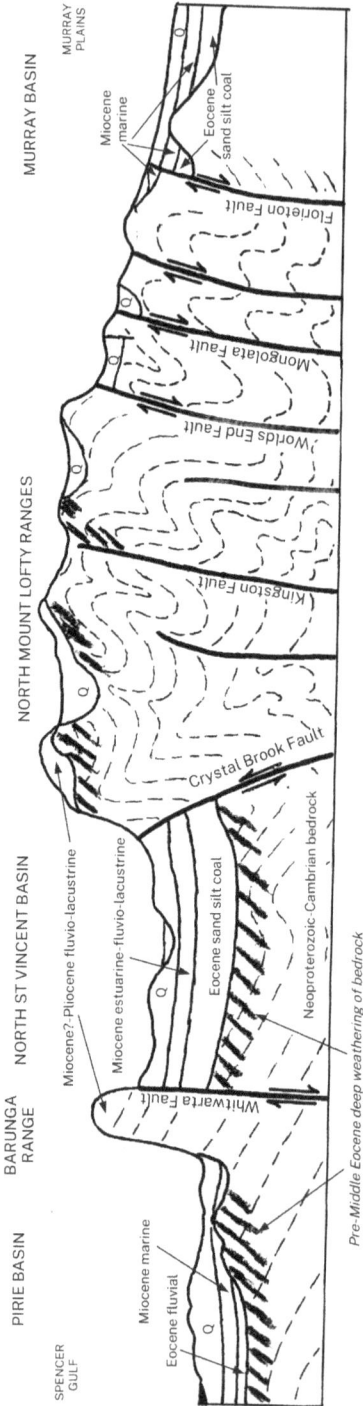

Figure 5.15. Squeezing: Preiss's section across the Mt Lofty Ranges.

Although the southern continental margin is presented as tensional and pull-apart, with normal faults dominating, the Cenozoic development of the Adelaide region, the Mt Lofty Ranges between Cenozoic basins along the northern flank of the Australo-Antarctic Gulf, is not the same in miniature, not a rift valley as long was thought. This is reverse faulting by squeezing, not stretching. And we regard this compression as inherited from the Early Palaeozoic Delamerian Orogeny, during which the schematically shown folding occurred along with the initial faulting. The fossil (biostratigraphic) dates of Eocene and Miocene for the strata in the flanking basins imply that the Mount Lofty Ranges Mark II are a Cenozoic product; elsewhere we can show that the hills originate in the later Eocene and Oligocene but their present form is very young, in the range of 10^5–10^4 years old and even younger. (The Coorong Shear Zone [Figure 5.11], marked by a string of granites at the eastern margin of the ranges, is a literally and fundamentally deeper structure.)

Source: Redrawn from Preiss and Cowley (2016).

Why does one cluster of ancient faults reactivate the better part of a billion years later when a neighbouring and ostensibly similar does not? Our best explanation is in heat—heat flowing through and out of the crustal heat, for the Sprigg domain is shown by repeated observation and investigation to be a high heat flow province. In such a province, deep heat might not in itself cause uplift but it does facilitate the reactivation of ancient faults, while otherwise similar ancient faults in the same horizontal compressional regime, but in the adjacent low heat province, slumber undisturbed.[8]

I conclude by expanding those why-questions to the southern margin of the continent. Why did Gondwanaland break up to develop the Australo-Antarctic Gulf, then the Southern Ocean, there and then? More generally there seem to be two schools of thought on this, one tending to ignore the inheritance, the ancient geological grain of supercontinents as irrelevant to oceanfloor spreading, whereas the other pays it close attention. We take inheritance seriously. The sudden jump in the spreading axis accompanying the sudden acceleration in spreading rate (Figure 5.11) is identified as a cluster of discontinuities, fracture zones—the George V, Tasman and Balleny fracture zones. The ancient rocks of southern Australia have several striking similarities with the rocks of Antarctica, but attention has focused progressively on two ancient structures in particular. The Coorong Shear Zone is matched with the Mertz Shear Zone and aligned with the George V; and the Avoca-Sorell Fault Zone is matched with the Lanterman Fault Zone and aligned with the Tasman. These fault zones cut right through the continental crust, they are much, much older than the oceanic crust, and they seem to have quite some influence over its growth. It becomes more interesting still: the Australo-Antarctic Gulf, growing very slowly for its first 50 million years, was terminated at what was to become the George V alignment. This implies that the Coorong Shear Zone stalled the eastwards propagation of oceanfloor spreading, and that the stalling was overcome only in the new, invigorated spreading regime.[9]

---

8   The 'Sprigg tectonic domain' refers to the Neoproterozoic and Cambrian rocks uplifted as the Flinders – Mt Lofty ranges in central southern Australia. It is a province of high rates of heat flowing up from the depths—significantly higher rates than in the surrounding regions. Holford et al. (2011) explained this with an informative cartoon which we reproduced in McGowran et al. (2016, Fig. 20e).

9   The influence of continental geological ('basement') structure on continental rifting and fracture zones in the growth of the Australo-Antarctic Gulf and the Southern Ocean are covered most comprehensively in two major studies by Gibson et al. (2012 and 2013). Progress in understanding breakup and Australia–Antarctica separation are discussed in Whittaker et al. (2007) and Williams et al. (2019). Breakup dates from 100-odd million years ago and separation dates from 50-odd million years. But with ancient continental crust (meaning an age of a billion years and more) influencing the lineaments of the modern oceanic crust (meaning the recent 50 million), my suggestion that Permian tectonism at about 300 million was foreshadowing the breakup (Fig. 5.9) still appears viable.

It's as if the Mammerickx pivot colluded with the Coorong shear in shaping the Southern Ocean.

The Khirthar Transgression: Fulcrum of the Palaeogene

In 1968 Murray Lindsay and I looked at some old drill cores from below the Nullarbor Plain seeking to sharpen our dating of the great limestone slab known as the Wilson Bluff Limestone, the onset of which turned out to be late Middle Eocene in age, about 42 Ma. And great slabs of shallow-water limestone began to accumulate in shallow seas on the western and northern Australian margins at the same time as our slab on the newly emerging southern margin (Figure 5.16).

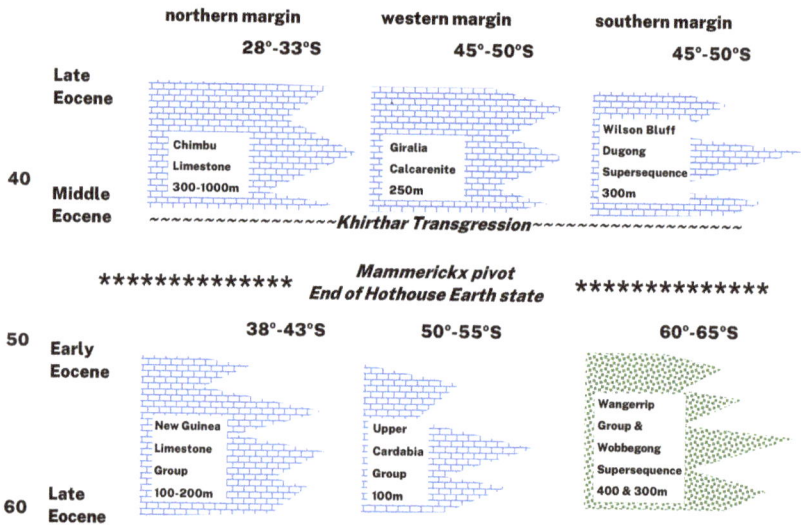

Figure 5.16. Big neritic carbonates on continental margins, same ages, different palaeolatitudes.

Big carbonates on two then three continental margins; big neritic limestones grew at the same time but not at the same palaeolatitudes around Australia. This synchronous stratigraphic pattern across tens of degrees latitude does not sit comfortably in a diachronous ruling paradigm.

Source: Redrawn and modified after McGowran, Li and Moss (1997).

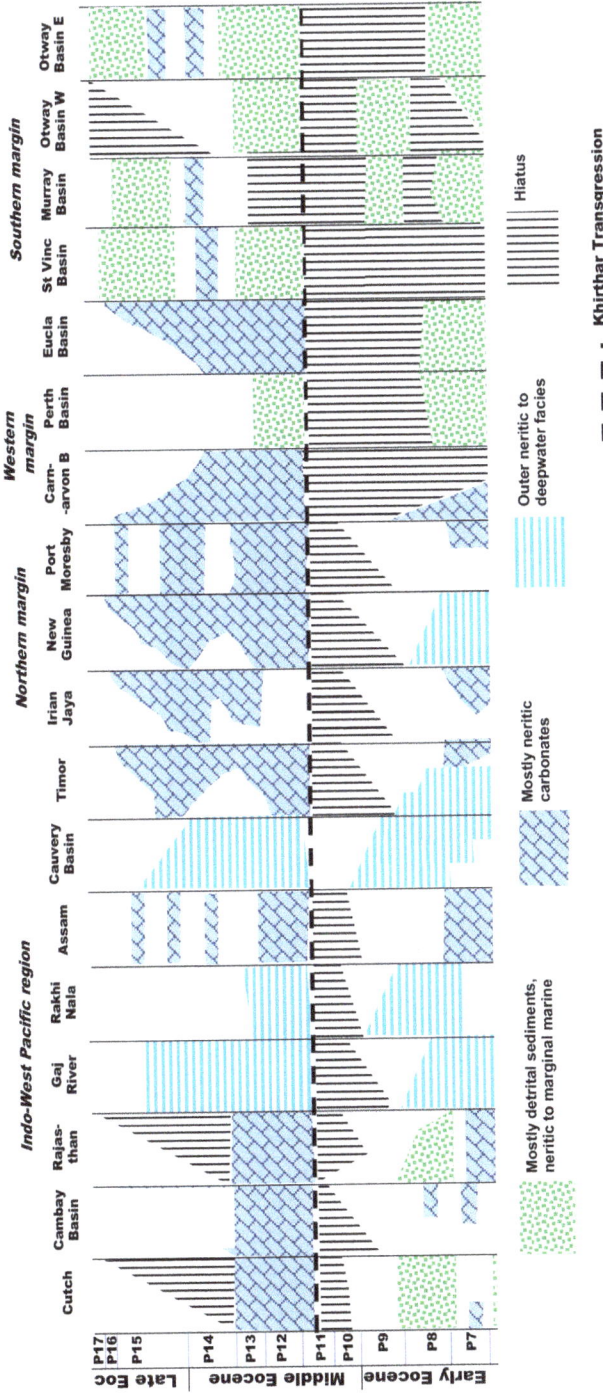

Figure 5.17. 1970s evidence for the Khirthar Transgression.

This is a correlation chart from the 1970s when (i) foraminiferal biostratigraphy was blossoming in the deep oceans and marginal basins, (ii) the plate-tectonic paradigm had taken hold, and (iii) we were beginning to realise that unconformities and hiatuses actually were biogeohistorical data, not simply the destroyers of data, to be blamed for the 'poor fossil record'. Notwithstanding the numerous gaps acknowledging tentativeness and uncertainty, the heavy line marks the Khirthar Transgression, terminating a remarkably widespread hiatus in the early Middle Eocene referred to in Chapter 7 as the Lutetian Gap.

Source: Redrawn after McGowran (1978).

We look still further afield—and recall that a far-sighted Indian petroleum micropalaeontologist stated in 1959 that this washing of warm and shallow seas across several continental margins at the same time, this coeval marine transgression, was the most significant stratigraphic event in the entire Indo-Pacific region. Yedatore Nagappa was working in the 1950s as planktonic foraminiferal biostratigraphy was taking hold globally. Working in the plate-tectonic paradigm, I followed his lead in reviewing the stratigraphy of districts more or less marginal to the Indian Ocean, and confirmed his discovery of the event, naming it the Khirthar transgression (Figure 5.17).[10]

Sedimentary strata accumulate where they do and when they do for a twofold reason. First, a sedimentary basin is formed or a pre-existing basin suddenly subsides (is 'rejuvenated') to provide accommodation space, meaning that strata can accumulate faster in this receptacle than they are removed by the erosive forces of nature. Second, sources are needed for those sediments. Since the eighteenth century we have realised that source and receptacle are renewed by uplift and subsidence, respectively—the guts of the so-called 'rock cycle'. When the eighteenth-century notion of the rock cycle intersects with the twentieth-century notion of the plate-tectonic cycle, it is not the smooth purring of these earthly machine-like processes that matters: it is their clash and jangle as a historical event. We have seen that a major transition in plate-tectonic regime was forced in Eocene times in response to the India–Asia collision. We can now see (on microfossil evidence for invasion by the sea and the accumulation of sediments) that several sedimentary basins formed or rejuvenated along the north flank of the Australo-Antarctic Gulf and in the southern Australian hinterland, about 42 million years ago, late in the Middle Eocene. And we can see that whatever is happening in our backyard is but part of a bigger picture. We will be returning to the Khirthar transgression.[11]

---

10   References are to Nagappa (1959) and McGowran (1978). The Khirthar Transgression and the Lutetian Gap are spelled out in McGowran et al. (2004).
11   Sauermilch et al. (2019) present a particularly ambitious synthesis of the history of the Australian-Antarctic Basin (meaning the sedimentary fill of the Australo-Antarctic Gulf). Unfortunately, weaknesses in their stratigraphy, such as ignoring the Khirthar Transgression and the Lutetian Gap, severely distort their account of the Palaeogene. See Chapter 8.

Stationary continents, mobile animals? Or mobile continents, stationary animals? Or a bet each way?

There are kangaroos and emus in southern Australia. Something of their story belongs in a chapter on tectonics.

As we have seen, the geographic distribution of terrestrial animals was central to both Darwin's and Wallace's theories of organic evolution by natural selection, and it was critically important to discover that the intercontinental contrasts could be found in Pleistocene fossil assemblages as clearly as in the living. And biogeography itself demanded some explanation. In the days when continental and oceanic barriers were considered to be stable and even permanent through deep time, the other significant control on the distribution of animals and plants seemed to be climate and climatic change. Conversely, our growing knowledge of biotic distributions seemed to reinforce the stability of continents and oceans, or at least the non-necessity of major tectonic transformation.

Or did it? And this is not the only basic question. Darwin and Hooker, in accord in due course about evolution by natural selection, disagreed about the biogeography of the southern continents. Hooker explained the striking biotic resemblances in terms of their evolution and migration by connections within the region. The Antarctic region was a centre of evolution. Darwin was more inclined to the notion of long-distance dispersal in waves from the north. This view prevailed. Primitive species were crowded southwards by their superior successors arising in the north; survivors were dubbed by Darwin living fossils, and Australia became perceived as a kind of sheltered workshop among the global biotas.

A clutch of three significant publications emerged within three years in the twentieth century. Arthur Holmes wrote *The age of the earth* (1913) about his radiometric time scale. He demonstrated that geological time, immense time, could be plausibly quantified and really was immense. Holmes also suggested that these stupendous amounts of time newly available to geology and earth history made seemingly unimaginable things seem possible. Right on cue, Alfred Wegener wrote *Die Entstehung der Kontinente und Ozeane (The origin of continents and oceans)* (1915). For Wegener, primarily a geographer, the disciplines palaeontology and stratigraphy delivered the most compelling evidence for continental drift. Also on cue, William

Diller Matthew wrote *Climate and evolution* (1915). Based in Matthew's unrivalled grasp of the global record of Cenozoic vertebrate animals, this was a global biogeography in the Wallace–Darwin paradigm, reaffirming stable continents and stable oceans and reaffirming the power of long-distance dispersal by plants and animals (Figure 5.18).

So palaeobiogeography was for continental drift; and palaeobiogeography was against continental drift.[12] This is about the matching of fossils and strata, about patterns, more than about the deep stuff of process, namely the mechanisms of stabilising continents, subsiding continents and drifting continents. In the next advance in the matter of patterns, the South African geologist Alexander du Toit made a detailed geological reconstruction of Africa against South America. He too wrote a book: *Our wandering continents* (1937). Du Toit's central point within a strongly collective or holistic argument was that the pattern of fossils and strata across the Atlantic Ocean was often closer than either district was to its immediate neighbours. The data were Palaeozoic fossils and strata; granites in the basements were also compared. The similarities were impressive but not quite compelling. For example, Wegener and du Toit were both impressed by the distribution of *Mesosaurus*, found only in South Africa and South America. This small, agile, fish-hunting reptile lived, on the balance of the evidence, in fresh and perhaps brackish water, and in the opinion of the palaeontologists competent to assess such things its traversing thousands of kilometres of deep ocean was a stretch too far. But doubts persisted, as they persisted for plants such as *Glossopteris*, logo for the Permian ice age. Palaeontology was plausible but not compelling, and not only among the geophysical and biological conservatives who demanded a convincing mechanism for continental drift before they—conservatives and continents—would budge. By the late 1960s, when continental drift and plate tectonics were securely established, the Triassic terrestrial reptile *Lystrosaurus* was discovered in an Antarctica surrounded by deep ocean, and its identifier Edwin H Colbert could produce a compelling diagram of Permian and Triassic fossil links across Gondwanaland (Figure 5.19).

---

12   Biogeography has lived through the revolutions in plate tectonics and palaeoceanography (to say nothing of molecular biology and evolutionary genetics) but for deep historical background to these topics, one cannot better Naomi Oreskes's *The rejection of continental drift: Theory and method in American science* (1999).

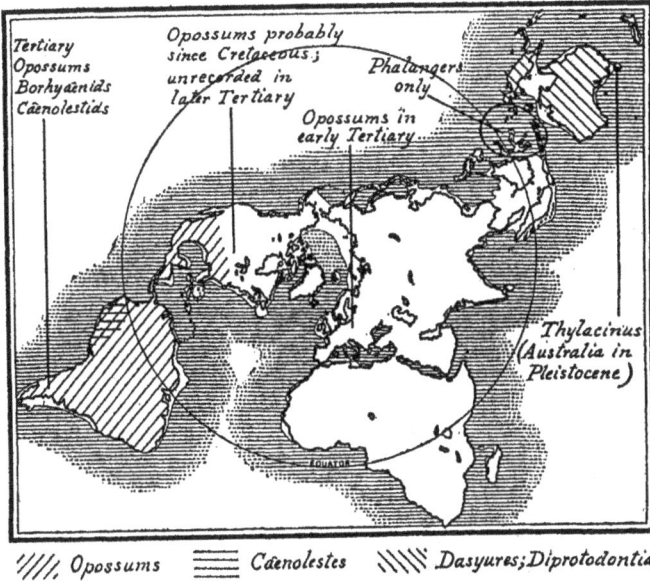

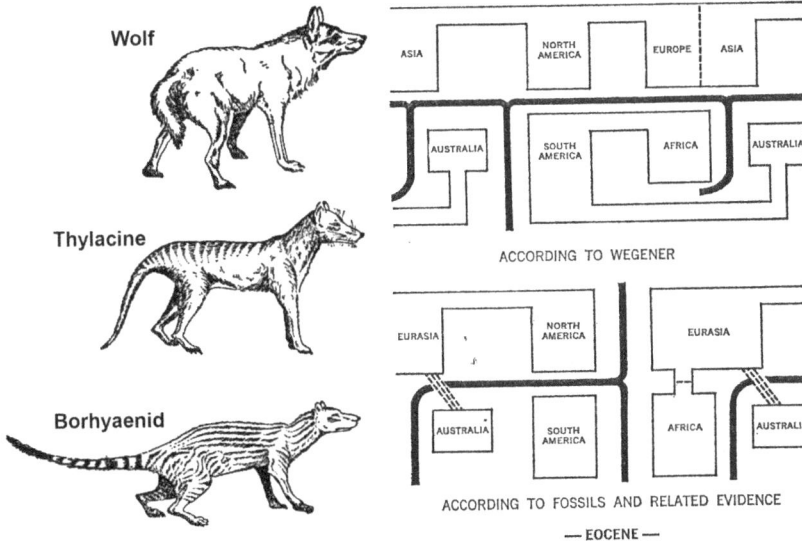

Figure 5.18. Matthew's geography of evolution: Cenozoic arguments against mobile continents.

Above, William Diller Matthew inherited the Wallace–Darwin view of an ancient northern centre of origin for animals and a radiating pattern of dispersal on stable continents. Dispersal in opposite directions to the ends of the earth during the Tertiary (Antarctica still absent) was accompanied by specialised parallel adaptations. Especially in the larger carnivorous mammals: if the Australian thylacine could come to resemble the placental wolf so closely, what's the problem with the South American borhyaenids separately and independently resembling the Australian thylacines?

Below, succeeding Matthew and Schuchert, Simpson took up the cause of continental stability: comparing Wegener's configuration of continents in the Eocene with the most likely configuration according to the available fossils and related evidence, he found radical discrepancy. (The Eurasian–Australian connection is a 'sweepstakes route'.) Put like that, Wegener had to be wrong. But here are two crucial points about Simpson's test: (i) being located in the Eocene it was anachronistic, and far too late to meet the Permian–Triassic challenge in our Figure 5.19; (ii) a warm, forested, eminently habitable Antarctica was duly discovered and could no longer be excluded from consideration (Chapter 7).

Sources: From Matthew (2015) and Simpson (1965).

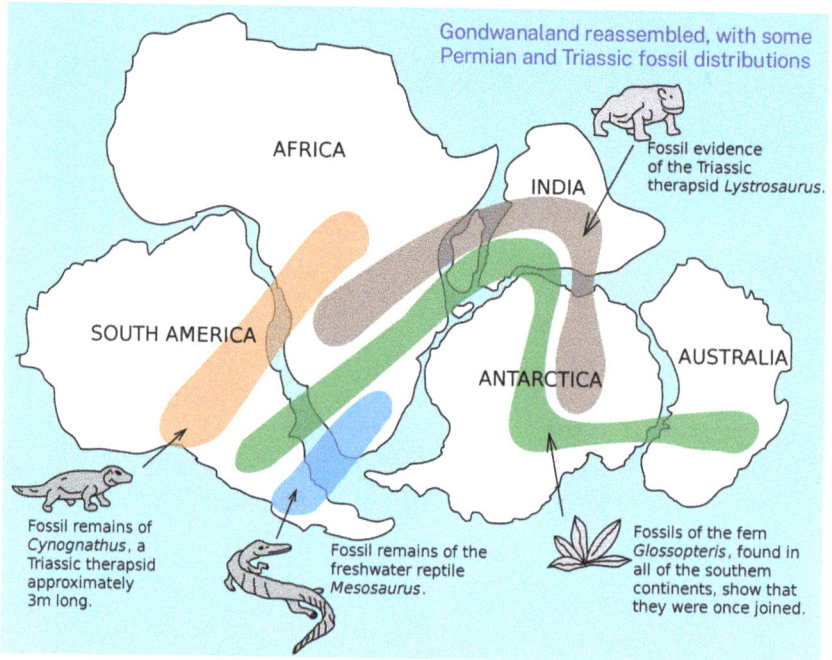

Gondwanaland reassembled, with some Permian and Triassic fossil distributions

AFRICA

INDIA

Fossil evidence of the Triassic therapsid *Lystrosaurus*.

SOUTH AMERICA

ANTARCTICA

AUSTRALIA

Fossil remains of *Cynognathus*, a Triassic therapsid approximately 3m long.

Fossil remains of the freshwater reptile *Mesosaurus*.

Fossils of the fern *Glossopteris*, found in all of the southern continents, show that they were once joined.

Figure 5.19. Colbert's evolutionary geography: Permo–Triassic arguments for mobile continents.

The notion of reassembling Gondwanaland arose when the early European geographers noticed a possible fit across the newly outlined Atlantic Ocean. The Permian ice ages boosted the case in the 1850s, but Hutton–Lyell uniformitarianism chilled it (pun unintended). That is, the process, whatever it was, had to be way outside observable experiences and therefore the theory was scientifically unacceptable. The important point about the pattern arguments as shown here, and especially the case assembled by du Toit in South America and Africa, is the age, Permian and Triassic. In the absence of a cogent drifting process, evidence of pattern was to fall short. That often is the way of science: people are reluctant to accept what happened and run with it until they know how it happened. This Wikipedia-renovated figure of Gondwanaland reassembled was first published by EH Colbert in *Wandering lands and animals* (1973, Fig. 31). Colbert had identified the robustly built terrestrial quadruped *Lystrosaurus* in Antarctic material after plate tectonics and continental drift were securely established. On Antarctica, deep ocean all around and not a land bridge in sight — surely *Lystrosaurus* would have tipped the balance towards continental drift in sceptical minds? No.

Source: Wikipedia-enhanced, after Colbert (1973).

Note that this is a Permian and Triassic configuration, a quarter of a billion years preceding our Cenozoic narrative. In his *Wandering lands and animals* (1973) Colbert made the point that palaeontological opinion about the drifting of continents was itself somewhat biogeographic. People working on older fossils and strata have tended to be more open to the notion of geographic mobility as an explanation; those working in the Cenozoic and later Mesozoic, less so: 'animals and plants move, not continents'. Perhaps this is an example of the infectiously ahistorical slogan to which palaeontologists and especially sedimentologists are prone, 'the present is the key to the past' (Chapter 10).

Matthew visualised land vertebrates as originating in and migrating out of the northern ('Holarctic') centres of dispersal; very little change in geography was required and the continents are generally permanent; and the evolution of land animals was in 'exact accord' with theories of climate change, such as moist and uniform climate alternating with arid and zonal climate: no transoceanic bridges, no sunken continents, continental drift quite unconvincing. Matthew presented maps of distribution centred on the North Pole and family trees with biogeographic distribution incorporated. Inspect Matthew's map for the marsupials. Cuvier's triumphant dissection of the lone Eocene animal aside (Chapter 3), there was nothing of a fossil record of marsupials in Africa or Eurasia. There was an expanding record of Cenozoic marsupials discovered in South America but only a Pleistocene fossil record in Australia. Where the remarkably similar marsupial fossils in the two continents were among Wegener's strongest evidence for a trans-Antarctic land connection, Matthew saw strong anatomical evidence for differences within the marsupials. He cogently rejected the inferred close and biogeographically significant relationship between the carnivorous thylacinids on one continent and the carnivorous borhyaenids on the other—it was an example of natural selection producing parallel evolution. Likewise, he showed that the South American genus *Caenolestes* was not a close relative of the Australian *Diprotodon*, marsupials though they both were.

And this parallelism is one key to Darwinian palaeontologists' unresponsiveness to bridges and drifts, for the fossil record is rich in examples of different organisms independently acquiring startlingly similar and felicitous adaptations. Then, palaeontologists believed that too much credence was given to Earth's crustal motion (of unknown cause) and not enough to the (known) mobility and powers of dispersal of organisms. Third, the Darwinians absorbed Holmes's injunction that the stupendous amounts of geological time now apparent radiometrically made extremely unlikely and unimaginable events possible. For example, odds of one-in-a-million, say, of an immigration happening in a given year, shorten dramatically if you have a window of opportunity open for several million years. And fourth, they asked, what does the record of marsupials in Australia actually tell us about their migrational history? It is widely agreed that our marsupials began with a few small animals migrating not from Asia but from the south, before 40 or more likely 50 million years ago, and that they founded our marsupial radiation in-house. The fossils documenting that great evolutionary event were discovered in the Pleistocene in Cuvier's time and are filling in a fossil record, back down through the Neogene, to 20–25 million years ago in the Late Oligocene Epoch. Below that, the strata are almost bare. Half a century's endeavours of filling out the botanical and marine fossil records, and hardly a bone or a tooth from the terrestrial vertebrates to be seen— absence of evidence, not evidence of absence, indeed.[13]

Matthew's spiritual heirs in biogeography and organic evolution, Simpson, Mayr and Darlington, similarly strong in advocating the powers of organic dispersal between stable continents, came to accept the new tectonic narrative for the southern continents. Compare Simpson's 1960s diagram of continents, falsifying Wegener's reconstruction, with Cracraft's a decade later (Figure 5.20).

Certainly the new global physical geography invigorated biogeography,[14] but still with us was the patchy fossil record. Antarctica was missing from Matthew's and Simpson's maps; the Antarctic record is still largely shielded from us by ice; there is still no evidence of marsupials in Asia. But the changing global geography and its history stimulated another look at the paradigm of northern origins, northern innovations and dispersal out of the north: a scrutiny of the birds.

---

13 Black et al. (2012) list one lonely fossil marsupial fauna, probably Early Eocene in age, from the rock record. And yet the four extant orders and one extinct order of Australia's marsupials, collectively the Australidelphia, seem to have diverged as long ago as the Palaeocene Epoch, roughly 60 million years ago. GG Simpson reminded us in Adelaide in 1968 that our fossil record was the most significant gap in the Cenozoic record of the mammals; and Neogene progress since then is sadly unmatched in the Palaeogene. It is not at all clear why we have discovered virtually no deep fossil record to keep the molecular clocks calibrated and honest; it can hardly be for want of trying. It is true that Australia has been rather too stable geologically for geologists' curiosity. The largest Eocene coals on the planet are in the Gippsland coalfields but they do not crop out. Spectacular though the limestone cliffs are in the Bight, the important stratigraphic anatomy of the Khirthar Transgression is deep down below.
14 Cracraft (1974).

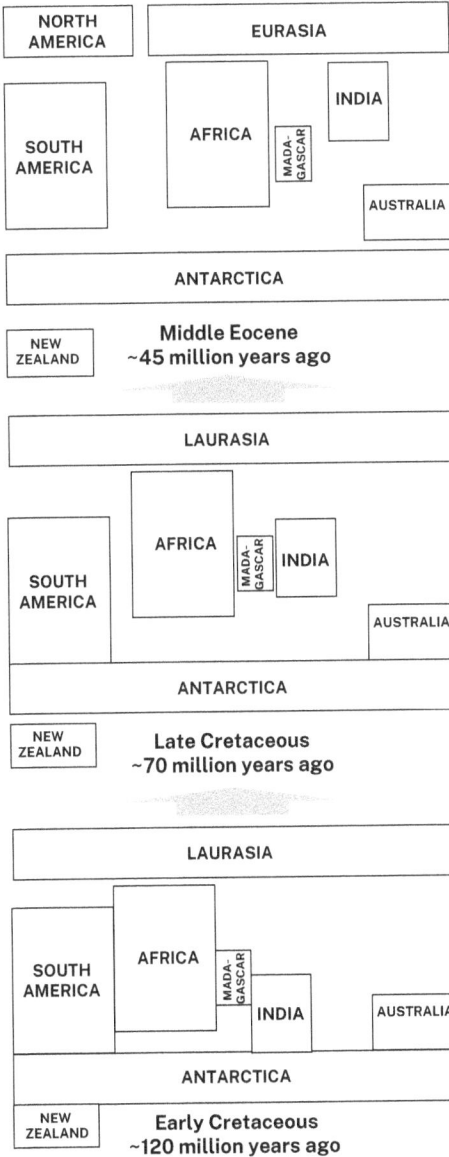

Figure 5.20. 'Dispersal versus vicariance' — do organisms move, or do continents?

'Dispersal versus vicariance' archly rephrases 'do organisms move, or do continents?' — which omits the shades of grey in complexity. When continental drift became acceptable, several biogeographers jumped on the newly mobile bandwagon. This tectonic succession of three scenarios (younging upwards) was Joel Cracraft's (1974) template for a reinvigorated historical biogeography, in which vicariance trumped dispersal.

Source: Redrawn after Cracraft (1974).

163

The song of Australia: Dead end no more

The ratites are a small group of birds, including some giants, and an interesting old problem. The ratites are the flightless birds of the southern continents, the emu and cassowary, the moa and kiwi, the ostrich and elephant bird, and the rhea. The ratites used to be regarded as primitive and predating flight (i.e. their common ancestor was flightless because feathers preceded flight and flying demanded higher organisation). Their geographic distribution was biological evidence for an ancient Gondwanaland. But Matthew reinterpreted the ratites like this:

> it appears certain that most, and possible that all of the existing ground-birds are readaptations to terrestrial habitat from flying ancestors, and their resemblances are due almost wholly to adaptive parallelism. (1915, p. 123)

In his 1974 review, in the glow of the new tectonic timetable, Cracraft argued that the ratites had a common southern ancestor which almost certainly was also flightless, so that barriers and land connections in the Cretaceous and Palaeogene 'were no doubt critical for their disposal'. Reasonably enough, the big birds sorted geographically on the fragments of Gondwanaland and their connections—moa with kiwi, elephant bird with ostrich, and emu with cassowary. It was very unlikely, wrote Cracraft, that the ancestors of the kiwi and the moas flew to New Zealand. However, ancient DNA did not sort that way (Figure 5.21).

Moa has hooked up with tinamou, and kiwi with elephant bird, making the dispersal of flightless ratites look very unlikely (and Gondwanaland was fragmented and the fragments dispersed long since). So, we are back to small flighted birds dispersing and only then evolving ratite-ness, not the other way around.[15] And what do these dispersals and independent adaptations forcefully remind us of? Why, none other than the adaptive parallelisms advocated by that old reactionary perched on the stationary continents, William Diller Matthew.

---

15   Mitchell et al. (2014).

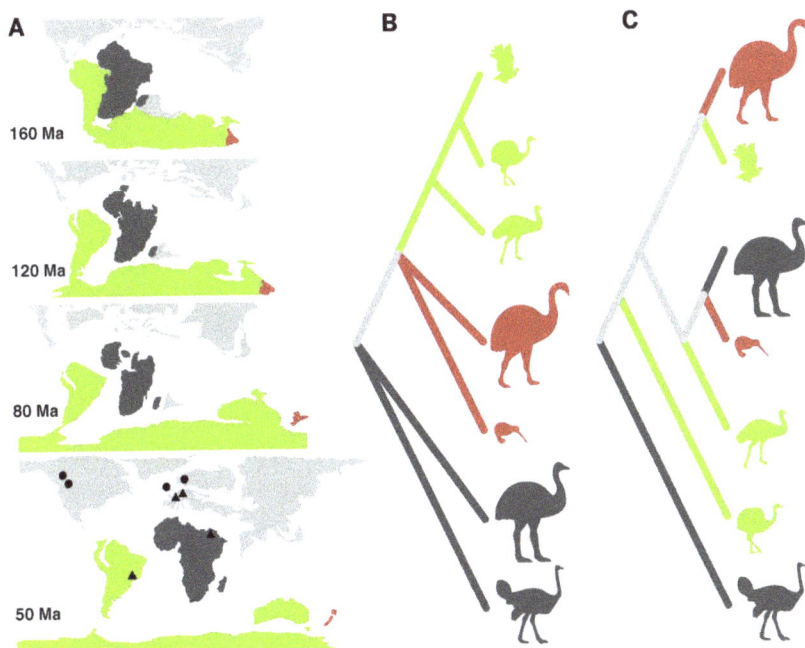

Figure 5.21. Mitchell's ratite patterns in evolutionary genetics.

The ancestors of the large flightless birds on the southern continents did not independently migrate southwards on fixed geography and then evolve convergently. Instead, the new tectonic timetable stimulated the theory of vicariant speciation on the fragments of a broken-up Gondwanaland. In an elegant test employing mitochondrial evolutionary genetics in 1914, *A* shows the order of lands' severances from a shrinking supercontinent: first dark grey, then red (both in the Cretaceous), then green (in the Palaeogene). (Circles are fossils of flighted Palaeogene birds and triangles flightless.) *B* shows the predicted relationships as governed by the timetable of tectonic breakup. But *C* shows the new phylogeny with the kiwi grouped with the elephant bird and the tinamou with the moa. On this evidence the birds dispersed as fliers, and only then converged to the state of flightlessness.

Source: From Mitchell et al. (2014).

Also recalled by this dispersing was the not-so-old reactionary, Ernst Mayr, the most forceful advocate of northern origin and innovation in birds. As recently as the early 1970s, Mayr could state clearly that the reality of continental drift was now firmly established, yet still it was certain 'that the long-standing thesis that Australia received nearly all of its bird life from south-eastern Asia through island-hopping is still fully valid'.[16] This

---

16   Mayr (1972). Much of Mayr's biogeography is assembled in *Evolution and the diversity of life*, his book of essays (1976). Richard Schodde (2005) has described the importance of May's ornithology in the Australian region for his theories of speciation, which of course included biogeography.

hypothesis of waves of immigration into Australia was based on such premises as (i) the stable earth's crust, with no sunken bridges and no continental drift; and (ii) the close resemblance of Australasian songbirds to Eurasian songbirds. DNA studies beginning in the 1980s have shown that all those resemblances were due to convergent evolution, not to direct relationship.

Ornithologist and biogeographer Richard Schodde chose a modern geographic configuration to depict most elegantly a scenario in the historical biogeography of the passerines or perching birds (Figure 5.22).

In the earlier vicariance (solid arrows), inherited from the disintegrating Gondwanaland, Australia acquired the oscines (the songbirds). Their dispersal (broken arrows) out of Australia, led by the corvids (crows, etc.) was later. This out-of-Australia event was well recognised by the turn of the twentieth–twenty-first century.

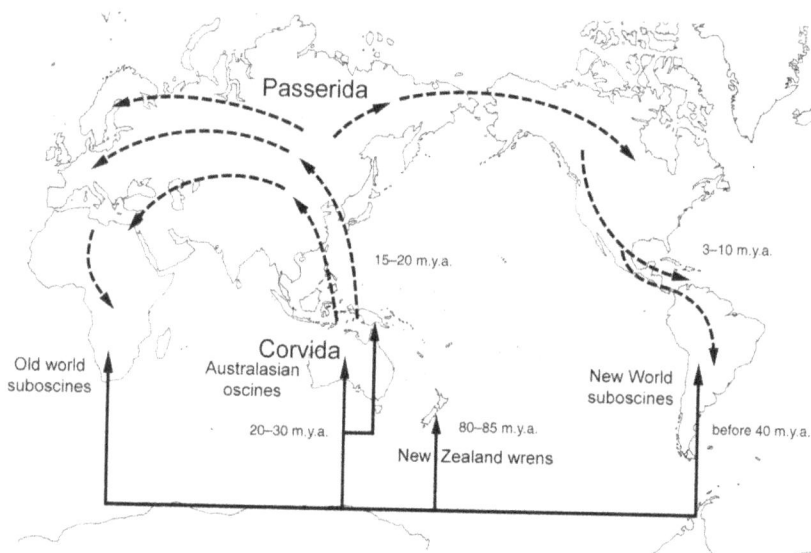

Figure 5.22. Richard Schodde's (2006) map of Australia's place in the biogeography of birds.

The arrival of the corvids (solid arrows) is now likely to be in the Early Palaeogene, perhaps 50–60 Ma. Some would agree with Schodde that the great dispersal (broken arrows) was during the Neogene, when dispersal was aided by the popping-up of Wallacea, the new tectonic lands between Australia–New Guinea and Southeast Asia. For other biogeographers, dispersal out of Australia was earlier, during the Late Palaeogene.

Source: From Schodde (2006).

Schodde's model of southern origins, southern speciations and out-of-the-south dispersals gave us the leading questions of recent decades: sorting out the tree-of-life genealogies of the songbirds and when the birds went north. One view has been that most of the action was during the later Palaeogene times, specifically in the Eocene. A second view is that it all happened 20-odd million years later. In this model Australia's tectonic collisions in the north were critical—the growth of a mountainous, climate-changing New Guinea and the emergence of all those tectonic islands known as Wallacea, between continental Southeast Asia and continental Australia – New Guinea.[17]

Songbirds originated in Australia! We did not import them after all; we exported them.[18] Just as Australian oak is actually Tasmanian Stringybark, and the 'Wollemi pine' (*Wollemia*) is not a pine and koala bears are not bears, so are our iconic magpies not magpies; and this statement applies not only to many haphazardly named Australian birds but to systematic research which, however careful, was unable to distinguish genealogical relationship from convergence. The underlying message is that the great majority of our birds (better: avian lineages) have been inherited from Gondwanaland and that the great songbird radiation began right here.

---

17   A title by a nine-author team puts it succinctly: 'Tectonic collision and uplift of Wallacea triggered the global songbird radiation' (Moyle et al., 2016). They are very clear in asserting that the diversification itself, not just dispersal into SE Asia and the rest of the planet, was due to tectonic collisions in the SW Pacific region.

18   Tim Low's *Where song began* (2014) is a rollicking tale of Australia's birds and how they changed the world. The tale is well told, supported by excellent plates, and deservedly popular; but it lacks informative figures, always useful in clarifying multifaceted biogeohistorical narratives; and it drifts into supersessionism, indulging in triumphalist putdowns (already looking premature) of the Darwinian titans Matthew and Mayr.

6

The great transformation and the last greenhouse

Contemplating *Acarinina mcgowrani*

When I was being immortalised in the naming of the planktonic foraminiferal species *Acarinina mcgowrani*, its christeners were confident of its age range including time of speciation and time of extinction, of its duration of about 11 million years and of who were its ancestors and who its descendants, all events potentially useful in biostratigraphic age determination. And now we have portraits of *A. mcgowrani*, hovering in the sunlit waters of the Eocene ocean while surrounded by its own cloud of 'microalgae', clearly its photosymbionts and not its salad lunch. So how do we know all this, given that *Acarinina* did not survive the Eocene Epoch and most likely was the last of its clade? How can we get to know a dead (extinct) clade? Is this portrait merely a somewhat insipid extrapolating of what we know of modern organisms, down into deep time? Or is there more to it? These questions are central to this chapter.

We begin with three grand statements of advances in recent decades. We have seen the rise and triumph of a theory of the mobile earth's crust producing a narrative of the making and breaking of supercontinents and the birth and death of oceans. We possess a comprehensive archive of microfossils for the pelagic realm throughout the Cenozoic Era. And new analytical techniques keep sprouting to satisfy the demands of new and ambitious questions about Cenozoic ocean history and earth history and life history.

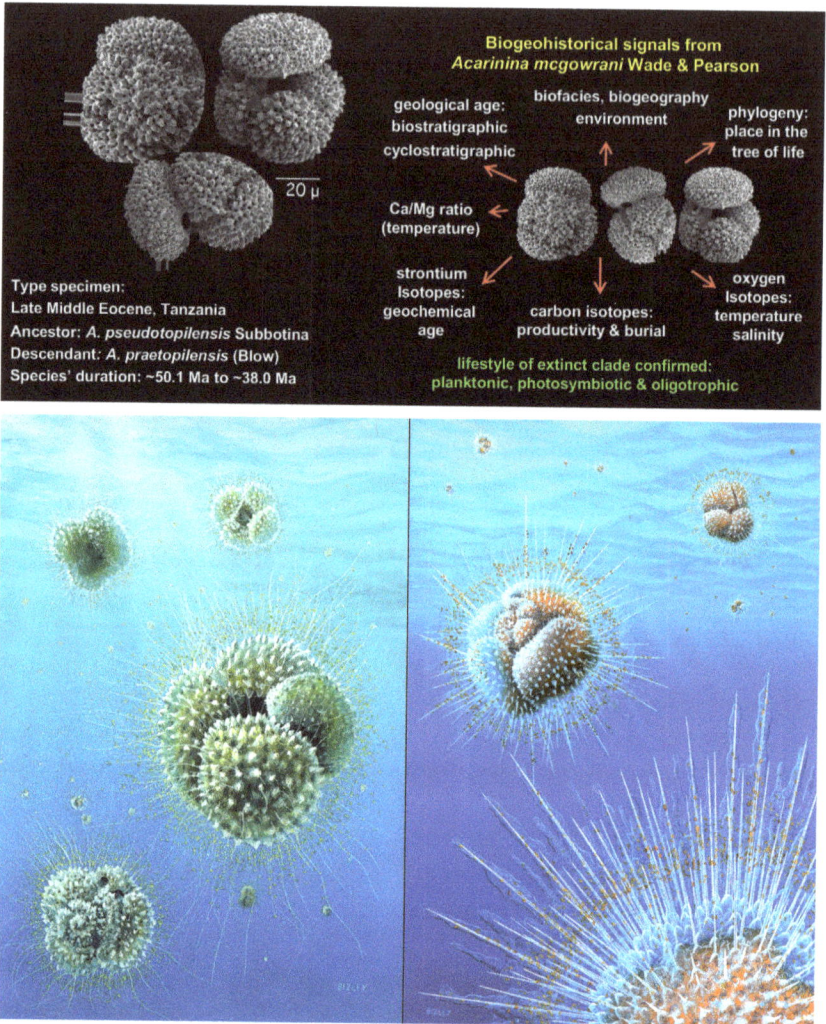

Figure 6.1. *Acarinina mcgowrani*: **Signals from an ancient ocean.**

6.1a: Portrait of an informative microfossil. *Left,* the type specimen (holotype) is the name bearer. Should this species be extensively revised and other names be required, as frequently happens with new reliable knowledge (i.e. scientific progress), then the new grouping including *this* specimen will retain *this* name. The three views of a trochospiral specimen are standard. Note the scale in microns: these shells are the size of sand grains. The species: (i) was planktonic, a denizen in the global ocean, but not a passive floater; (ii) lived in the brightly sunlit upper, warm, oligotrophic waters, the light and the warmth required by its halo of photosymbiotic 'algae' (single-celled eukaryotes, probably naked dinoflagellates); and (iii) arose from a known ancestor, gave rise to a known descendant and went extinct after its tenure on this planet of about 12 million years. We know all this from the lines of enquiry shown cryptically in the lower diagram (and several others, chemical). *Acarinina mcgowrani* is a powerfully informative messenger from an ancient ocean. It is not alone in that, but the genera *Acarinina* and *Morozovella* are particularly revealing about the Palaeocene and Eocene world.

6.1b: Portraits painted by Richard Bizley in 2011 and 2022, with scientific input from Paul Pearson. We once thought (Berggren, 1968; McGowran, 1968b) that *Acarinina* had spines like several modern species, but this idea has long been rejected. The prevailing view has been as shown on the left, where *A. mcgowrani* places its photosymbionts among filaments of soft tissue. Now, however, Pearson et al. (2022) present a compelling case for *A. mcgowrani* possessing spines of calcite very similar to, but evolved long before, independently of and convergently with, the spines evolved in modern species. Forward half a century, and the Berggren–McGowran notion arises again.

Source: (6.1a) Images of holotype from Berggren et al. (2006); (6.1b) Images © copyright Richard Bizley (bizleyart.com).

Consider *Acarinina mcgowrani* (Figures 6.1a and 6.1b). Certain other species would do as well; my choice is merely a mild case of geriatric vanity or identity politics.[1] The presence of *Acarinina mcgowrani* in a microfossil sample tells us that the age is constrained to a slice of Eocene time. Its association with other related species in samples from a wide range of sediments, from deep ocean to neritic, tells us that it was of planktonic habitat. And then there are the numerous chemical signals entombed within the crystalline calcite laid down during the growth of the shell, chamber-by-chamber. The best established and most versatile signals are the isotopic ratios of oxygen ($^{16}O/^{18}O$) and carbon ($^{12}C/^{13}C$). The delta numbers for the oxygen- and carbon-isotopic measurements, $\delta^{18}O(‰)$ and $\delta^{13}C(‰)$, are parts per thousand up (plus, 'heavier') or down (minus, 'lighter') from the agreed laboratory standard set at zero. The numbers are very small, they vary between species and their subtleties had no significant effect on the living organism.

The oxygen ratio reflects both the temperature of the oceanic water at the time of calcite crystallisation and its salinity, the latter being affected by the growth and decay of polar ice caps of significant size. Thus we have the problem of solving one equation with two variables. If we can assume no major icecap during greenhouse times (and the calcite has not been corrupted by groundwater after burial and fossilisation), then we have, in a succession of fossils in strata, a palaeothermometer. As the ocean cools, the $\delta^{18}O$ in successive samples increases ('gets heavier'), and if the polar icecap then grows, the salinity effect enters and $\delta^{18}O$ becomes heavier still. That is, the palaeothermometer becomes rubbery, but that can be compensated and the signal is still highly informative. In the planet's fully icehouse mode, growth of the icecaps resulted in sea level dropping by an estimated

---

1   *Acarinina mcgowrani* was described and named by Bridget Wade and Paul Pearson in the *Atlas of Eocene planktonic foraminifera* (Pearson et al. 2006), also the source of the holotype (type specimen) in Figure 6.1.

150 metres or so and the benthic $\delta^{18}O$ signal from the bottom of the deep ocean shifted by about +1.5‰. The calcium/magnesium ratio reflects water temperature and can be used to distinguish the temperature signal from the salinity signal in the oxygen number. The carbon isotopic ratio reflects the fact that photosynthesis preferentially fixes lighter carbon (again, in small numbers and physiologically ignored). If the production of organic carbon in the biosphere is high, and much is buried (e.g. as black muds, or peats, and perhaps becoming coal, oil or gas), then the carbonate carbon will be relatively heavy. Conversely, returning the organic carbon to the system as CO_2 will lighten up the ratio being fixed as carbonate skeletons. C_{org} shifts the reservoir numbers and C_{carb} records the shift.

These generalisations can hold at scales all the way from the microscopic and local to the megascopic and global. At the microscopic end, we can now justify the inferred cloud of symbionts enveloping *A. mcgowrani* in the brightly sunlit waters close to the surface of the Eocene ocean. The oxygen in *A. mcgowrani's* calcite is lighter than it is in several other contemporary planktonic species, indicating somewhat warmer and somewhat shallower water (in tens of metres). The carbon in its calcite is discernibly and consistently heavier, indicating that light carbon is being fixed photosynthetically in the microenvironment. Conclusion: *Acarinina mcgowrani* was tending its own garden in the bright light. We will be returning to the garden.

Oceanic palaeogeography and palaeoceanography

In the preceding chapter we brought the palaeogeography and tectonics of southern Australia almost up to date and showed its bearing on terrestrial southern biogeography. Now we return to the oceanic realm for the story of Cenozoic biogeohistory, and the natural starting point is the changing geography of the ocean basins (Figure 6.2.).

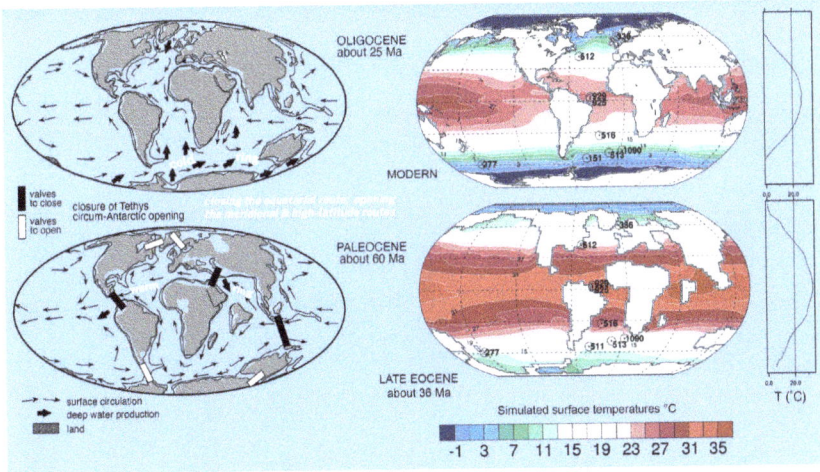

Figure 6.2 Two depictions of the making of the modern ocean.

Left, the plate-tectonic rearrangement centred on the Eocene was emphasised in the 1980s. The black and white bandages indicate the openings and closings that forced changes in oceanic circulation, especially the shift from warm-ring to cold-ring circulation. In the Palaeocene scenario, the deep water was dense saline water generated in shallow seas in warm latitudes. Warm brine drove halothermal circulation. Beginning after the Eocene transformation, the deep water was (is) cold water, its density enhanced in due course by brines under ice shelves. Cold brine drove thermohaline circulation.

Right, several decades later and the geographic patterns are well accepted, and carbon dioxide and palaeoclimatic modelling are prominent in two simulations of sea-surface temperatures. Locations are oceanic drilling sites; temperatures are contoured in 2°C intervals; latitudes are shown in 30° intervals and longitudes in 60° intervals. The Eocene world was warmer and temperature gradients were flatter.

Source: Maps at *left* from Haq (1983) and Seibold and Berger (1993). Maps at *right* from Pagani et al. (2011, supporting online material; copyright © AAAS).

Wolfgang Berger captured the fundamentals as a shift from the warm ring of the low-latitude ocean Tethys during the Palaeogene, inherited from the Cretaceous and about to die, crushed in the collisions of mobile continents, to the cold ring of the Neogene, especially the new, high-southern-latitude Southern Ocean. The Eocene was the critical time for the plate-tectonic opening or closing of 'valves', meaning the connections between the main ocean basins. Deep ocean water was produced in warm Tethys in the old times but is produced in the cold Southern Ocean today. Or so goes the narrative, rapidly produced in the early days, the late 1960s and 1970s, of the new science of palaeoceanography.[2]

---

2   For the beginnings of palaeoceanography, see Berger (2011, 2013), or the Seibold–Berger *The seafloor: An introduction to marine geology*, which ran to four editions (1982–2017), or Berger's tour de force *Ocean: Reflections on a century of exploration* (2009). No disrespect to leaders in a dynamic modern science, but Berger's thinking, writing and advocating are noteworthy for their brio.

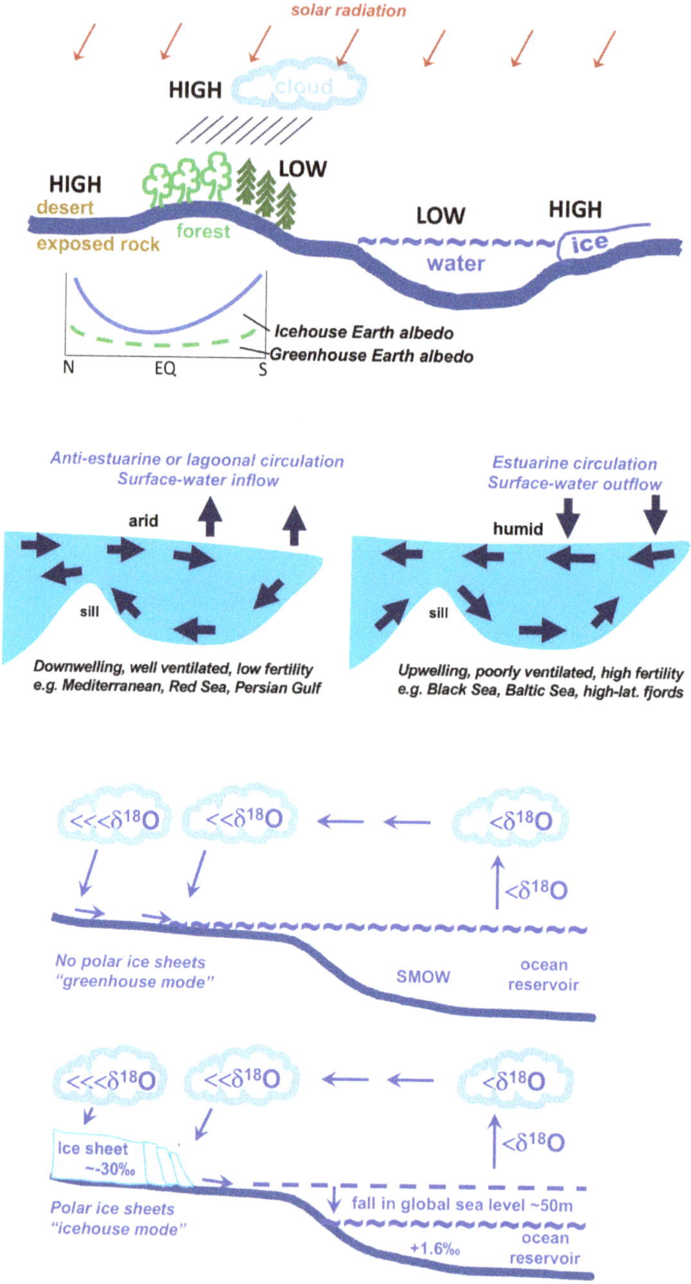

Figure 6.3. Aspects of climate and environment.

Above, albedo, reflectivity of solar radiation. Albedo is high on rocks and ice and clouds, low on water and vegetation. So, it was low in a Cretaceous world with no ice, less desert, more forest and vast shallow seas. (But cloud is the joker here.)

Middle, contrasting states in bodies of water in contrasting climates. The generalisations are useful at all scales, from a puddle to an ocean.

Below, why the oxygen-isotopic palaeothermometer is a rubbery instrument. Evaporation favours ^{16}O, raining favours ^{18}O, ice is isotopically very light. In a greenhouse world with no icecaps, we have a steady state. But in an icehouse world, locking up H_2O with a $\delta^{18}O$ signature of $-30‰$ leaves the ocean with $+1.6‰$ (SMOW is the chemists' standard mean ocean water) and a sea level lowered by some 150 m. So, a shell grown in that ocean preserves a single reading in the calcite of two variables, the temperature of the reservoir and the reservoir's isotopic signature at the time of calcifying.

Source: Author's depiction of textbook-type diagrams from the 1980s.

The internal driver of earth history manifests most obviously in plate tectonics—the restless earth. As we have seen. The external driver is solar radiation, which is absorbed unevenly, more at the equator than at the poles, more where reflective capacity is low (low albedo) such as water and vegetation, less where reflective capacity is high (high albedo). Again at all geographic scales, we note the lagoonal/estuarine contrast: between basins of water under arid and humid conditions; fresher and warmer water is less dense and saltier and colder is more dense. And we also note the hydrological cycle as seen through oxygen isotopes. In greenhouse mode we have a steady state, meaning lower contrasts in the oxygen signals in marine shells; in icehouse mode the waxing and waning of polar ice caps gives rise to stronger isotopic contrasts in planktonic and benthic shells (Figure 6.3.).

And so to the biosphere

In a fourfold and not unduly simplistic classification, we think of the hydrosphere, atmosphere and biosphere as underlain by the reactive lithosphere. The players in the interaction of these four 'spheres' are oxygen, carbon dioxide and water impacting on a small number of compound groups, namely minerals in the earth's crust ('calc-silicates'), organic carbon-based molecules ('carbohydrates') and carbonate carbon ('limestone'). I have needed to think about these matters for several decades but find it still useful to stare at four naive equations focusing on the peregrinations of carbon dioxide, in a table and a cartoon; and not forgetting silicon, as quartz and opal and high-temperature, rock-forming silicates (Table 6.1 and Figure 6.4.).

Table 6.1. Minimalist equations involving carbon dioxide in biogeohistory.

(i) Weathering of calcium-rich continental crust: $CaSiO_3 + 3H_2O + 2CO_2 \rightarrow Ca^{++} + 2HCO^-_3 + H_4SiO_4$ calc-silicate $\qquad\qquad\qquad\qquad\qquad$ dissolved silica
(ii) Photosynthesis; respiration or oxidation of biomatter: respiration and oxidation $CO_2 + 2H_2O \leftrightarrows O_2 + H_2O + CH_2O$ photosynthesis $\qquad\qquad\qquad$ 'carbohydrate'
(iii) [Bio]calcification; or (in reverse) dissolution of aragonite and calcite: $Ca^{++} + 2HCO^-_3 \leftrightarrows CaCO_3 + H_2O + CO_2$ bicarbonate $\qquad$ calcite/aragonite
(iv) Metamorphism: lime-rich and silica-rich oceanic sediments are returned to the crust at subduction zones: $CaCO_3 + SiO_2 \rightarrow CaSiO_3 + CO_2$ $\qquad\qquad\qquad$ calc-silicate

Source: Author's summary.

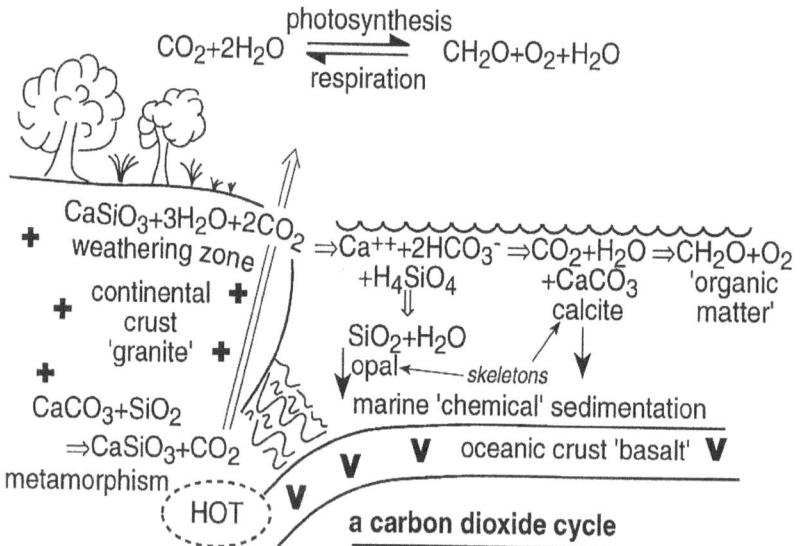

Figure 6.4. Equations in a carbon dioxide cycle.

The equations in Table 6.1 are displayed in a carbon dioxide cycle. Omitted are two important 'subcycles', namely deep ocean dissolving of calcareous skeletons in carbonic acid, and the interplay between the 'carbonate factories' in the open ocean (the pelagic realm) and the shallow seas (the neritic realm). C-org or C_{org} is organic carbon, fixed photosynthetically, and to be distinguished from carbon-carbonate, fixed in the shell or tooth or bone, or by mats of cyanobacteria.

Source: Author's depiction.

Equation (i) is boosted when metamorphic minerals, formed at high temperatures and pressures, become unstable when tectonic uplift and stripping exposes them to intensive weathering at the earth's surface. Equation (ii) runs in both directions in a steady state of photosynthesis versus consumption—until organic carbon is buried in sediments. It is here that the isotopic fractionation of carbon is so informative, but it is in equation (iii) that we see the signals of equation (ii) in fossil shells. It would be misleadingly incomplete to infer that a 'carbon dioxide cycle' is completed when the marine and biogenic calcite and opal are returned via equation (iv) to the infernal kitchen in the nether regions. For one thing, vast amounts of calcium carbonate are grown by molluscs and corals in the geologically unstable form of aragonite, which mineral is easily dissolved by groundwater, thereby supplying calcite cement in solid limestones.

More significant here is the fate of much of the vast amount of calcite comprising the '*Globigerina* ooze', now known as calcareous ooze because it is less the skeletons of the more visible planktonic foraminifera, more of the less visible planktonic coccolithophoric 'algae' generated in the photic zone. As the early oceanographers discovered, there is a lower limit to the pale calcareous ooze in the oceanic depths, a 'surface', a kind of analogue to the snow line in mountains. The rain of calcareous planktonic skeletons (much of it packaged as turds from the krill) meets dissolution in the cold, carbon-dioxide-charged abyssal waters known as the Antarctic bottom water. The depth at which this balance between supply and removal occurs at any given point in space and time is known as the calcite compensation depth (CCD). Above the CCD, calcareous ooze; below the CCD, brown clay. The CCD is another case of two variables but one solution. First, consider 'supply'. The rain of calcareous shells from the sunlit waters varies considerably in space and time, like all biological processes. The strength of aggressive acid attack in the deep bottom waters also varies, as a function of how much carbon dioxide is carbonic acid.

And now we bring things together in a consideration of two model oceans, cool and warm, as presented in a particularly prescient and heuristic discussion by the foraminiferologist Jere Lipps (Figure 6.5.).

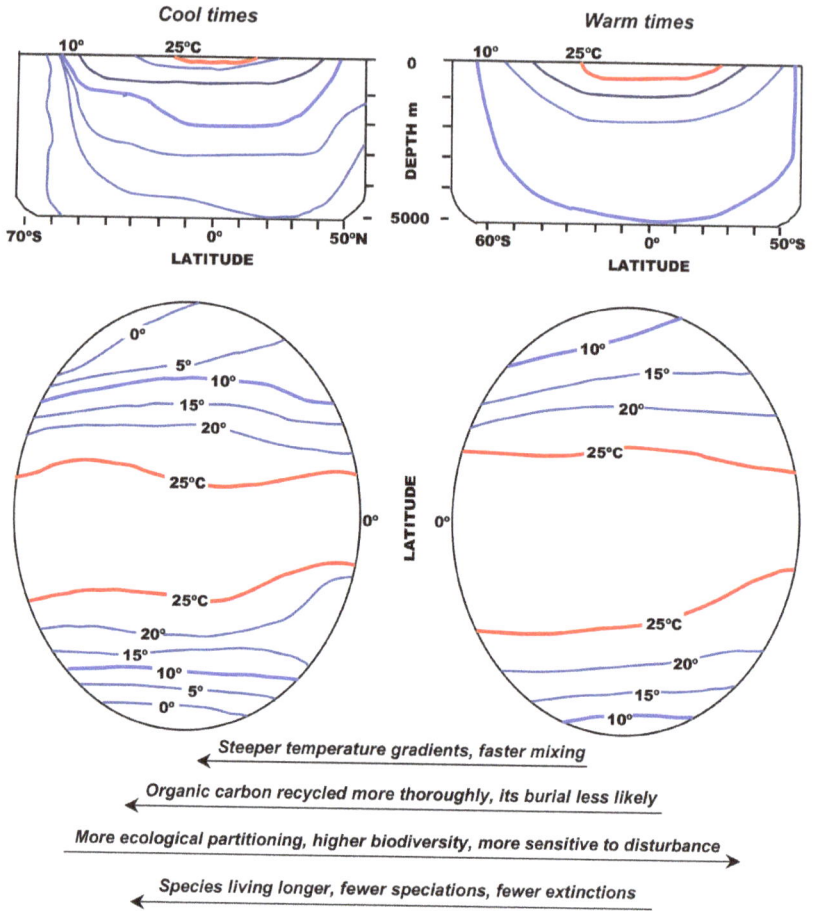

Figure 6.5. Lipps's early 1970s model oceans.

Jere Lipps's early 1970s model oceans, in cross-section and in plan view at two ends of a spectrum. The ecological and evolutionary generalisations are still heuristic in palaeoceanography and palaeobiology.

Source: From McGowran (2005a, Fig. 6.25) after Lipps (1970).

These models have implications for several disciplines. In warm times, equator–pole temperature gradients are flatter, so atmospheric gradients are flatter, oceanic mixing is less vigorous and the oceanic water column is more layered, or stratified. These are the times when organic matter is more likely to be buried before it is recycled: hence the great accumulations in the geological record of coal, oil and gas. Palaeogeographically, you will get shallow seas spilling more extensively across the continental margins. Ecologically, you will get a broader spectrum in the warm oceans, from oligotrophic (low fertility in the marine deserts) to eutrophic (high fertility in the great fishing grounds). Longer food chains will develop, from basic primary producers to top predators. There will be more specialising and more species (diversity), hand in hand with more fragility and greater sensitivity to physical perturbations. In cool times, vice versa all round in cooler oceans. Note too that the bottom waters of the deep oceans can be 10° and more warmer in a warm ocean. We have more insight into all of these matters today, but Lipps's ideas of 1970 stand up pretty well.

Miocene tutorial on environmentally informative isotopes — the Monterey theory of Miocene ice

For an illuminating example of the insights afforded by oxygen and carbon isotopes, we return to our Site 216 on the Ninetyeast Ridge to examine an integrated 1980s study of the Miocene section by Berger and Vincent (Figure 6.6.).

The argument runs as follows. The carbon signal in the succession of calcareous microfossils up though the strata at Site 216 displays a positive 'excursion', out in the heavy direction 18 to 17 million years ago and back about 5 million years later, all within the Miocene Epoch. Halfway through this interval the oxygen signal in the *same* succession of microfossils lurches in the heavy (positive) direction, suggesting that strong polar cooling and, indeed, growth of the Antarctic icecap cooled the oceanic bottom water. The inference is that the carbon excursion is recording the withdrawal of large amounts of organic carbon from the biosphere and its burial, and the resulting drawdown of carbon dioxide triggers 'reverse greenhouse', culminating in a major glaciation.

Although this outline is sufficient for our purposes, a perusal in some detail is rewarding because the study was so elegant and educational. The section of calcareous ooze at Site 216 is in metres, so the time is not linear. The dates in millions of years and the divisions of the Miocene come from the biostratigraphic subdivision using the coccoliths, the planktonic foraminifera and the radiolarians, always cross-checking and keeping each other honest. The study covers about 18 million years of Miocene time preserved in about 100 metres of calcareous ooze. Three species of foraminifera are analysed, one benthic species (BF) on the floor below kilometres of water, one deep-planktonic (DPF) and one shallow-planktonic species (SPF) perhaps a couple of hundred metres' depth apart in the upper waters.

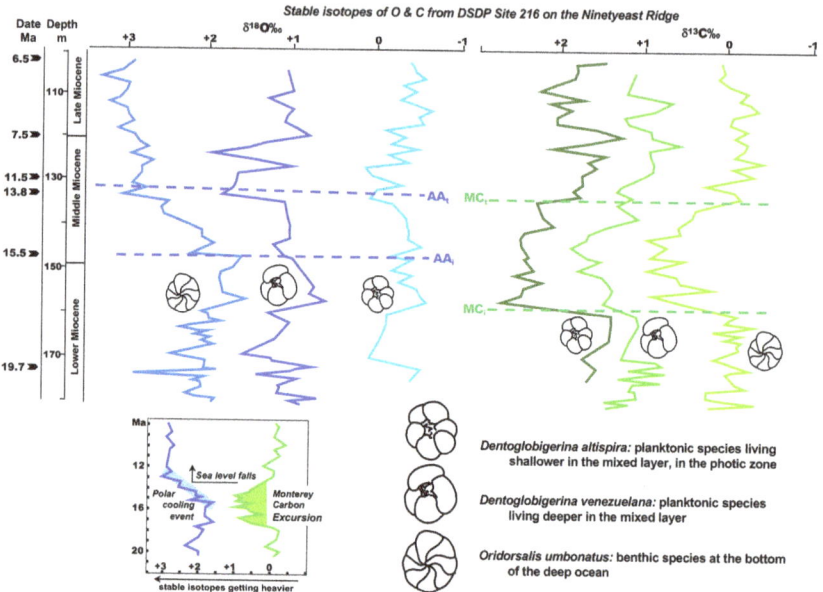

Figure 6.6. The Monterey event at Site 216 on the Ninetyeast Ridge.

BF, benthic foraminifer (*Oridorsalis umbonatus*), on the ocean floor on the Ninetyeast Ridge; DPF, deep planktonic foraminifer (*Globoquadrina venezuelana*) living in the deeper surface mixed layer; and SPF, shallow planktonic foraminifer (*Dentoglobigerina altispira*) living high in the surface mixed layer. They lived separate lives but they fossilised together. Note that the scale (*left*) is linear in metres' thickness of sediments, so the time in millions of years is not linear. Note that six sets of isotopic data (three carbon, three oxygen) run clear of each other, no entangling or overlapping. This shows that the ocean is being affected, top to bottom, to varying degree by the changes. MCi, Monterey carbon initiation, implying CO_2 drawdown, begins the story. AAi, Antarctic cooling initiation, is the reaction. MCt, Monterey carbon termination, is forced by the global chilling and expansion of the icecap. AAt, Antarctic cooling termination, when bottom waters remain cold, but the uppermost surface waters have remained warm.

Inset is the Monterey hypothesis as Berger simplified it. The oxygen signal of polar cooling and the (heavy) carbon signal of massive burial of (light) organic carbon are embedded in the same specimens in the same samples of calcareous ooze from our Site 216 on the Ninetyeast Ridge. So there are no problems of dubious correlation or false age determination. Burying carbon implies drawdown of CO_2. Drawdown implies cooling; cooling implies an icecap; an icecap implies a lowered global sea level; lowered sea level in a colder world implies both decreased accumulation of coal and other fossil fuels, and a return of carbon to circulation by erosion and oxidation; return of carbon will show up in a lighter carbon signal in oceanic shells. Which it does: QED. This is the Monterey hypothesis, matured to the Monterey theory.

Source: Berger and Vincent (1986), redrawn and simplified.

First, note the clean separation of the six curves throughout the section: no crossovers, no confusion between the deep ocean and the surface layer or within the latter. Next, the oxygen pattern: the benthic species is both heavier and getting distinctly heavier through time (+2 to +3 parts per thousand) than are the two planktonics, whose numbers wobble but with less pronounced trend. This suggests that the denser bottom waters were cooling. The lines AA_i and AA_t mark the initiation and termination of the spurt in growth of the Antarctic icecap, which is where the chilling of tropical deep waters comes from.

And now to the carbon. The heavier the carbon number in the shells, the more light carbon is being removed photosynthetically. Here, the highest numbers are the shallow planktonic species, which inhabited the sunlit uppermost waters where photosynthetic activity was most vigorous. The deep-planktonic species is lighter, indicating both less photosynthesis and recycling of organic materials from above. Lightest is the deep-ocean benthic species, due to the rain of organic materials generated in the sunlight far above. But inspect the positive isotopic excursion between the horizons labelled MC_i and MC_t, meaning the initiation of the Monterey carbon excursion and its termination, 5 million years later. Most importantly, the positive isotopic shift is very clear in all three profiles—that is, in the whole ocean, suggesting that light carbon was accumulated outside the ocean as a whole (the burial of future oil source rocks and the accumulation of brown coals on land both spring to mind), not simply redistributed between surface waters and deep waters. Berger and Vincent identified the culprit as being the source rocks for Monterey oil in the north-east Pacific margin; there are other options too.

So we have a temporal pattern here in six profiles, suggesting cause and effect like this: carbon shift throughout the water column in carbonate-carbon → burial of organic carbon → CO_2 drawdown → threshold in reversed greenhouse → chilling and icecap growth → global cooling → fall in sea level → return of light carbon to ocean. This is the Monterey hypothesis of rapid Antarctic icecap expansion by sequestering carbon dioxide. By the early 1980s there had been extensive scientific ferment in attacking the problem of the ice age—but that was a Pleistocene problem. Here we had a comprehensive theory for the recently discovered ice age forerunner in the Miocene Epoch, a theory of glaciation centred squarely on the biosphere, not merely upon plate tectonics and the behaviour of oceanic valves. I took this theory from the mid-Neogene glaciation theory back into the mid-Palaeogene by showing that the patterns of carbon vis-à-vis oxygen isotopes was remarkably anticipated in the decline of the hothouse-prone Early Eocene (see next chapter).

The farming photosynthesisers

There is a special category of the foraminifera, those lineages of microbes able to farm other, photosynthesising microbes. In this arrangement, photosymbiosis, housing accommodation and excreted materials including CO_2 are traded for carbohydrates and oxygen, and that trade pact has been achieved numerous times in the history of life.[3] It has been achieved frequently by the major groups of the foraminifera in the benthos and repeatedly by the planktonics (witness *Acarinina mcgowrani*).

The photosymbiotic benthics have repeatedly evolved species with large or very large shells with many rooms or cells and a large surface–volume ratio. All of this has to do with the basic requirements of their symbionts, namely shelter, warmth and light. Their partners are variously the four eukaryotic, single-celled groups of photosynthesising microalgae, namely the diatoms, dinoflagellates, chlorophytes (green algae) or rhodophytes (red algae); and also many kinds of prokaryote (cyanobacteria). The requirements and tolerances of the photosymbionts largely control the distribution of the foraminiferal species as to depth, temperature, water clarity and salinity.[4] We have a beautifully prepared profile from near the northern limits of large foraminifera in the north-west Pacific, thanks to sampling down the decades synthesised by Johann Hohenegger (Figure 6.7), who has also written a superb, highly accessible account of those same foraminifera (Hohenegger 2011).

---

3   Way back in the Middle Proterozoic Eon, the early eukaryotic microalgae underwent an evolutionary schism, into 'reds' (with chlorophyll c) and 'greens' (with chlorophyll b). The greens in due course gave rise to the land plants. The reds, in a Mesozoic revolution, gave rise to the phytoplankton dominating the modern ocean, namely the dinoflagellates (with organic skeletons), the coccoliths (with calcareous skeletons) and the diatoms (with siliceous—opal—skeletons). Their dominance extends widely, from their ecological importance to their contributing the bulk of the world's fossil fuels sourced in Mesozoic and Cenozoic seas (Falkowski et al., 2004; Katz et al., 2004), and to their partnerships with numerous lineages of foraminifera.

4   Prazeres and Renema (2018) have reviewed the evolutionary significance of the microbiome, which is the repeatedly invented partnership of the hosts, large benthic foraminifera, the symbionts, eukaryotic microalgae, and the endobionts, prokaryotes (bacteria). Diatoms are the most widely distributed microalgal symbionts and, requiring blue-green spectrum light, colonise the deepest levels of light penetration. The poster child of the dinoflagellates is the richly diverse ('speciose') genus *Symbiodinium*, partner of the photosymbiotic planktonic foraminifera (and most likely also those in extinct lineages) as well as corals and the giant clam. Prazeres and Renema observe that the large benthic foraminifera 'are essential ecosystem engineers and prolific carbonate producers, and the study of their microbiome should provide important information on their ability to respond to climate change' (2018, p. 16).

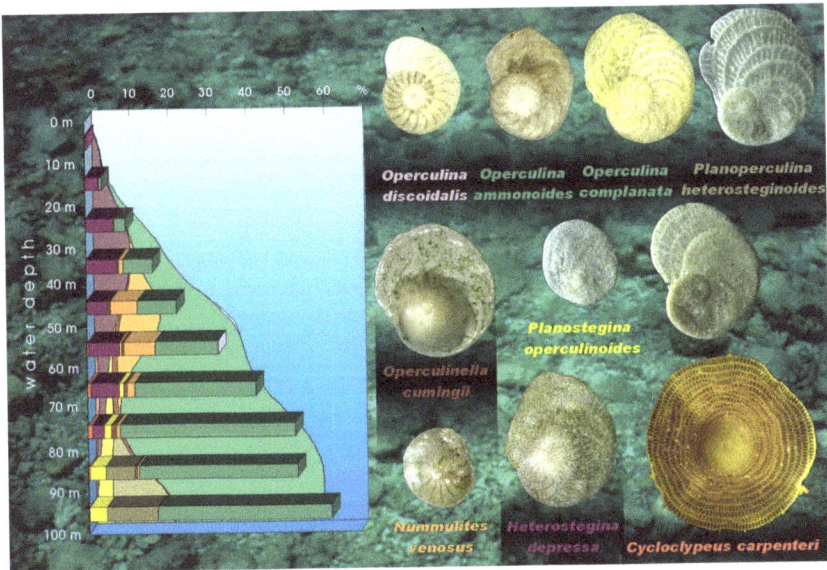

Figure 6.7. Hohenegger's large benthic foraminifera off Okinawa.

In the western Pacific Ocean, large benthic foraminifera have migrated northwards on warm currents, along with such tropical and subtropical organisms as corals. These single cells in multi-chambered shells are large (2 mm up to 13 cm), flat like a disc (with large surface–volume ratio), and with many chambers in the shell. They live in warm clear water in the photic zone (mostly the upper 100–130 m). This is about housing 'microalgae' (all of which are diatoms here), and bacteria as built-in gardens needing many photons, delivered reliably. It is a symbiotic arrangement whereby the foraminifer provides shelter and CO_2 in exchange for carbohydrate and oxygen. This profile off Okinawa was assembled by Johann Hohenegger. The colour-coding shows species distributed against depth, where the main variable is light (blues in the spectrum penetrate deeper than the reds).

Source: Courtesy of a Christmas card from Johann Hohenegger.

Recall the illustration from almost two centuries ago of *Nummulites* in the densely fossiliferous rock that built the pyramids (Figure 5.6). Just as the ammonite became the poster child for the Mesozoic Era and the trilobite for the Palaeozoic, so did Charles Lyell anoint *Nummulites* for the 'Tertiary or Cenozoic' on the frontispiece of his *Student's elements of geology* (1871). Why are they packed so densely? The easy answer is winnowing of the sediment post-mortem. The more significant answer is that they did not need much room. But these fossiliferous rocks become cemented by circulating bicarbonate and, not being susceptible to disaggregation freeing specimens in the laboratory, are studied as thin sections under transmitted light. The two most informative sections for restoring a three-dimensional coiled shell are the axial and the equatorial. And the new methods of X-ray tomography are revealing new levels of variation and complexity in these ancient shells, largely to be attributed to the needs of their algal and bacterial photosymbionts (Figure 6.8).

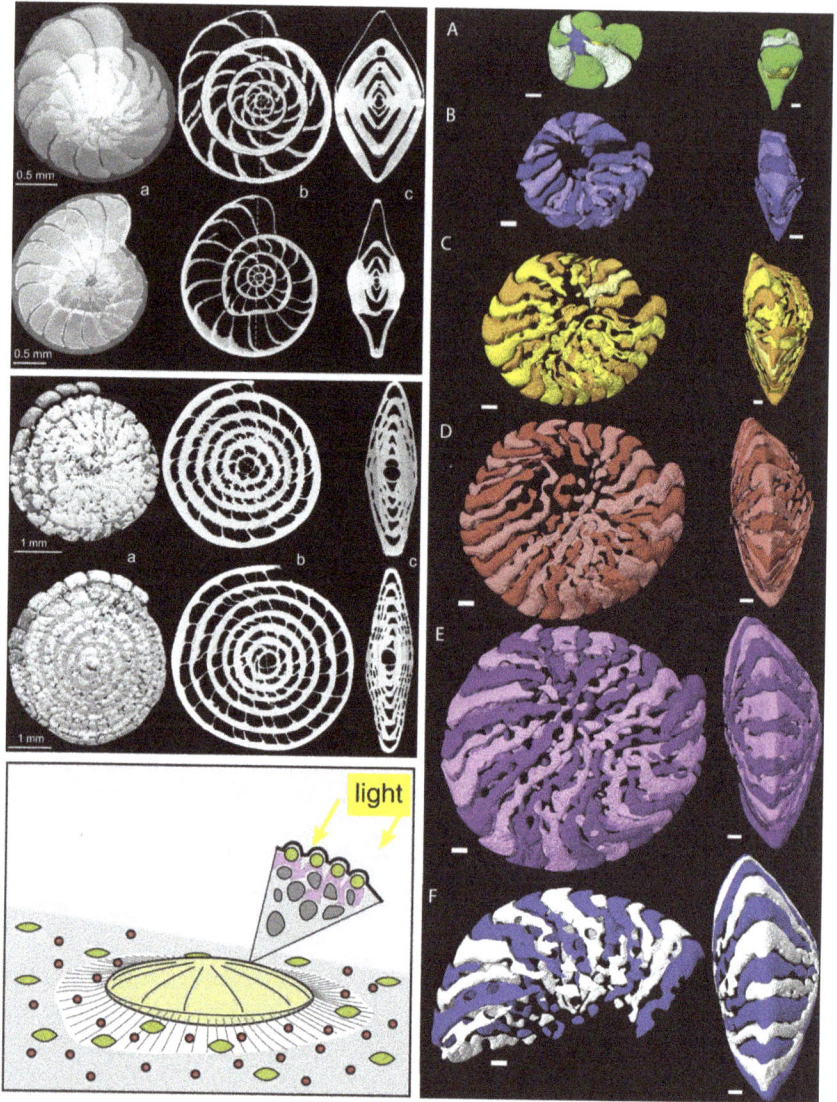

Figure 6.8. *Nummulites* and relatives illustrate the power and potential of X-ray tomography.

Top left, the modern species *Palaeonummulites venosus* and *Operculina ammonoides* (the same species are also in Figure 6.7) are shown as if the whole shell is removed (so we are seeing the chamber spaces as infilled, known as internal moulds), the equatorial section and the axial section. The shells look regular and uncomplicated (explaining the notion, falsified in the 1820s, that they were miniaturised versions of *Nautilus*). *Centre left*, the Palaeogene species *Nummulites fabianii* and *Nummulites fichteli* look rather more complicated. Those nummulites are from western Tethys whereas *Nummulites djokdjokartae* from eastern Tethys (*right*) shows distinctly more variation and apparent irregularity in its chamber spaces. Views *A* to *F* are of a single specimen showing (in equatorial and edge view) five and a half whorls in succession, again as internal moulds (alternating hues are for clarity). The spaces were packed with diatoms and bacteria. *Lower left*, a nummulite on a sunlit seafloor with phytosymbionts internally and externally (diatoms and bacteria).

Source: *Palaeonummulites venosus, Operculina ammonoides, Nummulites fabianii* and *Nummulites fichteli* are from Hohenegger and Briguglio (2012) and from Briguglio et al. (2013). *Nummulites djokdjokartae* is from Renema and Cotton (2015). The nummulite on the sunlit seafloor is from Pomar and Hallock (2008, Fig. 10E).

So we find these partnerships in clear sunlit waters on platforms, margins and atolls at tropical to subtropical latitudes, distributed through the photic zone according to the preferences or constraints of the symbionts for light of different wavelengths, protection from UV, and temperature. Together with warmth and light is the environmental factor of low nutrient—ecologically, at the oligotrophic end of the nutrient gradient. (The larger foraminifera have been compared in their recycling self-sufficiency to an Israeli kibbutz in the desert.) When nutrient levels rise or water temperatures fall, or oceanic mixing intensifies, the photosymbiotic strategy becomes less competitive.

Corals and coral reefs are conspicuous by their absence from Eocene strata in Tethys (in fact there is a worldwide dearth of coral reefs between the Cretaceous and the Oligocene). Instead, the large foraminifera have developed a striking fossil record, as displayed in the reconstructed transect in the low-latitude ocean, Tethys (Figure 6.9).[5]

As we have seen, Tethys was in the process of disappearing into what was to become the Alpine–Himalayan mountain chain, taking with it a kaleidoscope of marine environments preserved in the great limestones and other sediments of Palaeogene age. The environments ranged from shallow and over-saline, through sunlit carbonate platforms and ramps sloping through the photic zone into the deeper ocean. Geological surveying and petroleum exploration and production demanded geological understanding. This reconstruction was a major outcome.

---

5   The five thin sections in a reconstructed profile are from *Southern Tethys biofacies* (AGIP, 1988). The Southern Tethys domain comprised lands collapsed together in the Mediterranean and Middle East regions. The same biofacies are known from eastern Tethys—New Guinea and Timor-Leste ('Sahul') where David Haig et al. (2019) have presented a similar reconstructed profile for richly fossiliferous limestones.

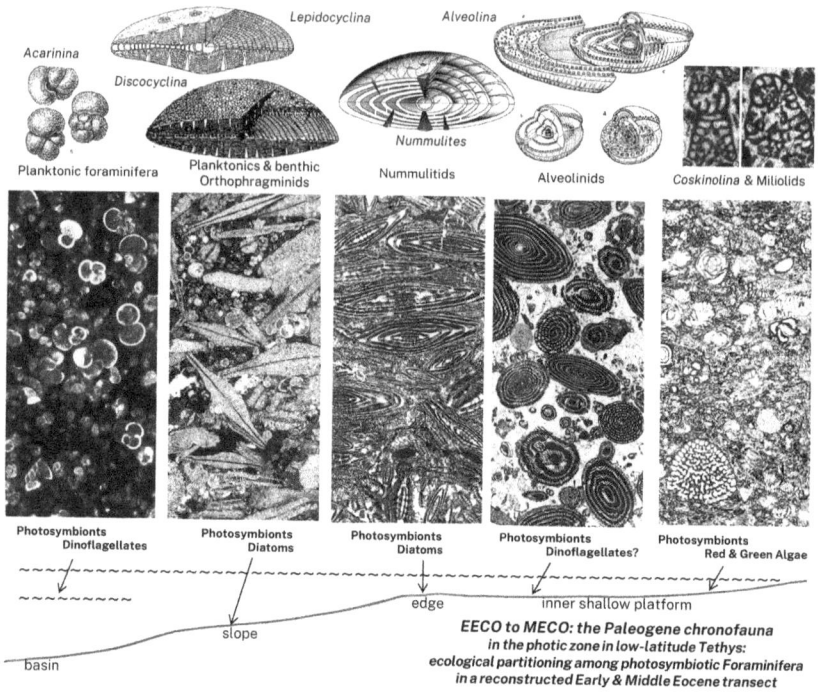

Figure 6.9. Eocene Tethys in the photic zone: Benthic foraminiferal partitioning. Thin sections in a reconstructed profile are from *Southern Tethys biofacies.*

The larger foraminifera peaked in the Early Palaeogene. In Tethys there were numerous shallow-water habitats — shelves, ramps, islands, atolls — and we find there the highest diversities and broadest range of communities. Thin sections of limestones show foraminiferal biofacies assembled in a reconstructed, idealised facies transect from inshore (shallow, may be very salty) to the open ocean, where the species were planktonic. The two main points are: (i) all are dominated by photosymbiotic species, and (ii) this spacing out of the distinct assemblages is an ecological partitioning. Hanging in the air is a question: where are the coral reefs? Third, the two on the right (inshore) are dark to transmitted light, whereas the others are glassy. The former needed protection as well as light! *Above*, the cutaway diagrams of shells reveal the internal structure, of many tiny chambers for housing the microalgae and bacteria. The disc-shaped *Discocyclina*, *Lepidocyclina* and *Nummulites* have wedges of calcite both strengthening the skeleton and illuminating the deep interiors (via their light-fast crystallographic axes). The grain-or spindle-shaped *Alveolina* are adapted to rolling in shallow water.

Source: Sartorio and Venturini (1988) ©Agip S.p.A.

Perhaps the most noticeable point is how densely the individuals are packed in a microscopic thin section. The individuals could live at very close quarters while surviving comfortably in their symbiotic arrangements. Instead of coral and algal 'reefs' we have foraminiferal 'banks'. Examining the cutaway diagrams of *Lepidocyclina*, *Discocyclina* and *Nummulites* skeletons (shells), we see high-density living in multistoreyed apartment blocks. And we also see wedges of calcite looking like pillars strengthening the shells. That too, but these wedges are calcite crystals with their light-fast axes penetrating the structures, persuasively analogous to light wells in hotels or office blocks. However, the Alveolinid and miliolid shells in the shallows do not have the glassy and transparent walls. Their calcite (known as 'microgranular') is at best translucent, frosted rather than glassy—probably they are prone to too much light and are shielded from UV.

This is a reasonably valid depiction of the photic zone in Tethys during the earlier part of the Eocene Epoch, the time given such labels as EECO (Early Eocene climatic optimum), and warm-ring ocean in a hothouse world. But we have seen that global tectonics was transforming that world. What happened to these diverse communities of large and photosymbiotic foraminifera? To begin with, the Nummulitids, Alveolinids, Orthophragminids and Miliolids on the seafloor and the planktonics in the surface waters never stopped evolving. Lineages split as species speciated and gave way to their descendants; the earth moved into hothouse mode in the Early Eocene and out again in the early Middle Eocene; there were extensive impacts on the five communities on the way in and on the way out—but still they held together, they stayed discernible for roughly 18 million years. This combination of incessant evolutionary change within a deep-time framework of overall stability is known as a chronofauna. This one we named the Early Palaeogene chronofauna,[6] and it came to an end in the turbulence of the Auversian Facies Shift. These concepts are developed in Chapter 10.

---

6   Lukas Hottinger (1997) recognised this long-term, overall stability of evolving communities of large benthic foraminifera in the Eocene shallow seas, which we named the Early Palaeogene Chronofauna (McGowran and Li 2000). See Chapter 10 for chronofaunas.

The temporal sweep of the Cenozoic ocean

In the past two decades the Cenozoic time scale has become a robust edifice reliably founded upon three main pillars. One is biostratigraphy, exploiting the speciations and extinctions in several major groups of fossils. A second is numerical ages, putting numbers on the ages of events and misleadingly called 'absolute ages'. Third is magnetostratigraphy, the well-dated succession of the reversals in earth's magnetic field and now the backbone of the Cenozoic time scale. And the robustness of this edifice has been raised to a new level by astrochronology, aka the Milankovitch cycles, the disentangling of rhythms recorded in sedimentary strata by the rhythms through time of the earth–moon–sun system.

Also in the past two decades, these advances have made it possible to track various indicators and phenomena of global environmental significance. There are the massive compilations of readings of the oxygen and carbon ratios, $\delta^{18}O$(‰) and $\delta^{13}C$(‰), signals in the calcite of benthic foraminifera from the bottom of the global deep ocean. The oxygen signal is a more or less trustworthy palaeothermometer for the first 30 million years (K-Pg to Oi-1), after which the ice effect gets stronger. Then there is a curve for 40 million years of atmospheric carbon dioxide. The reconstructed trajectory of the calcite compensation depth (CCD) is from the eastern-central Pacific Ocean. This assembly of multiple lines of evidence is a chronicling of the Cenozoic record. Coming to understand what the curves might mean would take us quite some distance forward in a history, a geohistory and biohistory of the Cenozoic Era, and would also serve as a template for what was happening in southern Australia (next chapter).

As of about 2010, that template of global environmental shifts during the Cenozoic Era could look as portrayed here in Figure 6.10.

Environmental history now, though, is framed more heuristically by the five natural divisions of the global climatic state as outlined in the review by Westerhold and his team (Figure 0.5).

The curves shown here (Figures 6.11, 6.12, 6.13) for oxygen and carbon isotopes, carbon dioxide and delta temperature are taken from that review, but I must point out that the running averages only are shown for clarity, omitting the fuzziness of the data as shown in the real curves of Figure 0.5. Stand back and look at the big picture, a two-part panel, Palaeogene Period below and before Neogene Period (Figures 6.10–6.13).[7] Overall, there are general trends from 50 million years ago to the present—cooling, lowered carbon dioxide, subsiding CCD, lowered sea level—but equally apparent is an 8-million-year punctuation in the middle, labelled the critical interval.

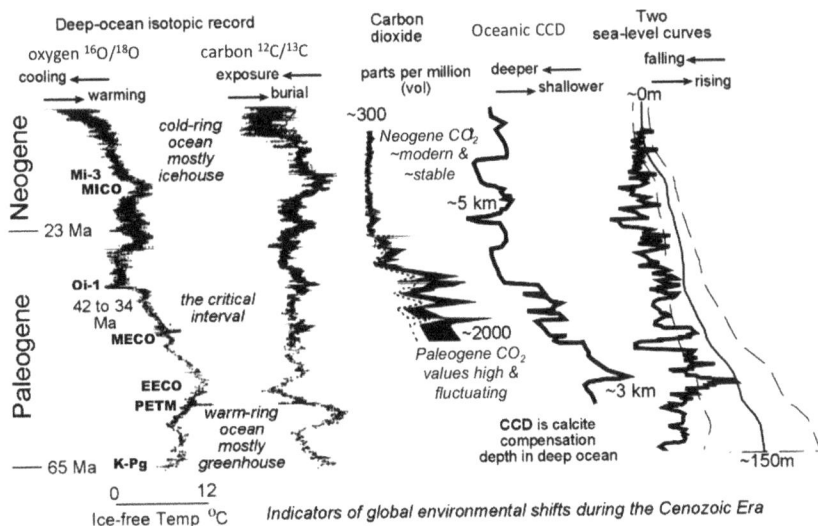

Indicators of global environmental shifts during the Cenozoic Era

Figure 6.10. Indicators of global environmental shifts during the Cenozoic Era.

Signals of Cenozoic global geohistory as of about 2010. The temporal sweep of the global ocean is captured in a grand parade at the grand time scale. We see in one gulp the trajectories of the ocean cooling, of 'organic' carbon in the biosphere, of carbon dioxide, of 'carbonate' carbon in shells and of reconstructions of sea level (by two methods). Oi-1 and Mi-3, isotopic signals of glaciations. EECO, MECO MICO, Early Eocene, Middle Eocene and Miocene climatic optima, respectively. K-Pg, Cretaceous/Palaeogene boundary. PETM, Palaeocene–Eocene thermal maximum.

Source: From McGowran (2012a, Fig 3): carbon and oxygen, Zachos et al. (2001, 2008); carbon dioxide, Pagani et al. (2005); calcite compensation depth, equatorial Pacific, Lyle et al. (2010); sea level, Kominz et al. (2008).

---

7   The photos of massive neritic carbonates from our region point to the correlation of neritic limestone deposition with geologically extremely rapid shoaling (shallowing) of the oceanic CCD at just the time of the Khirthar transgression and the accumulation of large—huge!—neritic limestones (Figures 5.15 and 5.16).

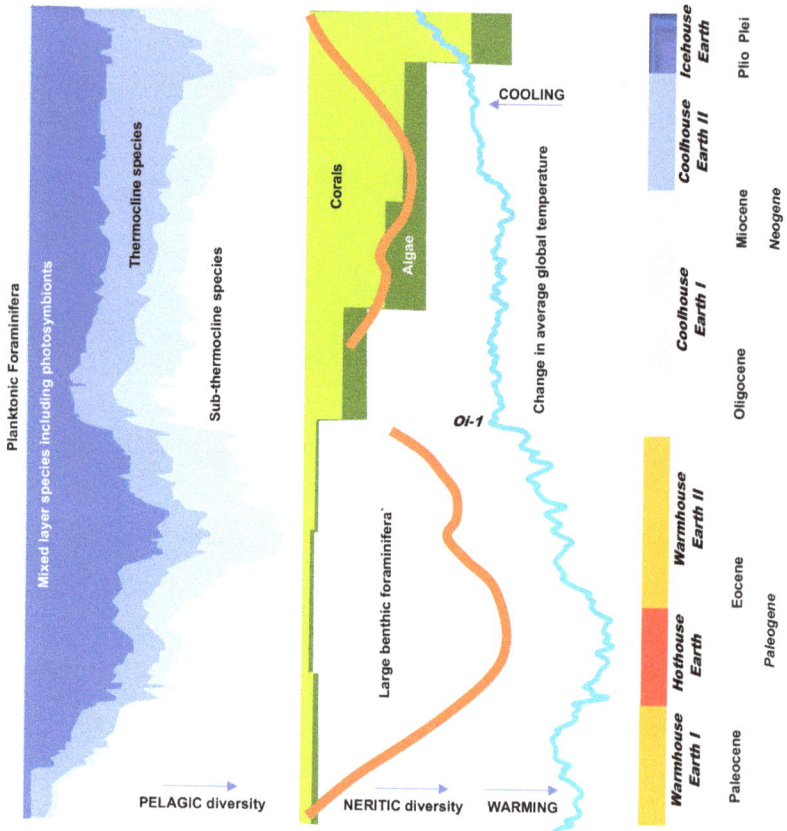

Figure 6.11. Foraminifera and corals in a natural two-part Cenozoic Era.

Calcareous skeletons making fossil records, good (neritic algae and corals), better (neritic large foraminifera) and best (planktonic foraminifera), reinforce the natural two-part or binary Cenozoic, so far as major components of the biosphere in the neritic and pelagic realms are concerned. That is the main point of this diagram. The planktonic foraminifera are ecologically sorted three ways, into inhabitants of the mixed layer of the ocean, in the vicinity of the thermocline and just below the thermocline. (The maximum diversity attained was 45–50 species, based on their shells.) They recovered from an almost complete wipe-out at the end of the Mesozoic Era; they peaked in the Middle Eocene *after* the hothouse; they suffered extinctions particularly at the end of the Eocene; they recovered gradually in the Neogene. At this very broad perspective, the large foraminifera in the neritic have strong similarities and some differences with the plankton in the pelagial; and they have stronger differences with the corals and algae in the neritic. Pamela Hallock and her colleagues explain the contrast in terms of different living conditions in the warm shallow seas. The Eocene seas had weaker thermal gradients at first, therefore less vigorous circulation, less mixing and lower nutrient levels, suiting the lifestyles of the foraminifera. Neogene seas were mixed more vigorously under steeper thermal gradients, promoting the washing of nutrient-laden water across the sessile corals and the flourishing of the algae.

Source: Author's combination of: *left*, planktonics plot from Ezard et al.(2011); *middle*, large foraminifera, corals and algae from Pomar et al. (2017); *right*, global temperature curve and earth's climatic states from Westerhold et al. (2020).

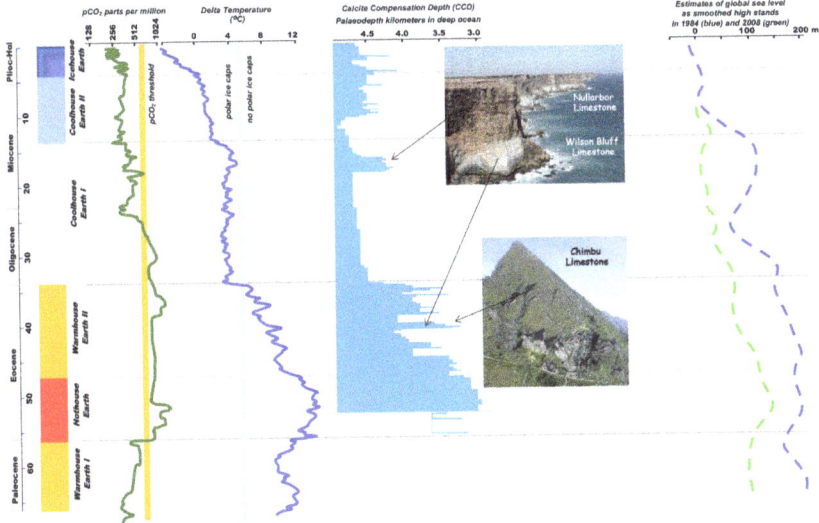

Figure 6.12. Environmental shifts during the Cenozoic Era, updated.

Signals of Cenozoic geohistory. The strongest message is twofold. First, compare the overall parallel slope in the curves for atmospheric carbon dioxide (decreasing), oxygen isotopes (getting colder and icier), and CCD (deepening), all obtained from deep-ocean drilling. Second, the 'big picture' is a two-part picture. The Palaeogene ocean was a warm ring ocean in a mostly greenhouse world; the Neogene ocean is a cold ring ocean in a mostly icehouse world.

Source: Westerhold et al. (2020), with CCD from Pälicke et al. (2012, Fig. 2) and sea level curves from Kominz et al. (2008, Fig. 12).

Before this, during the Eocene, oceans in warm-ring configuration were warmer than today (10° and more warmer, in the bottom waters of the global ocean), carbon dioxide levels were several times higher than preindustrial levels, and calcite compensation depths were 2 kilometres shallower. The hothouse world was a 'greenhouse world'. We have known since the days of Charles Lyell that a vast Eocene sea covered large parts of what is now Europe, that extensive limestones swarming with *Nummulites* gave the French name *La Nummulitique* to the Lower Tertiary or Palaeogene, and that terrestrial floras and faunas with palms and crocodiles, well north of the Arctic Circle, were to be compared with, say, the modern Florida Everglades. From 50–55 Ma to 40–45 Ma the $\delta^{18}O$ curve implied cooling and the CCD curve deepened during the Eocene. Perhaps the reason for the deepening CCD was twofold. Cooling the ocean would have steepened the gradients, invigorating circulation and mixing and promoting the exchange of oceanic carbon dioxide with atmospheric oxygen, thereby reducing the corrosive carbonic acid in the bottom waters and lowering the CCD; warming the ocean would reverse the processes. Meanwhile the depositing of extensive tracts of limestone in the neritic realm would raise the oceanic CCD to shallower levels.

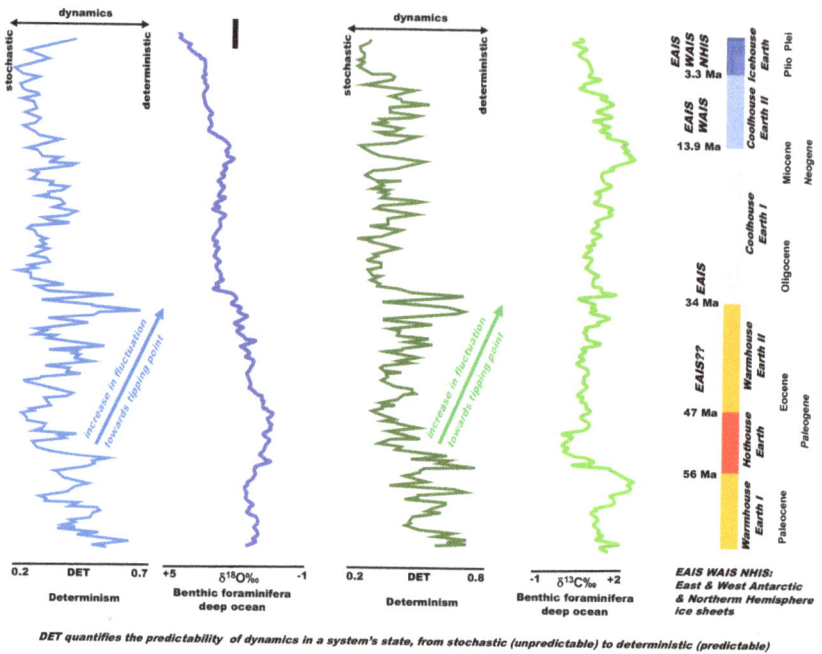

DET quantifies the predictability of dynamics in a system's state, from stochastic (unpredictable) to deterministic (predictable)

Figure 6.13. Westerhold et al. (2020) in determinism in Cenozoic climatic states.

Westerhold and 23 colleagues extended the carbon and oxygen database for the Cenozoic Era, significantly boosting the Palaeogene sector. Recurrence analysis of determinism (DET) shows that climate is more predictable (deterministic) in the warmhouse state than it is in the hothouse, coolhouse or icehouse states. Westerhold et al. point out that dynamic changes are rising in amplitude on a 13-million-year trajectory during the warmhouse times of the Middle and Late Eocene. The dynamics attain a threshold, namely the tipping point at the end of the Eocene that turns out to be the true tipping point in Cenozoic biogeohistory.

Source: Westerhold et al. (2020).

After this, during the Oligocene, things were never the same again. The oxygen signal of cooling went down and stayed down below Middle Eocene levels, fluctuations notwithstanding. The carbon dioxide curve went down likewise and the CCD curve went down likewise. The world became icecap-prone, an icehouse world. It still is. What happened? Staring at this chart, yes, I see the grand sweep from the greenhouse to the icehouse in these grand global metrics, but critically I see two turning points, or pivots. One is at about 42 Ma, just before MECO (the Middle Eocene climatic optimum) interrupts the cooling trend and the CCD does a truly spectacular halt, beginning a series of geologically rapid fluctuations of half a kilometre and more in amplitude in the water column. The second turning point at about 38 Ma marks the onset of the icecap-prone world, the massive CCD fluctuations cease, and it is the beginning of the end for the greenhouse levels of carbon dioxide in the atmosphere. It is this 42–34 Ma slice of time, slightly adjusted, that Wolfgang Berger and Gerold Wefer named the Auversian Facies Shift.[8] Initially a perception of fundamental palaeoceanographic change—in the pelagic realm—the AFS actually encompasses the neritic and terrestrial as well. A lot happened during the AFS, as we shall see.

First, though, we have the double question of why the Eocene hothouse? And why did it end? Recall the tectonic scenario for the Indian Ocean region including India's rapid flight northwards and plate-tectonic reorganisation (Chapter 5). These were times of high ocean-crust production rates at spreading ridges, and such times have long been linked to high sea levels, high carbon dioxide levels and warm global temperatures. Increased spreading rates means lengthened and tumescent ridges which displace water, spilling it across continental margins in tectonoeustatic transgressions. On the other side of the world the ridge in the Norwegian–Greenland sea was lengthening, a million cubic kilometres and more of basaltic lava was extruding, and there was a huge hydrothermal outgassing of carbon dioxide. The CO_2 flush brought on EECO, the warmest episode in the Cenozoic Era. However, these warm and wet conditions encouraged increased chemical weathering, soaking up a lot of the CO_2. Beginning with decreased crustal spreading rates, things tended to reverse across the board, culminating in a cooling of the deep oceans by about 45 million years ago, in the Middle Eocene. And the warm ring was being closed as the cold ring was being opened.

---

8   Berger and Wefer (1996). Figures 6.11, 6.12 and 6.13 display prominently the critical transformational interval between the warm Palaeogene ocean with its expanded shallow seas and the cool Neogene ocean with its contracted shallow seas. Berger and Wefer called it the Auversian Facies Shift, referring to the changes, revealed by deep-ocean drilling, from mixed, variable sediments under a shallow CCD to cleaner chalks etc. under a deeper CCD; the Auversian was a later-named Eocene stage, now redundant. I (2005a, 2009) expanded the concept of the Auversian Facies Shift to mark the global zone of change and instability in biogeohistory between the Khirthar Transgression and the first seriously large ice sheets in the Early Oligocene. But the term has not caught on, and Berger (2011) cooled off too.

Concluding ruminations. And what about the 'last greenhouse'?

Consider Western Civilisation as a succession of temporal slabs at scales of 10^3 to 10^4 years. The Greeks and Romans. The Dark Ages. The Renaissance. The Reformation. The Enlightenment. Terms with fuzzy boundaries, highly contested terms, but highly useful terms, even so. The Cenozoic Era is like that—but at 10^6 to 10^7 years' scale. Clearly the era has two parts, two periods, in an overall trend from hothouse to icehouse. Equally clearly, there may be sharp and clear breaks in the various parameters but there is a critical interval of changeover. That is, the Oligocene Epoch is a fuzzy interval between the upper and the lower 'natural' parts of the Cenozoic Era.

We see the two-part pattern on this grand scale when we inspect the fossil records of the limestone fossils, the large benthic foraminifera and the planktonic foraminifera, and their contrasts with the corals and the calcareous algae. The main contrast is between the two benthic groups of photosymbiont bearers, the large benthic foraminifera and the corals. Pamela Hallock and her colleagues explain the contrast as due to rather different living conditions in the warm shallow seas. The Eocene seas at first had weaker thermal gradients, therefore less vigorous circulation and lower nutrient levels, suiting the lifestyles of the foraminifera. Neogene seas were more vigorously mixed under steeper thermal gradients, promoting the washing of nutrient-laden water across the sessile corals and the flourishing of the algae.

Which begs the question in the title: what is the last greenhouse? Recall the rhetorical question from Chapter 1: when was the last ice age? It does not have a straight answer because it depends on the scale. Likewise here. Global cooling was a prolonged but episodic, 50-million-year affair. It occurred in four main steps, each stronger than the last, and truncated by warmings, each weaker than the last. Looking back from the present, we are warned that global warming is driving us toward the environments of 125,000 years ago and then the still warmer conditions at times in the Pliocene Epoch, between 3 and 5 million years ago. 'Greenhouse' is applied to each of these in its respective context. But in the 10^6 to 10^7 years' window the Miocene interval known as MICO (Miocene climatic optimum), round about 15 million years ago, stands out as the last gasp of the Eocene hothouse.

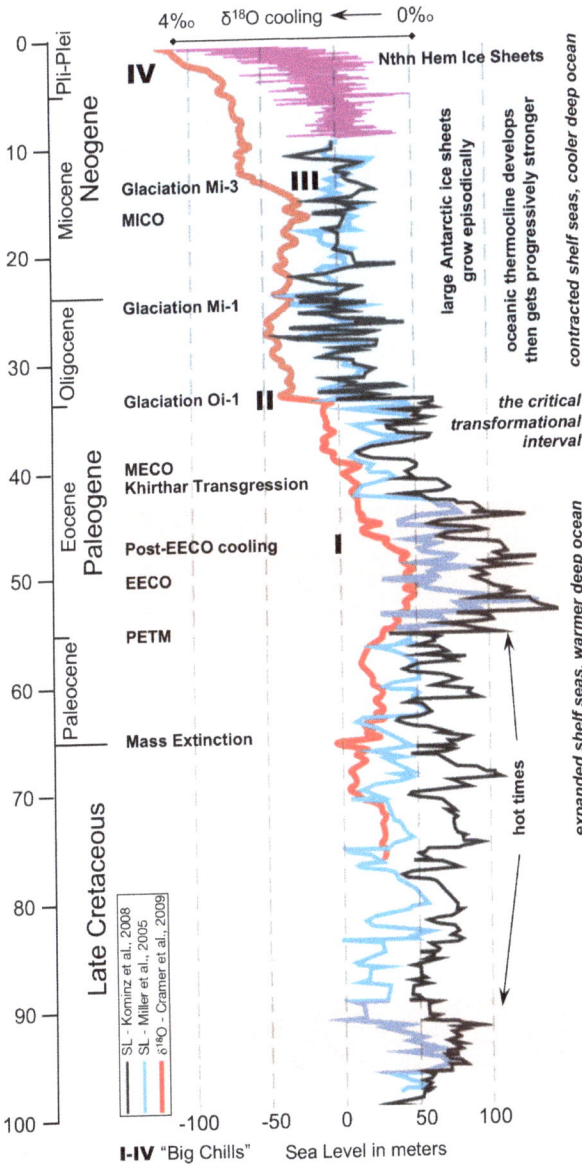

Figure 6.14. Sea levels reconstructed over 100 million years.

This puts the Cenozoic Era into a 10^8 years' perspective, using two curves of reconstructed sea level from the Atlantic margin of North America and a deep-ocean, oxygen-isotopic curve. At the grandest scale global sea level parallels global cooling, which is fair enough. But looking at the black and blue curves, we see amplitudes of sea level in hothouse times of more than 50 metres and comparable with amplitudes in icehouse times. How can that that be? We lack a cogent and consensual answer.

Source: Adapted from Pagani et al. (2014, Fig. 6; sources as shown).

But wait. Let's step up to 10^8 years' scale. We have long known that the Mesozoic included times of extensive, really extensive, transcontinental seas and that the dinosaurs mostly had it pretty warm.

Here is a chart (Figure 6.14) displaying the present state of play for the past 100 million years. The first 50 seem to display a kind of steady state, strong fluctuations in the reconstructed sea levels notwithstanding. The second 50, in contrast, are on a long downward trend, punctuated by four 'chills', I–IV. At this scale the Eocene hothouse labelled EECO might well be identified as the last greenhouse.

It's all in the scaling.

7

The Palaeogene Australo-Antarctic Gulf: Tropical swamps in winter darkness

Cenozoic deep time in four acts

Southern Australia is facing Wilkes Land across the Australo-Antarctic Gulf (AAG) at 60–70°S. The dark winters are becoming warmer. Any connection at the head of the gulf with the south-west Pacific Ocean is still narrow and shallow. There is no Southern Ocean, not yet. Among the numerous strange features of global geography, the most striking (and to become the most preoccupying among the community of the still-young science of palaeoceanography) are Antarctica's connections with South America and Australia. But global plate tectonics and oceanfloor spreading are being reorganised at the planetary scale, thanks to India's flight terminating in collisions, and the AAG is about to expire into the Southern Ocean. The faunas and floras are becoming, in the minds of many, relics of the Cretaceous Period, living fossils, evolutionary losers, a kind of living museum.

The small snapshot in the summary diagram (Figure 0.3) is taken at 38 Ma, above and after the informal but highly meaningful boundary between the Early Palaeogene and the Late Palaeogene at about 43 million years ago. That the Cenozoic Era falls naturally into four parts has been a staple of my intellectual furniture since Pat Quilty and I spelt it out independently in the 1970s. That is how southern Australian biogeohistory is constructed on one page and more or less the framework for the narrative in this and the next chapter.[1] The Early Palaeogene saw the rise and fall of the Eocene hothouse, the warmest times of the Cenozoic (Figure 7.1).[2]

Figure 7.1. Eyre Formation type section on Coopers Creek.

In the region of Lake Eyre, the Eyre Formation is an extensive blanket of sand sitting on a widespread unconformity marking a hiatus of millions of years. In this photo by Heli Wopfner, the unconformity is on the bench just above Coopers Creek and dipping gently to the right, at the base of the Coolabah. Heli's sketch of a locality nearby (about

---

1   The fourfold depositional episodes in southern Australia are adopted also by James and Bone (2021) in their book describing biogenic sedimentary rocks.
2   The photograph was taken by Wopfner in 1961 and published in 2020. His sketch in Wopfner et al. (1974) was done in the grand European tradition of field sketching. Deposited on floodplains and fluviatile settings (Figure 0.4) under rich vegetation during ultra-wet Hothouse Earth, the sands were bleached during wet-tropical weathering and fossils are found only in those drilled strata that escaped the worst of the weathering. Sporomorphs indicated to Harris, who worked with Wopfner on the 1974 paper, a two-part Eyre Formation, respectively Late Palaeocene – Early Eocene and Middle Eocene in age.

12 m wide) illustrates deposition, erosion and more deposition as channels switched in a high-energy, fluviatile system. *1*, *2* and *3* indicate fine, medium-coarse and very coarse to pebbly sands.

Source: Photo by Heli Wopfner in 1961, published in Wopfner (2020), courtesy Geological Survey of South Australia. Sketch from Wopfner et al. (1974).

The mysterious descent into the icehouse began in the Late Palaeogene. As we struggle to get our ages of strata and their correlations in order, we see that these global trajectories are imprinted upon southern Australia. And southern Australia has something to contribute to the grand global narrative.

Continent drifts north into new environment? Or new environment sloshes south over continent?

Capturing the scientific and popular imagination, Australia's continental drift became a universal explanatory solvent. Here at last was the scenario accounting compellingly for our distinctive biomes on land and at sea. Making its stately way through 25° into the lower latitudes, the continent entered an environmental zone diachronously, encouraging the accumulation of limestones in its shallow marginal seas, at first in the north, in due course in the south. Time advances, and tropical-type shells have moved into the shallow southern seas. Entering the latitudes of the belt of atmospheric high-pressure cells, the Australian continent dries out, its floras tracking that process towards what we are today, an arid land with damp fringes. Such scientific disciplines as sedimentology, palaeobotany, biogeography, geomorphology, palaeopedology and the modern, efflorescing environmental sciences have frolicked in the paradigm of continental drift and plate tectonics. These days you can hear about mobile Australia from tour guides, nature lovers and animal-welfare activists. It is colourful and plausible, even majestic, this narrative of an isolated continent sailing from one global environmental zone to the next, traversing roughly 2,800 kilometres in 40-odd million years at some 7 centimetres per year. And the narrative is partly correct.

Missing from the above is the counter-notion out of deep time, namely that the global environmental zones might move to meet the continent. And move a great deal faster. It is not only the endogenic crust and hot subcrust that are mobile; the exogenic systems of ocean, atmosphere, land surface and biosphere are ever restless. Within two decades of the *Glomar Challenger* putting to sea in 1968, we could confirm with real evidence that the Cenozoic earth's surface was characterised by overall cooling, that that cooling occurred in four main steps, successively stronger, and that the four coolings were each preceded by a warming, successively weaker. If southern Australia fits that global pattern, then its drift equatorwards is but part of the story.

But here are two more specific tests based in stratigraphic patterns—specifically, that is, based in space and time. First, recall the pattern of Eocene limestones, the massive shallow-water carbonates marking the Khirthar Transgression in our corner of the Indo-Pacific region (Figure 5.16). Their seriously important characteristic is that they accumulated not diachronously but at the same time, isochronously, geologically speaking, while spread across three Australian continental margins and across perhaps 30° latitude. That pattern does not fit the stately progress of continent encountering new environment, and the reason is to be found in the second test. A global plot of isotopic temperatures from the surface waters of the ocean against geological time and palaeolatitude yields isotherms, lines marking 10, 15 and 20°C, and the isotherms are swinging through tens of degrees latitude (Figure 7.2).

Thus the environment indeed was moving to meet the continent, considerably faster than the continent was moving to meet the environment. At about the time of the Khirthar Transgression about 40 million years ago, the 20°C isotherm had swung a very long way to the south, and that observation is consistent with the latitudinal spread of neritic carbonates at that time.[3]

---

3   The diagram was adapted by McGowran and Li (1998) from Frakes et al. (1994) and Frakes (1999), who produced 'grossplots', marine isotherms, of oxygen-isotopic ratios from planktonic foraminifera located in their time and space in the oceans (using data from both hemispheres). Swings in the oceanic isotherms have been better predictors of events in Cenozoic biogeohistory than Australia's trajectory of continental 'drift'.

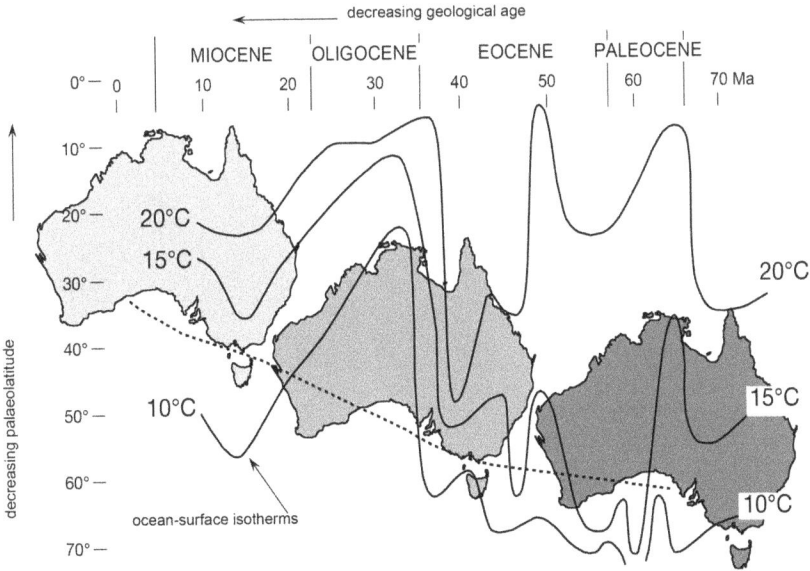

Figure 7.2. Australia's continental drift and two speeds in earth history.
Plotting palaeolatitude against geological age reveals two speed zones in earth history. The dotted line shows Australia's trajectory with the dogleg marking the fundamental change in the Middle Eocene (the three images of the continent are snapshots along the way). The continent took roughly 50 million years to traverse 20° into lower latitudes. Compare that with a contouring of 1980–1990s oxygen-isotopic data on oceanic surface temperatures. An isotherm could swing tens of degrees in a very small fraction of that time, especially across the Eocene–Oligocene boundary.
Source: Adapted by McGowran and Li (1998); Frakes et al. (1994) and Frakes (1999).

No limestones and not many shells

It became apparent in Chapter 5 that the first third of the Cenozoic strata in southern Australia did not have as many shelly fossils as did the younger strata. There were hardly any rocks of that age now exposed, the only known strata being near Cape Otway, including my Pebble Point foraminifera, and there seemed to be virtually no limestones. Instead, we have piles of clays, muds and sands, collectively siliciclastics, known from the matching of geophysical patterns called seismic stratigraphy. The best illustration we have of this pile is Guy Holdgate's reconstructed section in Victoria in the south-east (Figure 7.3).

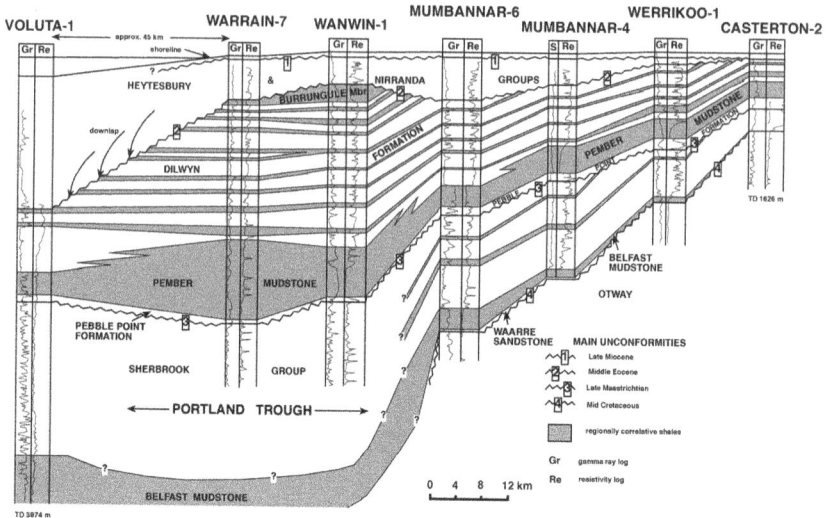

Figure 7.3. Early Palaeogene pile in western Victoria.

Holdgate's cross-section in south-western Victoria, reconstructing the Early Palaeogene pile of muds, clays and sands continuing the story of the Late Cretaceous in the Portland trough, part of the Otway Basin under the AAG. Exposed only near Cape Otway (Figure 4.17), these rocks are accessed by drilling. Each drillhole has two geophysical logs, gamma ray and resistivity. Note the four main unconformities and, especially in our context, unconformity #2 in the Middle Eocene, which is the location of the Lutetian Gap and concluded by the Khirthar Transgression. Unconformity #1 is seen at Point Margaret (Figure 4.18). As always, vertical scale is greatly exaggerated, as seen by the total depths of the Casterton and Voluta drillholes. The Middle Eocene unconformity #2 has cut into the Early Eocene strata, which were slightly tilted. The time-equivalent of the unconformity is the Lutetian Gap and here it actually conflates two well-marked unconformities in south-east Australia, the Latrobe and the Marlin unconformities.

Source: Assembled by Guy Holdgate and from McGowran et al. (2004).

The pattern of thin clays (regionally correlative shales) is very consistent, not only in this cross-section of the trough and its flanks but also to the west and the east. Rhythmic consistency over many kilometres is the theme, not lateral facies change.

These rocks are the last burst of siliciclastics delivered into the narrow basins and ridges of the rift valley underlying the AAG. They became prime targets for the explorers looking for fossil fuels. But it was a frustrating business looking for microfossils to guide the drilling. After I had extracted reasonable ages from the foraminifera on the Otway coast it soon became clear that we needed much more resolution and precision in our ages than the sporadic foraminifera could provide for those sands, muds and clays, a thick set of strata being targeted in the exploration for fossil fuels. Except for a few thin horizons with shells, like the Pebble Point, the foraminifera either were absent or consisted only of the agglutinated kind, or none. That implied mostly hostile conditions, the usual suspects being brackish water, low oxygen or general instability. We knew that the sea was there, in the gulf, but it was a very strange and narrow sea. And why no limestones? Enter, palynology.

Sporomorphs and dinocysts: Enter palynology

Palynology is the enquiry into fossil spores and pollen grains. Dissolve a sedimentary rock in strong acids, hydrochloric acid removing all the carbonates and hydrofluoric acid removing all the quartz, clays and rock-forming minerals, and what, if anything, is left? If the sediment had escaped oxidation by deep weathering or groundwater, what might be left is a concentrated sludge of tough, acid-resistant 'organic material', such as cellulose of plant origin, which could be smeared and preserved on a microscopic slide and scrutinised under a powerful light microscope. The targets would be two main classes of microscopic objects together known as palynomorphs. One lot are the sporomorphs, which are the spores and pollen grains, evolved to be distributed by water, by insects and birds and by blowin' in the wind (Figure 7.4).

The second lot are the dinocysts, the cyst stage in the life-cycle of the single-celled eukaryotes, microalgae known as dinoflagellates, mostly of marine and marginal-marine habitat. Very small in size and very large in numbers, the assemblage of specimens in a palynological sample may contain information simultaneously on one, the marine environment (Figure 7.5), and two, the neighbouring terrestrial environment, and three, the geological ages in both, and four, the biogeography, as environment expands into region.

Like us foraminiferal micropalaeontologists, the palynologist in glass-half-empty mode can feel suspended between biology and geology, between animal, vegetable and mineral with their different cultures, curricula and administrative structures; but in glass-half-full mode there await her the exhilarations of further vistas from the edge, deeper insights and greater possibilities of cross-fertilisation and heuristic integration.

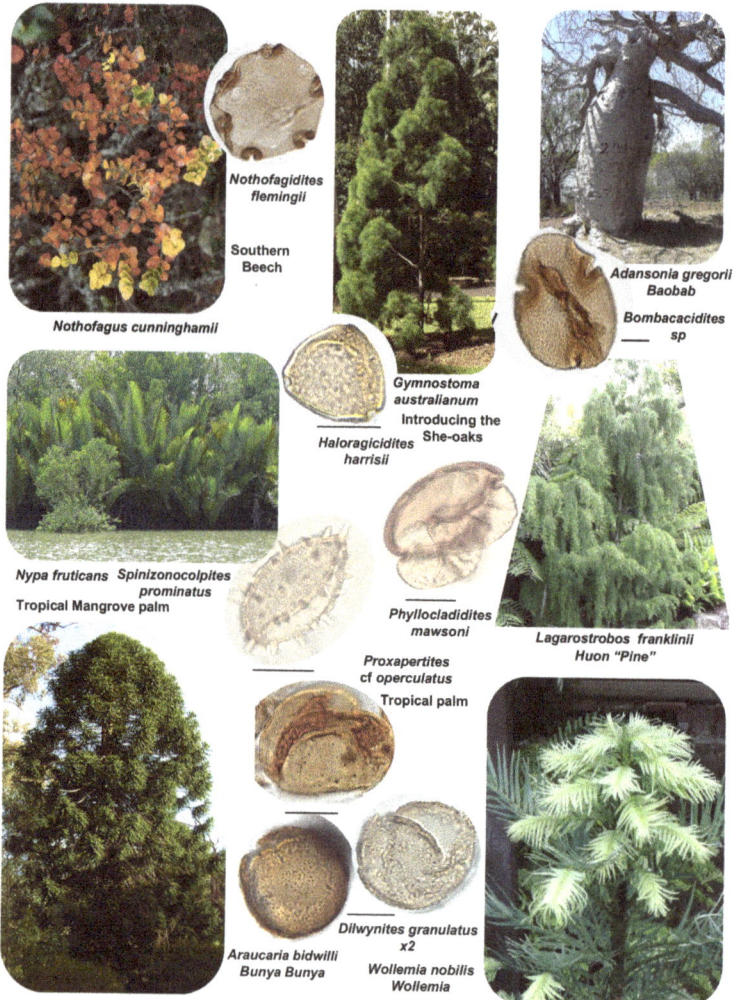

Figure 7.4. Modern trees and ancient pollens.

Modern trees give significance as nearest living relative to ancient pollens in this selection. *Phyllocladus, Wollemia* and *Araucaria* are living reminders of the conifer forests ringing much of the AAG in Early Palaeogene times. The nearest living relative of the plant that shed *Dilwynites* pollens might be the recently discovered and newsworthy *Wollemia*. (Or tropical species of the tree *Agathis*.) The conifer pollens in the Eocene coals in Gippsland are dominated by *Phyllocladidites mawsoni*. The presence of *Spinizonocolpites*, pollen of the mangrove palm pollen *Nypa*, in strata on the flanks of the AAG is the strongest terrestrial persuader of a tropical climate so far south. However, it is in good company with, for example, the tropical palm pollen *Proxapertites*. Likewise with *Bombacacidites* pollens off Wilkes Land: the baobab and virtually all its other living relatives are tropical or near-tropical. *Nothofagidites*, the prolific pollens of *Nothofagus*, the southern beech, tended to expand in the stratigraphic record at the expense of the conifer pollens during somewhat cooler times, beginning with the end of the hothouse. *Gymnostoma* and its pollen *Haloragicidites*, present in Hothouse Earth, herald *Casuarina* and *Allocasuarina*, the she-oak saga in the drying and leaching soils of Gondwanaland and Australia.

Source: *Nothofagidites flemingii* and *Bombacacidites sp.*, Eocene, off Wilkes Land (Contreras et al., 2013); *Nothofagus cunninghamii*, south-east Australia, I, KuresH, CC BY-SA 3.0, via Wikimedia Commons; *G. australianum*, northern Queensland, Battsv, CC BY-SA 4.0, via Wikimedia Commons; *Adansonia gregorii*, northern Queensland, Reise-Line, CC BY-SA 4.0 via Wikimedia Commons; *H. harrisii, D. granulatus, P. mawsoni* and *S. prominatus*, Palaeocene–Eocene boundary, south-east Australia (Frieling et al., 2018; Huurdeman et al., 2020); *Nypa fruticans*, Sarawak, Bernard Dupont CC BY-SA via Wikimedia Commons; *Lagarostrobos franklinii*, Tasmania, Secungar CC BY-SA 3.0, via Wikimedia Commons; *Araucaria bidwilli* and *W. nobilis* (B McGowran); *Proxapertites* cf *operculatus*, Early Eocene, Tasmania (Macphail and Jordan 2015).

Before deep-ocean drilling, before we had any clear conception of the AAG, before we had any clear notion of high levels of carbon dioxide or their sheer environmental impact, we were beginning to glimpse the very different environments of southern Australia. It was very wet then where it is very dry now. There were widespread conifer (gymnosperm) rainforests where there are no forests today. In the 1960s my colleague and palynologist Wayne Harris demonstrated strong changes through time on the Otway coast in the relative numbers of the three big plant groups seen in palynological preparations, namely the angiosperms, the gymnosperms and the pteridophytes (mostly ferns); and he inferred a strong warming from the Late Palaeocene into the Early Eocene.[4] Within Eocene times, forests of angiosperms (flowering plants) expanded at the expense of the conifers in a change identified as the '*Nothofagus* revolution'.

An awareness of Eocene warming in the northern hemisphere had eighteenth-century roots and was well established when Lyell came to write his *Principles of geology* (1831). Fossils from corals to cycads to crocodiles were demonstrating that tropical and subtropical conditions, on land and at sea, reached higher northern latitudes than they do today. Assemblages of fossil plants being unearthed from the London Clay showed that south-east England had an Eocene climate like south-east Asia's today. The presence of the mangrove palm *Nypa* was decisive.[5] In the southern hemisphere the evidence was more elusive at high latitudes, most Antarctic fossils being under the ice. But new discoveries are adding to progress, such as finding *Nypa* in the Eocene in western Tasmania.[6]

---

4   Harris (1965, 1971).
5   See the specimen of *Nipadites umbonatus* in the nineteenth-century plate of Lower Eocene fossils in Figure 10.13.
6   Pole and Macphail (1996). *Nypa*'s sporomorph, *Spinizonocolpites prominatus*, is sufficiently well known in Gippsland palynology for its incoming to be listed as a key biostratigraphic event at the Palaeocene–Eocene boundary (Partridge, 2006). This observation, together with Harris's (1965) inferring the strong warming event in what became known as the Australo-Antarctic Gulf, foreshadows discovery of the Palaeocene–Eocene thermal maximum (PETM) in southern Australia.

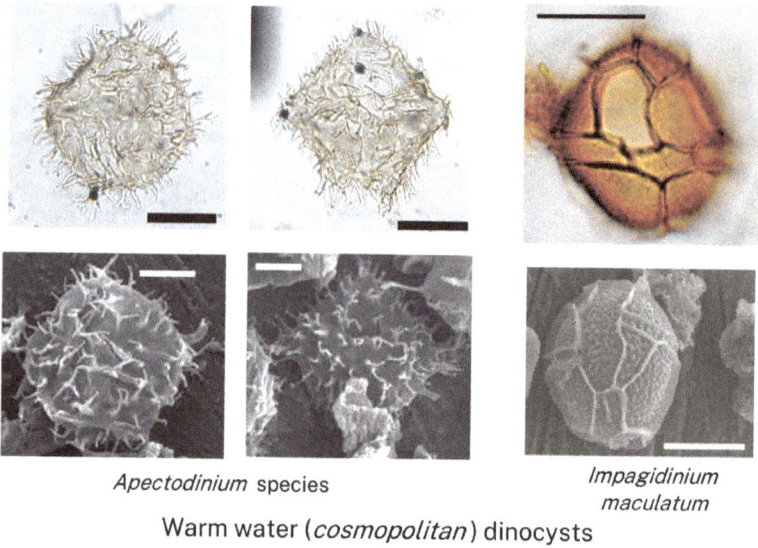

Apectodinium species

Impagidinium maculatum

Warm water (*cosmopolitan*) dinocysts

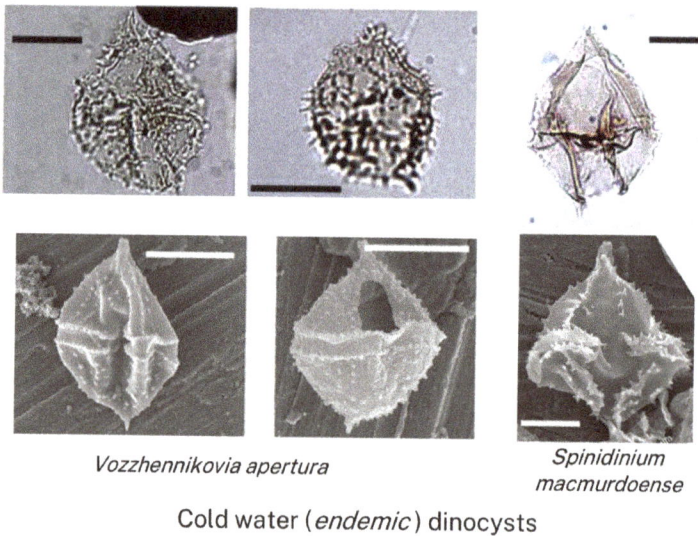

Vozzhennikovia apertura

Spinidinium macmurdoense

Cold water (*endemic*) dinocysts

Figure 7.5. Dinocysts warm and cold in Eocene biogeography.

Dinocysts are the cysts of dinoflagellates, single-celled eukaryotes. Dinocysts survive and concentrate along with pollens and spores in the strong-acid digestion of sediment and sedimentary rocks. They are similarly susceptible to oxidation atmospherically or by groundwater. These species are shown in parallel images from transmitted-light microscopes and scanning electron microscopes. They represent distinct, climatically based provinces in the southern oceans and marginal seas of the Middle and Late Eocene — the Antarctic (endemic) and cosmopolitan (warm) provinces. Scale bars are 25 μ.

Source: From the synthesis by Bijl et al. (2013) and Peter Bijl (pers. comm.).

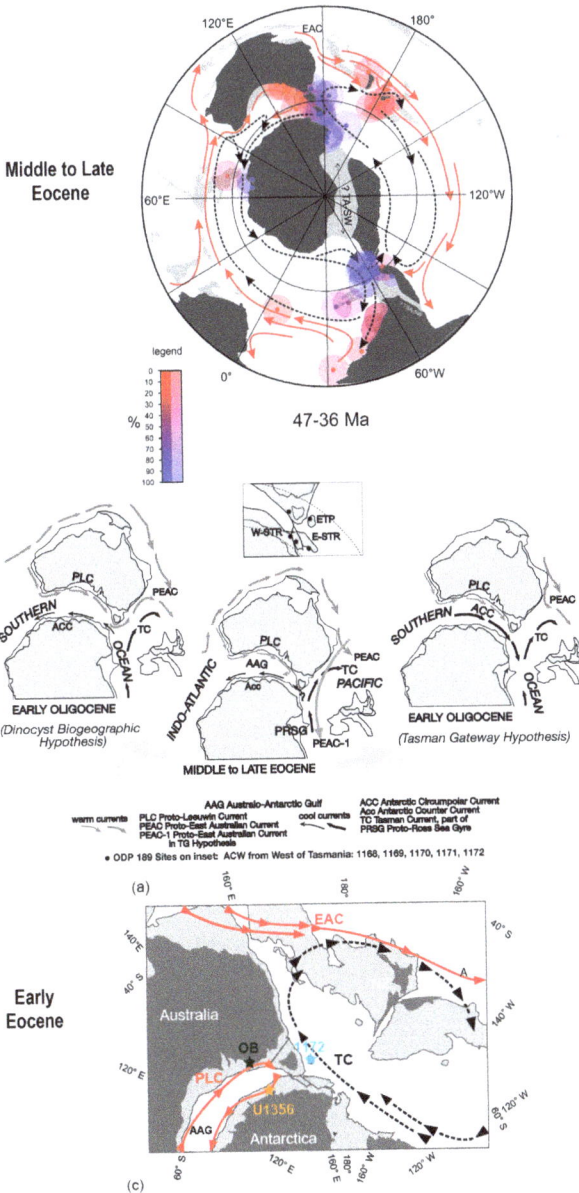

Figure 7.6. Dinocyst biogeography and southern reconstructions.

Geographic changes from Hothouse Earth to Warmhouse Earth II, as the AAG is absorbed into the nascent Southern Ocean.

Above, Peter Bijl's iconic map of dinocysts, counted on a spectrum from endemic to cosmopolitan, with currents and speculative palaeogeography. The key point is the strong contrast between the eastern AAG and the south-west Pacific as the warm (cosmopolitan) marine microfloras pile up against the Tasmanian barrier. For demonstrating the growth and decay in this episode of endemism through time, see Figure 8.9.

Middle, two palaeogeographic scenarios of death of the AAG. *Central map,* later Eocene, showing clockwise circulation in the AAG with a significant proto–Leeuwin Current (PLC), and two opposing views of a south-west Pacific, either a strong Tasman Current (TC, black) as part of a cool proto–Ross Sea Gyre (PRSG), or a strong proto–East Australian Current (EAC) delivering heat virtually to Antarctica itself (PEAC-1, grey). *Map on the right,* Early Oligocene, the Tasman gateway hypothesis (Kennett et al., 1974; Kennett and Exon, 2004). *Map on the left,* Early Oligocene, the alternative, acknowledging the strong influence of high-latitude provincialism and biofacies in dinocysts and diatoms, here labelled the 'Dinocyst biogeographic hypothesis' (e.g. Huber et al., 2004; Warnaar, 2006). *Inset,* ETP—East Tasman Plateau, W-STR and E-STR—South Tasman Rise, east and west blocks.

Below, the AAG is very warm on all evidence, marine and terrestrial. Marine connections south of Tasmania, if any, are very narrow and shallow.

Source: The map (*above*) is after Bijl (2011). The two hypotheses (*middle*) are from McGowran (2009). The map (*below*) is from Frieling et al. (2018).

Meanwhile, drilling during the exploration and development of fossil fuel resources was building a pattern of dinocyst succession through the Palaeocene and Eocene. One particular biogeographic discovery aroused wide interest—that many of the dinocyst species in the high southern seas in the Early Cenozoic Era were distinctive and localised: they were endemic. This discovery gave rise to the theory of the Transantarctic Flora, perhaps held together by a Transantarctic Seaway postulated across an ice-free Antarctica. Oceanic drilling at these high southern latitudes brought about the next major advance by a Dutch invasion, perhaps the most significant advance since the days of the European marine explorers, including Tasman and several other Dutch captains. The new sailors and marine and nonmarine palynologists were a cosmopolitan lot, but the main scientific drive erupted from the University of Utrecht. They scrutinised the timing of the death of the AAG as it disappeared into the new Southern Ocean and they produced data much-needed in the modelling of ancient atmospheres and oceans. They strengthened the succession of microfloras from the abundant microfossils coming out of the cores of drilled sediment—the pollens and spores blowing or washing off the lands mixed in with the dinocysts reflecting the oceanic waters changing in space and time. The oceanic drilling confirmed the two-part dinocyst distributions, the one assemblage apparently preferring warmer waters and the other apparently preferring cooler or cold waters (Figure 7.6).

This classifying as cosmopolitan vis-à-vis endemic could be quantified, permitting us to see very clearly the contrasts already being perceived within and outside the AAG, especially the impact of a proto–Leeuwin Current.[7]

Into the hothouse in the Australo-Antarctic Gulf

Words such as 'tropical', 'subtropical' and 'temperate' are familiar enough, if not meaning quite the same thing to all people. But we need to venture a little further into jargon with this plain statement: *during the Early Eocene the coastal lowlands of Wilkes Land in Antarctica, on the southern side of the AAG at 70°S and experiencing up to two months' winter darkness, had a pantropical rainforest.* 'Pantropical' meant an evergreen rainforest living under mean annual temperatures between 20°C and 25°C, but also with a mix of species that could tolerate frosts (up to certain levels of severity and frequency, at any rate) with species that could not. But what kinds of evidence support that kind of palaeoclimatic statement? The answer is broadly threefold: first, our insights into the ecology and physiology of living forests, woodlands, grasslands and deserts; second, recognising the nearest living relatives of ancient plants as useful analogues; and third, independent lines of evidence from the marine realm.

The AAG in Palaeocene and Early Eocene times was surrounded by forests. It was very wet, being in, as the scientists would say, 'high-precipitation regimes', and the forests were rainforests. There were three kinds. The warm-temperate rainforests were dominated by the group of conifers

---

7   Unlike the old captains, the modern sailors were venturing into *terra* hardly *incognita* when Ocean Drilling Project Leg 189 took place in 2000—meaning, problems in palaeontology and stratigraphy. Bob Hill had assembled a rich series of chapters on *History of the Australian vegetation: Cretaceous to recent* (1994), and the notion of the trans-Antarctic flora had been clarified (Truswell, 1997). Biostratigraphic palynology—terrestrial sporomorphs and marine dinocysts—had been developing for decades in economic exploration, understandably geological in its emphasis and too much of it hidden away 'commercial and in-confidence' in the mantra of modern capitalism. But Alan Partridge produced a prime reference, *Late Cretaceous-Cenozoic palynology zonations, Gippsland Basin* (2006) (the evidence remaining unpublished). Emerging from ODP Leg 189 was *The Cenozoic Southern Ocean: Tectonics, sedimentation, and climate change between Australia and Antarctica* (Exon et al., 2004). The early Dutch work on southern dinocyst biogeography was included in *From greenhouse to icehouse—the Eocene/Oligocene in Antarctica* (Francis et al., 2009). It includes a version of our Figure 7.6; our version is from Bijl (2011, Ch. 3, Fig. 3). The next oceanic drilling leg in our region was covered in *From greenhouse to icehouse at the Wilkes Land Antarctic margin: IODP Expedition 318 synthesis of results* (Escutia et al., 2014). The all-important controls on ages, datings and correlations were drawn together in *A magneto- and chemostratigraphically calibrated dinoflagellate cyst zonation of the early Palaeogene South Pacific Ocean* (Bijl et al., 2013).

known as podocarps. The cool-temperate rainforests were dominated by other conifers, the araucarians, and by the southern beeches, *Nothofagus*. The paratropical rainforests were dominated by ferns including tree ferns (Cyatheaceae). But it is the presence from time to time in the Early Eocene strata of the sporomorph *Spinizonocolpites prominatus* that signalled the peak 'hothouse' conditions in the swampy and estuarine margins of the Early Eocene AAG, because the parent of this sporomorph is very closely related to the modern mangrove palm, *Nypa*, the quintessential 'mangrove megathermal' indicator, requiring mean annual temperatures above ~24°C.

The Palaeocene forests flanking the AAG were rich in the podocarps and the araucarians, and among the abundant pollens shed by the araucarians were those that Harris named *Dilwynites*, seemingly close to the later discovered modern survivor, *Wollemia*. Abundant ferns indicated mild, frost-free conditions, but the climate varied in degrees of warmth, for there were times when palms and cycads flourished and times when they did not. Towards the end of the Palaeocene Epoch there was a strong warming shift throughout the south-west Pacific region and in the AAG. Paratropical forests replaced the temperate forests, ferns increased and angiosperms foreshadowed an Eocene expansion and diversification at the expense of the gymnosperms such as the podocarps.

This heating in the southern terrestrial environments matches closely the Palaeocene–Eocene thermal maximum (PETM), a very strong and sharp carbon- and oxygen-isotopic signal at about 56 Ma. We have seen the PETM signal ushering in the hothouse times in Figures 6.10, 6.14 and others. Workers in every subdiscipline pertinent to Palaeogene palaeoceanography and biogeohistory have swarmed over the PETM since Kennett and Stott discovered it in ocean drilling in the far south in 1990. It has been studied intensively for three decades, not only as a sharp and significant event in biogeohistory, but for what its speed and intensity and impact upon the carbon cycle and the biosphere might tell us about the impending crisis in modern (post-industrial) global warming.[8]

---

8 Kennett and Stott (1991). We knew that things were happening in the oceanic, neritic and terrestrial realms at the end of the Palaeocene Epoch. For example, the mass extinction at the end of the Cretaceous Period, the K-Pg event, did not capture many of the deep-ocean benthic foraminifera; their faunal turnover was delayed until the end of the Palaeocene (see Chapter 10). The PETM was the real beginning of the spectacular ecological spread of the large foraminifera in the warm shallows of Tethys, as displayed in Figure 6.9. The Kennett/Stott discovery sharpened focus and invigorated research. Two particularly interesting reviews of the PETM are by Sluijs et al. (2007) and McInerney and Wing (2011).

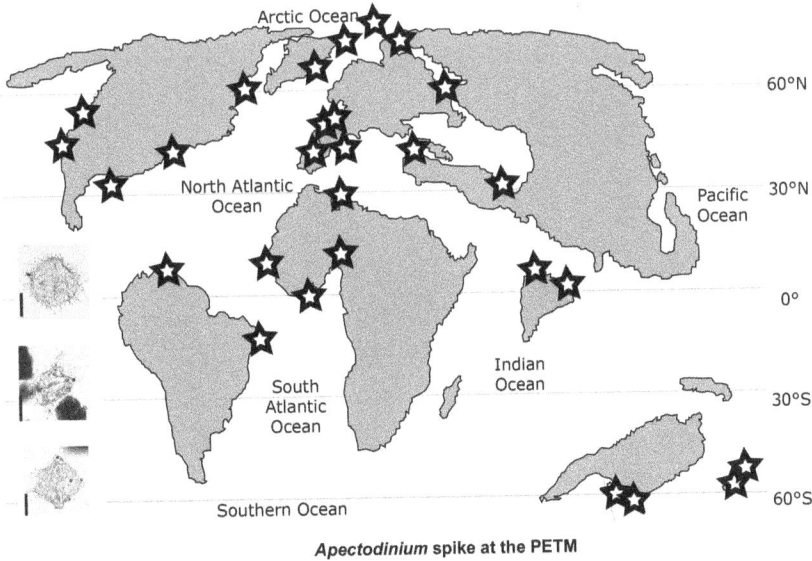

Apectodinium spike at the PETM

Figure 7.7. Dinocyst herald of the hothouse: *Apectodinium* spread globally at the PETM.

The dinocyst *Apectodinium*, prime indicator of a very warm ocean, spiked spectacularly to become a prime indicator of the PETM. This map is the biogeographic situation as known, in 2007, modified to include subsequent records from the AAG and the South Tasman Rise. Every palynological sample containing marine dinocysts, from the PETM and from the far north of the planet to the far south, was found to include *Apectodinium*. The sheer spread of this hothouse genus to high northern and southern latitudes urges the conjecture that this hothouse time was also a time of seriously flattened longitudinal gradients in temperature.

Source: Sluijs et al. (2007).

The marine dinocysts illustrate the PETM. Marine palynologists rapidly established the awesome geographic spread of the spike in numbers of the genus *Apectodinium* (Figure 7.7).[9]

In the Gippsland Basin in south-eastern Australia facing the south-west Pacific, species of *Apectodinium* are prominent in the dinocyst zonation across the Palaeocene–Eocene boundary; they were found at Site 1172; and they were found on the other side of the Tasmanian barrier in the Otway Basin flanking the AAG. Also on both sides we have the indicator of the (robustly inferred) very warm-water species, *Florentinia reichartii*.

9   Figure 7.8a is from Frieling and Sluijs (2018). Figure 7.8b is simplified from a principal component analysis by Houben, Quaijtaal et al. (2019); their data are from Late Eocene samples but their three robust trends, confirming the value of the three dinocyst 'complexes' as palaeoecological indicators, hold true through the climatic shifts in Eocene times.

At PETM time the biogeographic contrasts exemplified by the distinctness of cosmopolitan and endemic microfloras were very muted. However, environmental contrasts are another matter (Figures 7.8a and 7.8b).

This illuminating reconstruction can be applied to the northern margin of the AAG. (No matter the sketch lacks the lush southern rainforest of those times; there is abundant runoff.) Increased rainfall and runoff might advance a brackish, lower-density lid across the local sea with a pulse of nutrient off the land, signalled by a spike in *Senegalinium*. Upwelling brings nutrient up into the photic zone while estuarine stratification suppresses it; such environmental differences can be inferred from the microfloras.

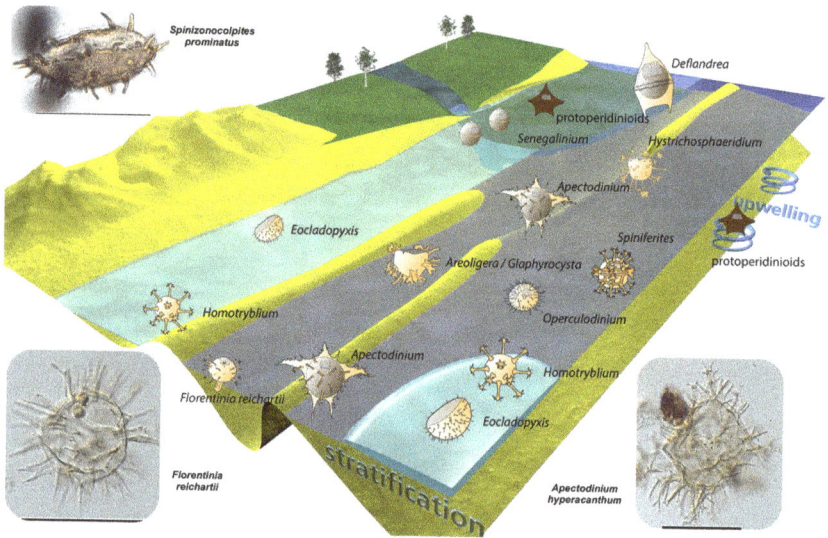

Figure 7.8a. Dinocysts in a reconstructed PETM sea.

The distribution of fossil dinocysts in space and time is revealing environmental preferences, permitting environmental reconstructions, as in this sketch abstracted from painstaking and consilient inferences about the PETM. *Apectodinium* required sea-surface temperatures of 20–25°C and *Florentinia reichartii*, as high as 30–35°C. *Senegalinium* preferred high nutrients and low salinities. The Protoperidinioids in high numbers indicated high levels of nutrients, either from runoff from the land or from deep waters upwelling into the photic zone. Or both! *Spiniferites* in contrast preferred the low-nutrient (oligotrophic) habitats of the open ocean. *Homotryblium* indicates unstable conditions such as lowered salinities, not only inshore but also in a stratified ocean such as the AAG in hothouse times. Scale bars for the three hothouse indicators, two microbes and the pollen of the mangrove palm *Nypa* are 50 µ.

Source: Frieling and Sluijs (2018), with dinocysts and sporomorph from Frieling et al. (2018).

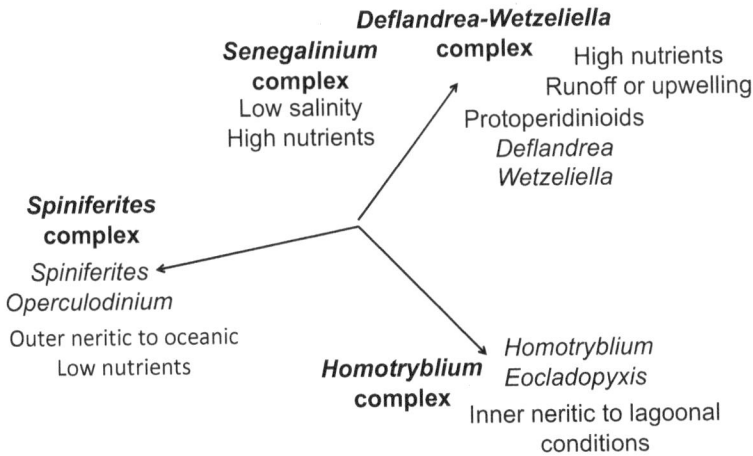

Figure 7.8b. Dinocyst complexes indicate ecological preferences.

The genera and species of dinoflagellates tend to clump together statistically as 'complexes', the binding material being similar ecological preferences. Principal component analysis of Eocene microfloras produces three ecological trends on two axes (omitted) accounting for most variability.

Source: After Houben, Quaijtaal et al. (2019); their data are from Late Eocene samples but their three robust trends, confirming the value of the three dinocyst 'complexes' as palaeoecological indicators, hold true through the climatic shifts in Eocene times.

The PETM foreshadowed the hothouse warming of the Early Eocene climatic optimum (EECO) climaxing at 52–50 Ma, and the evidence in the terrestrial biomes around the AAG for the existence of EECO is conclusive. At sea level in western Tasmania, independent lines of evidence from plant fossils gave a mean annual temperature of 24°C at about 65°S latitude, and the mangrove palm *Nypa*, together with other palms and cycads and other plants, implied that temperatures remained well above freezing even during several months of winter darkness. On Wilkes Land the mean summer temperature was ~21°C and the mean winter temperature was ~11°C, and frost-free. The coastal forests on Wilkes Land had a great deal in common with the paratropical forests of modern-day New Caledonia.

How come? Answering, the authors of the Wilkes Land research:

> suggest that the high atmospheric CO_2 levels of the early Eocene greenhouse climate were a decisive factor in the physiological ecology of high-latitude forests, most probably through causing a reduction in carbon respiration during the polar winter and an increase in photosynthetic carbon gain during the growing season.

> Our new data from the peak early Eocene greenhouse world indicate that a highly diverse forest vegetation containing evergreen elements can successfully colonize high-latitude, warm-winter environments when atmospheric CO_2 levels are high.[10]

This evidence from palynology and palaeobotany for high temperatures in the deep south is strongly consilient with outside evidence from marine organic chemistry, climatic modelling and marine palynology. Comparing a dinocyst with a sporomorph illustrates this point forcefully. The marine dinocyst *Apectodinium* comes and goes in bursts in the warm-water microbiotas, thus indicating climatic fluctuations in the marine realm. *Apectodinium* 'acmes' signal both PETM and EECO, identified and delineated by oxygen-isotopic and carbon-isotopic spikes in the stratigraphic succession of foraminiferal shells. Meanwhile, in the mangroves at the edge of the sea, the tropical mangrove palm *Nypa* sheds the sporomorph *Spinizonocolpites*. Dinocyst and sporomorph send the same message of heating at the same times in the same residue of acid-resistant microfossils.

These strands come together in the discovery of the carbon-isotopic PETM at Point Margaret in western Victoria (Figure 7.9).

Two metres of dense and dark claystone and mudrock contains a remarkable, multipronged story of the extreme climatic spike heralding the Eocene hothouse. There was a rapid, massive and sustained turnover in the vegetation. The rainforests similar to modern rainforests in north-eastern Queensland transitioned abruptly from mesothermal to meso-megathermal, at 60°S palaeolatitude in winter darkness. The marginal marine mangrove palm *Nypa* was prominently in residence, having immigrated on the proto–Leeuwin Current. The dinocysts are similarly informative in the marine realm: the dark muds below poorly ventilated water have abundant *Senegalinium*, followed by a spike in *Apectodinium* at the zenith, then somewhat clearer water with more *Spiniferites*.

---

10   Pross et al. (2012).

Figure 7.9. Sporomorphs capture the PETM at Point Margaret.

Two metres of solidified dark mud from Point Margaret in south-west Victoria have captured the PETM from the north flank of the AAG (white bandage near top). The Pember Mudstone was trenched for dense palynological sampling. The Late Palaeocene Pebble Point foraminiferal and molluscan faunas come from the lower, strongly banded strata. (Early Eocene shells have not been found at this locality.) Sporomorphs from ferns, conifers and angiosperms were grouped as mesothermal (14–20°C), meso-megathermal (20–24°C), and megathermal (>24°C) to demonstrate a very powerful shift in the climate — and powerful confirmation of the onset of the global hothouse. The vegetational succession is consonant with two estimates of shifts in mean annual air temperature (not shown): a formula derived from nearest living relatives (NLR); and the organic-chemical biomarker molecules of branched glycerol dialkyl glycerol tetraether (brGDGT). And it is all consilient with the strong carbon-isotopic signal (CIE) of the PETM. The early arrival and constant presence in 15/15 samples of *Spinizonocolpites*, the sporomorph of the mangrove palm *Nypa*, attest to the persisting influence of the (proto-) Leeuwin Current flowing from the northern and western margins of the continent.

Source: Palynological studies of the PETM section on the Otway coast are in Frieling et al. (2018), Huurdeman (2017) and Huurdeman et al. (2020). The pollen counts are selected and simplified from the latter, which has the mentioned tables pertaining to nearest living relative. The main photo and the inset image of Joost Frieling (*left*) and Peter Bijl (*right*) sampling in harness are by Steven Bohaty. For scale, notice them at the bottom of the trench in the main photo.

As we have noted, 'nearest living relative' is a touchstone in palaeobotany.[11] What we know about living organisms can be applied to the fossil record. Thus *Dilwynites granulatus/tuberculatus* are identified as the sporomorphs of Araucarian conifers such as living *Agathis* or the recently celebrated living fossil *Wollemia*. Therefore, we can talk usefully about conifer rainforests 50–60 million years ago. Ask, what does a species do? Answer: it exists, it reproduces successfully for some time, it responds to an unfamiliar environmental crisis by migrating or retreating into 'refugia', or it speciates with the new species acquiring the new adaptation for surviving, or it goes extinct without issue. What the species seems not to do so much is to evolve gradualistically (Chapter 10). In this study of the PETM at Point Margaret about 50 sporomorphs were tabulated, each with its nearest living relative, and in a second table the nearest living relative was listed with its climatic requirements including mean annual air temperature. This could all be quantified and condensed into a plot through time of mean annual air temperature. The temperature rose in the deep south by perhaps 4–5°C, which exceeds beyond our comprehension the grimmest of scenarios for our present global crisis.

When independent lines of evidence converge on high temperatures and very high rainfalls, on land and at sea, at latitudes of 65–70°S, we have to contemplate a very different world from today's at 50-plus million years ago. At this point I recall the two pioneering studies of oxygen isotopes which showed a cooling through Cenozoic time, in the deep ocean overall and in surface waters at high southern latitudes.

Putting the two studies together (Figure 7.10), Wolfgang Berger demonstrated the overlap in surface waters in the Early Eocene, implying no latitudinal gradient at all. Berger thought that this could not be, not at any rate on this planet. Thus there must be a very strong low-salinity component in the oxygen signal. That inference implied, in its turn, very wet conditions at high southern latitudes, so wet indeed that large tracts of the far south-west Pacific Ocean were virtually estuarine—that is, with a somewhat brackish, lower density lid above the normal salt water.[12]

---

11 Nowadays, analysis of nearest living relatives is but part of the artillery trained upon a fossilised leaf, together with climate leaf analysis multivariate program, leaf area analysis and leaf margin analysis. Using these proxies on a dozen fossil floras, Reichgelt et al. (2022) add to the accreting evidence that southern Australia in the Early and Middle Eocene was considerably warmer and wetter than it is today.
12 There is an excellent review by Savin (1977) of the early stable isotope research at Cenozoic time scales.

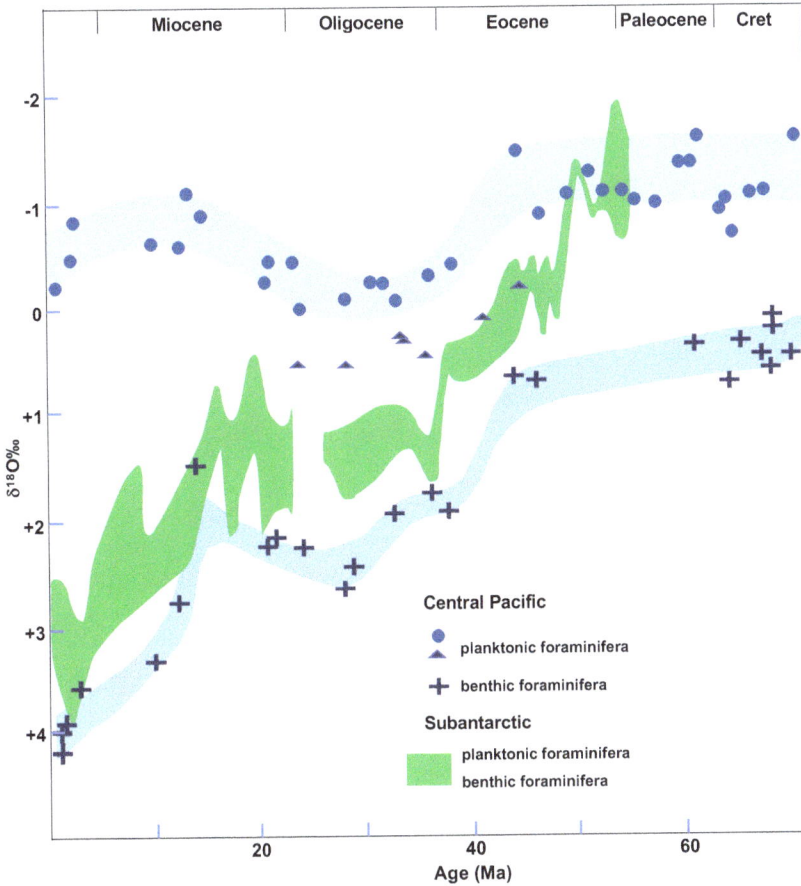

Figure 7.10. Berger's theory of the brackish lid on the south-west Pacific.

Wolfgang Berger put together the two pioneering plots of oxygen isotopes at Cenozoic scales, from planktonic and benthic foraminifera (Shackleton and Kennett, 1975; Douglas and Savin, 1978). In the equatorial Pacific curves, we see the two sets diverge increasingly through time as the $\delta^{18}O$ signal from the deep ocean gets heavier, implying colder, whereas the surface waters stay warm. In the Subantarctic case they both get colder. So far, so good; this all fits the scenario of a cooling ocean, cooling from the poles into the deep. But inspect the earlier part of the Eocene: the Subantarctic deep benthos is giving the same signal as the surface-equatorial plankton (but there is no reading for equatorial benthos). That implies either a very low planetary temperature gradient, or a $\delta^{18}O$ signal distorted by lots of fresh water, that is, a south-west Pacific Ocean with a brackish mixed layer. Or, most likely, both.

Source: Berger (1979), republished (2009); redrawn.

217

Well, the paratropical forests skirting the very warm AAG seem to be well established as reality, four decades later, and so too does year-round rain; and I wouldn't dismiss the possibility of a transient oceanic lid either.[13] But what happens then to thermohaline circulation, depending as it does on cold, dense water, generated in polar regions and delivered to the world's ocean basins, thereby producing the surface–bottom gradient in temperatures? A plausible answer might be that there was no cold, dense water. Carrying this situation further spawned the theory of halothermal circulation, replacing the thermohaline ocean from time to time. (Refer back to the thermohaline/halothermal contrast in Figure 6.2.) In a halothermal ocean there were shallow basins at low latitudes acting as giant evaporating dishes producing dense brines, which escaped from time to time into the oceanic bottom waters.[14] The most direct and accessible implications for very low latitudinal gradients should be very low mixed-layer–bottom-water oceanic gradients—and there does exist such evidence. Drilled sections beneath the tropical Pacific Ocean indicate separation during the Palaeocene in isotopic temperatures between the mixed layer and the bottom waters, kilometres below—that is, a robust gradient. But for a time in the Early Eocene, as the bottom waters warmed strongly but the mixed layer stayed constant, the isotopic numbers overlapped—implying that the gradient collapsed.[15]

Beginning with a sluggish circulation, a radically flattened longitudinal gradient in oceanic temperatures has challenging implications for the global environment. Here is a question from a couple of decades ago: why are corals and their reefs, plentiful in the Late Cretaceous and the Palaeocene and again in the Oligocene, so rare in the Early Eocene? Putting it another way: why no corals in the Tethyan transect where there are no less than four flourishing communities of large tropical-type foraminifera? Contrast

---

13   And I didn't. An extensive brackish-water lid might explain quite a few things about the AAG in Early Palaeogene times, such as the paucity of calcareous microfossils and macrofossils. The big three variables pushing the biosphere in its wet winter darkness would be high temperature, low salinity and low oxygen.

14   Halothermal and thermohaline: for Woodruff and Savin (1991) the warm saline waters generated in Tethys' shallow marginal seas and platforms influenced the circulation of the Palaeogene global ocean. The evaporating dishes disappeared when the oceans Tethys and Paratethys disappeared into mountain belts in the Miocene. An isotopic reversal in the far south (Weddell Sea) in oceanic Eocene sections implied warmer water *below* colder in the water column. Discovering this reversal in the foraminiferal shells led Kennett and Stott (1991) to push the halothermal narrative a stage further. Thus Eocene *Proteus* (halothermal circulation) was succeeded by Oligocene *Proto-Oceanus* (mixed halothermal and thermohaline circulation) then Modern *Oceanus* (thermohaline circulation). Sadly, as too often at the anti-romantic cutting edge of science and its funding, the names have not prospered. But salinity, temperature and density powerfully affect circulation at all scales from puddle up to ocean.

15   See Dutton et al. (2005).

the lifestyles of the corals and the large foraminifera, the two great groups of photosymbiotic calcifiers (i.e. producers of skeletons, or 'carbonate factories') living in the tropics and subtropics. To go with their symbiotic farming arrangements, corals are plankton catchers, also called suspension feeders, meaning that their food comes to them on currents—for them, turbulent mixing in the sea is highly beneficial. A cooling ocean with steepening temperature gradients is prescribed. The large foraminifera also farm symbionts, but they are pseudopodial grazers too, and they graze on microbes on the surfaces on which they live. An increasing global temperature with flattening temperature gradients implies a decrease in local turbulence—to the forams' overall benefit at the expense of the corals. Hence their splendidly arrayed communities on the warm platforms in Eocene Tethys. (These communities reinvented themselves to some degree in the Miocene Indo-Pacific region.)

But sluggishness had other effects in the far-southern tropics. The large foraminifera never arrived during the Palaeocene or the Early Eocene hothouse. Carbonates are virtually absent from the AAG and the poor and sporadic record of marine shells indicates hostile environments. Conditions being extremely wet, excess runoff caused two problems in this enclosed sea by forming a brackish lid. First, lowering the salinity severely inhibited marine organisms' physiological mechanisms for calcifying—for constructing their shells. Second, the brackish lid inhibited gas exchange, by which waters expel their respired carbon dioxide and replenish their oxygen. Brackish above and foetid below, the AAG was not a friendly place for many organisms.[16] The upshot was that the forests around the warmhouse and hothouse gulf never had the marine counterparts that were flourishing in seas elsewhere, neither diverse coral communities nor the spread of the phytosymbiotic foraminifera across the neritic seas.[17]

---

16   But clean up the place and they will come. Darragh (1994, 1997) and Stilwell (2003) have described the shelly fossils mainly of molluscs that found conditions to be tolerable sporadically and all too briefly.

17   Our references to sluggish circulation in hothouse oceans and atmospheres imply low temperature gradients. That is, that the high latitudes warm and cool during climatic changes more than do the tropics. The evidence supports this, but the modellers have had difficulty in producing the effect, known as polar amplification. A recent study has supported polar amplification to the extent that the Early Eocene latitudinal gradient was reduced by at least 32±10 per cent compared to the modern latitudinal gradient (Evans et al., 2018).

Hothouse biogeography outside the Australo-Antarctic Gulf

EECO, the global warming event lasting more than 4 million years (53.3–49.1 Ma), included several especially warm events identified by sharp spikes in carbon-isotopic profiles ($\delta^{13}C$ becoming negative) and known as hyperthermals with durations of 40–200 thousand years (the PETM is the first and best-known). A hyperthermal known as the J event defines the onset of EECO, and at or very close to the J event the planktonic foraminiferal faunas of the open ocean suffered a major impact, a switch in dominance between the dominant genera *Morozovella* and *Acarinina*, at first near the equator, then at higher latitudes north and south. Why? It might have been too hot or too acid in the photic zone, perhaps bleaching by inhibiting the photosymbionts. The central point here is that oceanic microfossils in their hundreds of species and thousands upon thousands of specimens can be microchemically analysed and dated to within thousands of years, to pose and answer questions from 50 million years ago.

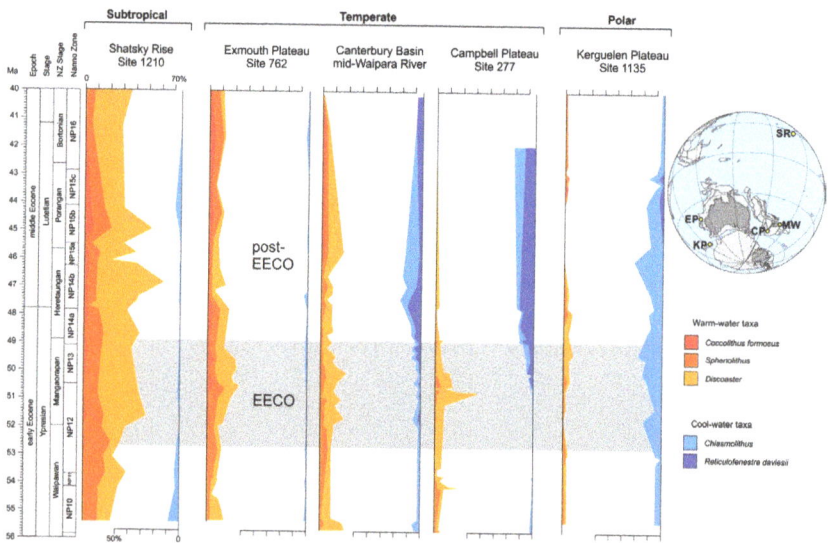

Figure 7.11. Southern calcareous phytoplankton out of the hothouse.

Biogeography reflects both global environmental gradients and global environmental change in deep time. Temperature-sensitive taxa among the calcareous phytoplankton — coccoliths and discoasters — display the end of Hothouse Earth around the Antarctic, Australian and Zealandian continental masses at southern latitudes but are absent from the AAG. These microfossils display a latitudinal gradient during EECO and the gradient steepened discernibly after EECO.

Source: Adapted from Crouch et al. (2020).

Zealandia has a good example of all this in addressing the notion of extreme temperature and flattened temperature gradients on Hothouse Earth (Figure 7.11).

The coccoliths and discoasters are distributed biogeographically in a pattern that can, reasonably, be called subtropical, temperate and polar. On this evidence the oceanic temperature gradients are not extremely flattened in EECO. But the coccoliths and discoasters distributed biostratigraphically from Hothouse Earth into Warmhouse Earth displaying strong response to the global climatic shift.[18]

Before the wide brown land

The igneous and metamorphic rocks of the earth's crust, the lithosphere, comprise minerals formed at high temperatures and pressures. These minerals are the silicates—the olivines, pyroxenes, amphiboles, feldspars and micas. Bring them to the earth's surface, and you bring them out of their stability zone. The lithosphere interacts with the hydrosphere and the biosphere. It becomes the reactive lithosphere. At 'our' temperatures and pressures, subjected to expansion and contraction and wetting and drying, penetrated by plants and soaked in carbonic acid and rotting vegetation, the high-temperature minerals are intrinsically unstable. They break down physically and chemically during what is known as weathering; the more or less undisturbed residue of weathering is the saprolite; the saprolite plus any materials from 'outside' make up the regolith, and the regolith includes the soils. Advanced chemical weathering produces the clay mineral kaolinite, the ferric iron minerals such as goethite in laterites, and the end member, bauxite, rich in aluminium minerals especially gibbsite. Laterite but especially bauxite have long been taken as signs of wet-tropical weathering, so their presence at high latitudes indicates expansion of tropical conditions at times past, especially the Early Eocene in south-east Australia. But such inferences have been challenged, and loudly.

---

18   The demise of *Morozovella* at the onset of EECO is described by d'Onofrio and four co-authors in (2020). *Morozovella's* mean percentage abundance dropped from 32 per cent to below 7 per cent, an ecologically huge shift for a dominant oceanic taxon. The coccolith studies in Zealandia and the south-west Pacific Ocean are in Shepherd et al. (2021) and Crouch et al. (2020), from whom Figure 7.11 is adapted. This Zealandian research is a strong biogeohistorical program. The leader, Chris Hollis, asked in 2014, *Was the early Eocene ocean unbearably warm or are the proxies unbelievably wrong?* The proxies, palaeobiological, geochemical and modelling, are still some way from achieving consensus and consilience in their answers.

Australia has long been called an old continent—too old for oil, for example—displaying very large areas that have been exposed to weathering for hundreds of millions of years and even more. Weathering in some places has penetrated hundreds of metres to change rock-forming minerals. Given all that time, it is not difficult to believe that the processes can be so slow as to be imperceptible. It is also noteworthy that Australia mostly missed the enormously rejuvenating effects of the Alpine mountain-building episodes which occurred in Zealandia and at the northern edge of geological Australia (Sahul), in Papua New Guinea and Indonesia. In a second 'miss', we experienced little of the cleaning and scraping and refreshening effects of the Pleistocene ice sheets, so prominent on the northern continents. One outcome has been a particularly thick and relatively undisturbed regolith. At the same time, though, we mostly lack the great piles of sediment shed by the rising Alpine chains and deeply eroded in their turn, for those strata in, say, North America or China archive not only leaves and bones but also ancient soil horizons exposed to geological eyes.

Biogeohistory is based on succession, the sum of untold numbers of superpositions, and on dating—what is older, what is younger and by how much time? But consider the outcrop and typical profile shown here (Figure 7.12).[19]

The Painted Desert is in rocks mostly of Cretaceous age. They have never been deeply buried then exhumed; they have not been deformed beyond some gentle bending and breaking. They have been bleached in deep weathering to the white of the clay mineral kaolinite, the main residue (plus quartz, the great survivor among common minerals), with some red colouring from the relatively insoluble rusts of ferric iron. All that we can say about the timing of the weathering event or events is that it must have been at latest Cretaceous or Cenozoic in age, and that it might have been long and slow or short and sharp, or a bit of both. We don't have tangible evidence superimposed by stacking, such as might be offered in a succession of fossils. The dominating processes of soaking, rearranging and removal might be successional, but the outcome is not superpositional in a way that Arduino, Cuvier or Smith might have recognised two centuries ago.

---

19 Johannes Walther (1860–1937), whose weathering profile appears in Figure 7.12, was one of the very best, much-travelled, highly curious and productive natural historians, his head seething with ideas about geology, palaeontology and marine biology, best known in biogeohistory for Walther's law of facies, stating that sedimentary facies stacked conformably in outcrops were also formed alongside each other. His field work in Western Australia was cut short by the outbreak of World War I. See Gischler (2011).

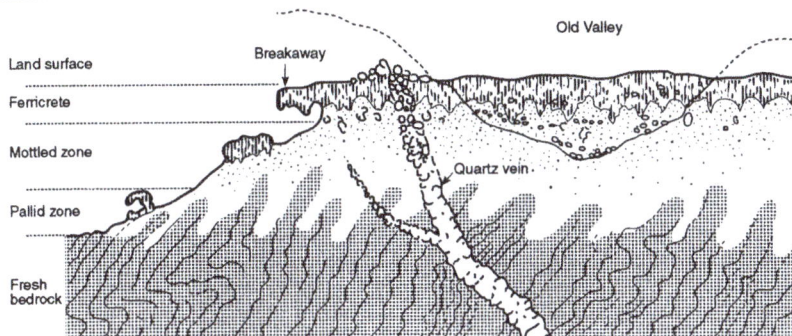

Figure 7.12. Mt Arckaringa and Walther's classical arid weathering profile.

Above, Mount Arckaringa, in the 'Painted Desert' in South Australia. These are sedimentary rocks of Cretaceous age, bleached and ferruginised during intensive weathering extending to depths of tens to hundreds of metres, in environments very different from the arid zone shaping the distinctive modern desert landscape.

Below, the ferricrete (ironstone) profile, or typical cross-section, through the arid landscape of Western Australia, as published by Johannes Walther in 1915. Note the outcropping quartz vein, indicating that the rock being weathered is in situ, unlike the ancient valley (palaeochannel), also being weathered. The fresh bedrock here is deformed, metamorphosed and of Proterozoic or Archaean age.

Source: Mt Arckaringa image, David Olsen (via Wikimedia Commons: commons. wikimedia.org/wiki/File:Arkaringa.jpg). Walther's profile (1915) from Bourman and Ollier (2002).

Inspect in Walther's figure, however, the generalisation called a weathering profile or a typical cross-section, constructed from outcrops on the craton in arid Western Australia. There is evidence here of three processes: (i) 'physical' erosion is occurring, laterally, by undercutting the resistant cap of ironstone which falls off in chunks from the distinctive 'breakaway'; (ii) 'chemical' erosion is/was occurring vertically, bleaching below and concentrating ferric iron above (and retaining the quartz vein intact until now); and (iii) the profile was cut at some stage by a stream, long gone, leaving a choked channel ('old valley').

Any evidence of when the channels were cut, such as plants or their sporomorphs preserved in the channel fill (against the odds), would also be evidence for a minimum age for the weathering profile and for the profound change of climate implied by a vigorously cutting watercourse. Except that Walther drew his ferricrete, the resistant cap, across the palaeochannel with no discernible interruption. Which leaves us with no evidence against the grand null hypothesis of Lyellian geology, namely that nothing much changes in earth history until you can convince us with modern examples of the conjectured process that something has indeed changed. For the ferricrete must have been forming before and after the cutting of the palaeochannel.[20]

There have been two perceptions of deep weathering on Australia: gradualist and episodic. The gradualist school points to evidence that laterites can form in wildly varying environments. All that you need for deep weathering is plenty of time and plenty of water. If it is wet, then slow and cool can be as effective in the long run as fast and hot. Some proponents look to a favourable parent rock; others, efficient drainage; still others point to tropical-type bauxites associated with fossil floras of cool aspect; probably most believe that weathering intensity has not fluctuated greatly through deep time. The gradualist view encompasses a very wide range of workers— probably every discipline pertaining to landscape evolution is represented. If they have one thing in common, it is that there is no cogent evidence of cycles or rhythms in the record of deep and deep-time weathering.[21] No jerky patterns.

---

20   I disagree with a significant detail in Walther's splendid diagram displaying the ferricrete on the old rock as coeval with the ferricrete on the geologically very young palaeochannel. The ferricrete on the old rock, it has been widely believed, was a residue, a concentrate remaining when all else has been removed. If so, then it could not be the same residue atop the intact and unshrunken fill of the palaeochannel. Instead, the latter would have been imported as part of the fill and become the cement for ironstone sandstones and conglomerates at the base of the old valley, not at its top. 'Laterite', 'ferricrete', 'ironstone'—all have been criticised for loose use. All we need here is to distinguish between residual ferric crusts and imported ferric cements (McGowran et al., 2016). The cemented ironstones are widespread in central southern Australia where Middle Eocene channels were cut into the deeply weathered EECO profiles and the abundant ferric iron was remobilised to produce large red-brown sedimentary clots within the accumulating strata, mostly sands. The metre-scale clots are popular as ornamental stones.

21   Taylor (1994) assembled ('without critical review') a plot through time of weathering events, and indeed there is no cyclicity or episodic pattern to behold among three dozen sources in the literature. Taylor and Shirtliff (2003) assembled a still wider array and still found no fluctuations in weathering intensity. For Bourman (1995, 2007; see also Milnes et al., 1985), a committed ultra-gradualist, the 'standard laterite profile' like Walther's is usually referred to in scare quotes.

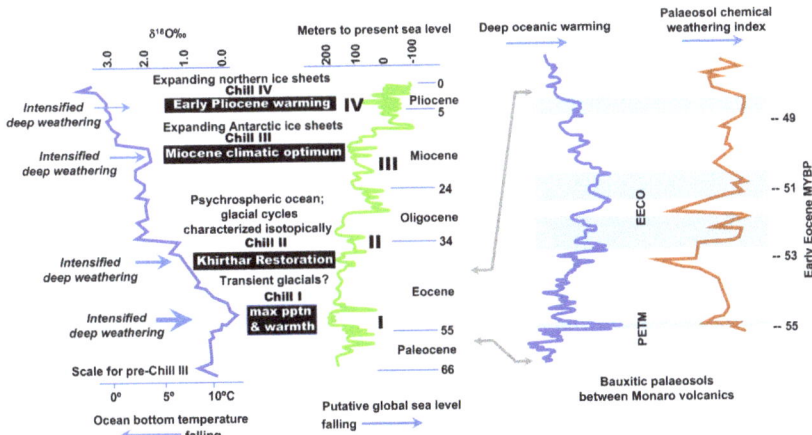

Figure 7.13. Argument and test of episodic deep weathering in the hothouse.

Left, the central argument in the 1980s for episodic deep weathering. That the Cenozoic world changes episodically was becoming apparent in the pelagic, neritic and terrestrial domains, in climatic change and in the history of the biosphere. The proposal was that times of warmth and wetness would be times of intensified deep weathering. It seemed extraordinary that the anatomy of the landscape, the regolith, should be outside of episodic biogeohistory in all its forms and essentially unknowable as to its history. The size of the arrows suggests the degree of intensity predicted for deep chemical weathering.

Right, a test of the theory of episodic deep weathering by Greg Retallack: 53 palaeosols within an Early Eocene succession of volcanic rocks in the Monaro in south-east NSW included several with an advanced chemical weathering index, signalled by the mineral bauxite. They revealed that local spikes in warmth and precipitation at high southern latitudes coincided with global spikes in warmth, precipitation and high atmospheric levels of CO_2. The four spikes are spikes in nature, and not mere remnants of incomplete preservation. The pattern is entirely consistent with the transient hyperthermals, an intrinsic feature of the Eocene Hothouse as displayed in the patterns of oxygen and especially carbon isotopes.

Source: *Left*, adapted from McGowran and Li (1998). *Right*, adapted from Retallack (2008).

By the late 1970s I was well-marinaded in the global exogenic system, meaning the interaction between the hydrosphere, biosphere and reactive lithosphere and their episodic, stop-and-start patterns through millions of years and more. Episodic patterns were widely apparent, such as in oceanfloor spreading and orogenic ('mountain-building') pulses, palaeoceanographic shifts and global climatic events, and in reading the fossil record as punctuated organic evolution. If much of known biogeohistory were like that, and not to be shelved as gradualism masked as a very imperfect record of what happened in deep time, then what about the lands and their weathering? It was apparent that dating the landforms and their regolith was fraught with problems. It was also apparent that the gradualistic paradigm held sway, so I could hardly help wondering why, when gradualism was melting away everywhere else, the regolith was exempted.[22] My idea was merely (word used advisedly) that Cenozoic biogeohistory was resolving into a four-part pattern—and perhaps deep weathering was, too?

During the global cooling after the Early Eocene, there were three main reversals truncated by renewed cooling. The 'Chills I–IV' got stronger, the episodes of deep weathering got weaker? Simple, no?

That notion of episodic weathering lingered in some limbo between plausible and compelling. Palaeomagnetic dating applied to ferruginisations of the Australian regolith produced a clear two-part answer, a big signal at 60±10 Ma and a second signal at 10 Ma ±5.[23] The clearest evidence came from Retallack's detailed palaeopedological study of a series of soil horizons among volcanic flows in the Monaro Tableland in south-east NSW. Retallack developed a chemical weathering index, the end member being the mineral bauxite. The basalts gave a control on the ages. Thus the bauxitic high points in the weathering could be correlated with high points in the deep-oceanic oxygen-isotopic curve including PETM and EECO. Previous investigations had found that cool-upland macro- and microfloras implied bauxitic weathering that was less-than tropical, but these tight correlations revealed wet-tropical episodes of deep weathering.[24] Local spikes in warmth and precipitation on land at high southern latitudes coincided temporally with global spikes in deep-oceanic warmth, in precipitation and in high atmospheric levels of CO_2.

The spikes between 49 Ma and 56 Ma are real geohistorical spikes in a real succession and not mere remnants of patchy preservation. And deep weathering is an intrinsic component of the fourfold division of Cenozoic rocks and time which is a central theme of this book.

---

22 McGowran (1979a, b). Regolith studies were boosted in the ensuing decades because the Australian regolith masks so much potential mineral wealth, and an informative snapshot is to be found in Eggleton's *The state of the regolith* (1998). Figure 7.13 (left) is redrawn and adapted from McGowran and Li (1998) in that volume.

23 Pillans (2002). Encouraging, but no cigar.

24 Retallack (2008) gets the cigar. The deep weathering was proved here to be short, sharp and, importantly, variable in its intensity among the soils in the series through a drill core. Bega #7 drilled through almost 200 metres of the Monaro volcanics encountering 53 palaeosols, several of them bauxitic, on basalt flow surfaces, each capped abruptly by the chilled surface of the next flow. Retallack successfully falsified (in this situation, at any rate) an array of arguments against episodic deep weathering—such as cool bauxites, or deep weathering, long and slow and cool (Taylor et al., 1990). Zhou et al. (2015) described another thick gibbsite–kaolinite palaeosol from Bridle Creek in the Monaro, reinforcing the wet-tropical climate at about 60°S at ~52 Ma, in the middle of EECO. (It is unimportant here, but the bracketing of the bauxitic episodes is now from about 56 Ma to just before 49 Ma, spanning the Eocene Hothouse).

8

Farewell, hothouse and farewell, Australo-Antarctic Gulf

An isotopic profile of the long trajectory of Warmhouse Australia

The clade of *Casuarina*, the she-oaks (Figure 8.1), flourished during the Early Eocene climatic optimum (EECO) and flourished again when EECO and Hothouse Earth closed. In those days the clade comprised rainforest forms, of which the genus *Gymnostoma* still survives. *Casuarina*'s ongoing narrative is one of adapting to soils of low fertility (especially lacking phosphorus), to the drying-out of the waterlogged EECO environments, and to the disappearance of the winter darkness. It is also noteworthy in cultural background that *Allocasuarina verticillata* (Lamarck) was described in 1786 by the same great biologist who described *Trigonia margaritacea* and many other shells, meanwhile discovering organic evolution.

We need a chronological framework to hang the turbulent story of the Late Palaeogene, with its punctuated cooling and the still-mysterious, early growth and decay of polar ice.

Figure 8.1. Tom Roberts, *Sheoak and sunlight* (1888).

The species is *Allocasuarina verticillata* (Lamarck), the drooping she-oak, identified thanks to Ian Sluiter.

Source: Courtesy of the National Gallery of Victoria, with permission.

Figure 8.2 is a reference for 14 million years of global trajectory and its acronyms and for the regional names and correlations which will become familiar in this chapter. Particularly welcome is the clear depiction in the oxygen and carbon signals of the transition from the Bartonian to the Priabonian Stage, because this transition is hazier in the literature than are MECO (Middle Eocene climatic optimum) and E-OT (Eocene–Oligocene transition).

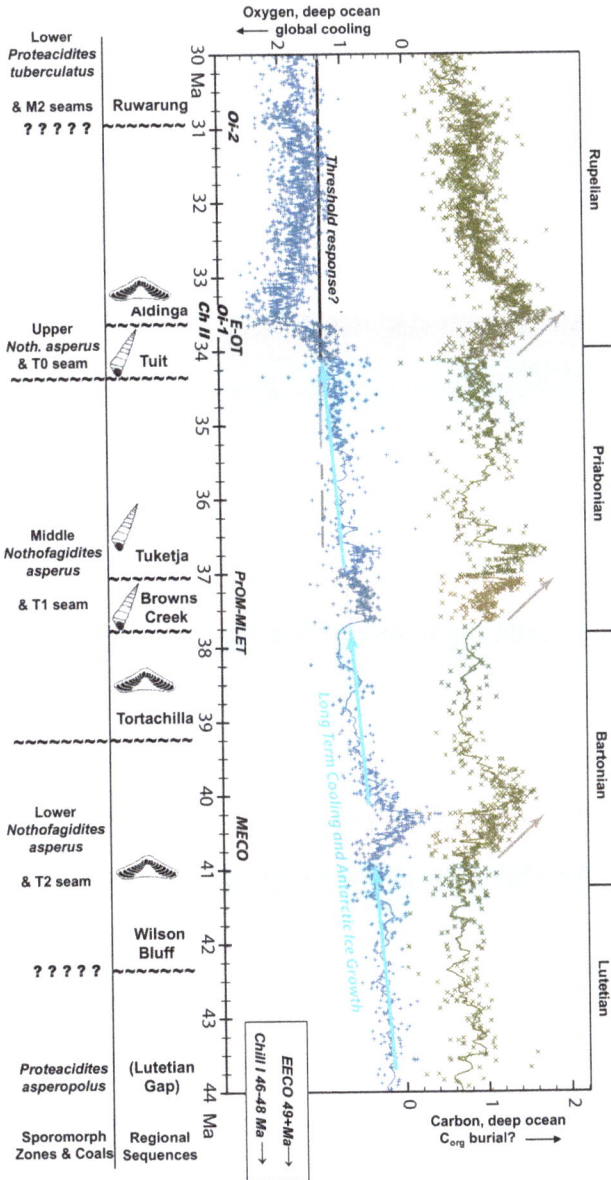

Figure 8.2. Lutetian–Rupelian global cooling.

This 15-million-year trajectory of global cooling shows three successively weaker warming reversals during the Middle and Late Eocene, as compiled in the deep-oceanic, oxygen-isotopic record in benthic foraminifera (Henehan et al., 2020). The three reversals are accompanied by three pronounced positive carbon excursions, suggesting that photosynthetically fixed light carbon has been buried somewhere — whether in the pelagic, neritic or terrestrial realm. The repetitive arrows hint at some kind of cyclicity at 3± million years. To emphasise fluctuations in fertility in the Australo-Antarctic

Gulf, the snail *Spirocolpus* signals eutrophic tendencies and the benthic foraminifer *Halkyardia* signals oligotrophic tendencies. MECO, Middle Eocene climatic optimum. PrOM, Priabonian oxygen maximum. MLET, Middle–Late Eocene planktonic transition. EOT, Eocene–Oligocene transition. Oi-1, Oi-2, isotopically signalled glaciations. Threshold response?, predicted critical level for southern-polar icecaps to grow and remain stable. See the text for the regional sequences in the rundown to the icehouse, the sporomorph zones and the Gippsland coal seams.

Source: Isotopic curves: Henehan et al. (2020). Pollen zones and coal seams: Partridge (2006), Holdgate and Sluiter (2021), Korasidis et al. (2019).

Biogeographic digression on the Leeuwin Current

We are about to reintroduce the large, photosymbiotic foraminifera to the story. Marine biogeography in the southern hemisphere is largely controlled by the anticlockwise oceanic gyres, so that extratropical excursions by tropical-type organisms can and could occur on the western margins of southern continents.

The palaeocirculation sketches (Figure 8.3) display a warm, clockwise current in the Australo-Antarctic Gulf (AAG), labelled Proto–Leeuwin Current. It would seem to be a deflection of the warm, anticlockwise gyral systems in all the oceans south of the equator in the warm Early Palaeogene world. However, the name comes from the modern, regionally influential Leeuwin Current which runs along the western and southern margins of the continent—against the anticlockwise Indian Ocean gyre in the modern icehouse world. This is a Neogene topic in the next chapter, but we see a biogeographic pattern like the Leeuwin Current pattern much deeper in time—indeed, all the way back to the Late Cretaceous, in a different ocean in a different world. Note, on the limestones diagram (Figure 5.15) and on our minimalist maps for EECO and MECO (Figure 8.3), that the large photosymbiotics came down from the north to 60°S in the Middle Eocene but not in the Early Eocene—not during the hothouse times of EECO when there were no carbonates in our gulf and no large photosymbiotic foraminifera. We attributed this contrast to the accelerated opening of the AAG, causing the first serious deflection of Indian oceanic circulation in that direction; that is, to the south of Australia. Sustaining the inelegance in contrasting with the squirting theory, I called this the sucking theory. While we pointed to the contrasts between the Palaeogene and the Neogene oceans, we emphasised the recurring of the Leeuwin Current under changing geographies and controls, whereas we might better have distinguished instead between proto–Leeuwin Currents and 'the' subsequent Leeuwin Current. As we saw in Figure 7.11, the currents and water masses in the neighbourhood are distinguished from their modern counterparts by 'proto-'.

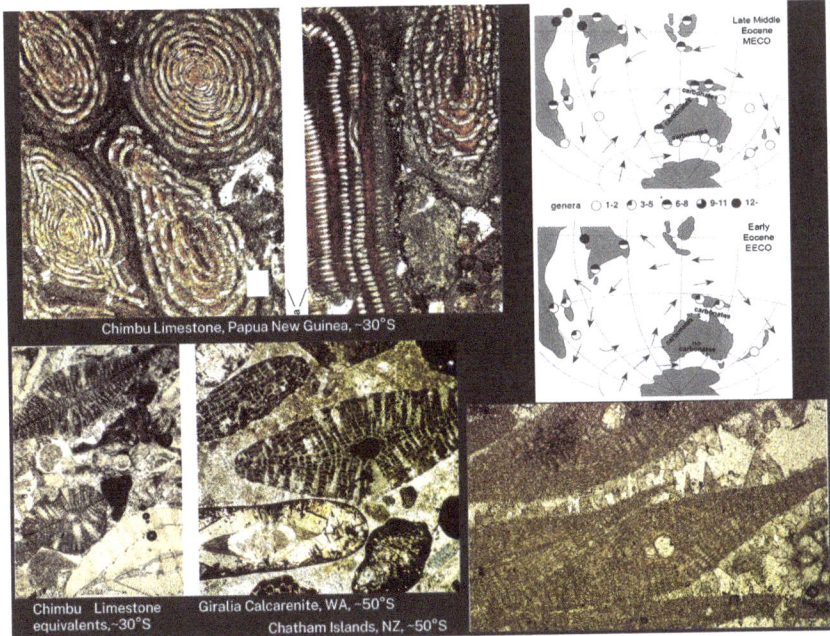

Figure 8.3. Eocene large forams expand to the deep south.

Shells of foraminifer cemented into a solid limestone are studied in thin sections under transmitted light. The Chimbu specimens from New Guinea are Alveolinids; the others are Orthophragminids and Nummulitids (recall Figure 6.9). The point of the diagram was that these big photosymbionts could come to the far south during EECO with the anticlockwise oceanic gyres (Africa, New Zealand) but not against the gyres (Western Australia) — until, that is, the Leeuwin Current was running (MECO). We now know that there was a proto-Leeuwin or Leeuwin Current in the latest Cretaceous and at least episodically throughout the Cenozoic Era. The critical point now is that the Australo-Antarctic Gulf (AAG) was mostly hostile to the big calcareous photosymbionts along with animals with calcareous skeletons; marine palynology is providing another perspective. Then again, dinocyst biogeography is largely an oceanic and planktonic story, whereas this foraminiferal story is benthic and neritic.

Specimens from New Guinea and the western Australian margin are of late Middle Eocene age (MECO time); from the southern margin AAG they are slightly younger in the Tortachilla sequence; from New Zealand specimens of *Asterocyclina* are Early Eocene (EECO time). Note the MECO record in the middle of the Indian Ocean (recall Figure 5.2). This is *Asterocyclina* and *Discocyclina*, in the photic zone at DSDP (Deep Sea Drilling Project) Site 253 on the Ninetyeast Ridge (and now beneath 2 km of water), a spectacular example of dispersal from the north, presumably by island-hopping or atoll-hopping along the ridge.

Source: From McGowran (1990). Sketch maps for EECO and MECO, adapted from McGowran, Li, Cann et al. (1997).

We have the sporomorph *Spinizonocolpites prominatus* proving that the tropical mangrove palm *Nypa* inhabited the western continental margin and was well established in the AAG during Late Palaeocene and early Early Eocene times. Therefore, the lack of carbonates informs us of hostile conditions in the AAG; it does not point to the absence of the proto–Leeuwin Current.

Into the Late Palaeogene: Khirthar Transgression, gigantic bryozoan reefs and MECO

The hothouse state of the Early Eocene was terminated. Chapter 5 told how the India–Asia collision forced a global rearrangement of seafloor spreading patterns after the Mammerickx pivot at 47.3 Ma in magnetochronological Chron C21. In the new tectonic regime, Australia–Antarctica separation accelerated. In Chapter 6 we saw three trends broadly in parallel within the time bracket, 49–48 Ma to 46–45 Ma, across the boundary between the Early Eocene and the Middle Eocene. There is the first sustained oceanic cooling (variously labelled Chill I and post-EECO cooling, now the onset of Warmhouse Earth II). There is a sustained fall in global sea level, a sustained contraction of shelf seas, that is, of the neritic realm, showing through the short-term variations. And third is a plunge in the calcite compensation depth (CCD) in the equatorial Pacific Ocean—plunging about a kilometre, which is carbonate chemistry at the global scale.

Here is a précis for the transition from the Lower Eocene world to the Middle Eocene:

i. The collision of India with Asia, two thick slabs of crust, brought on gridlock in global oceanfloor spreading. One outcome was decreased injection of CO_2 into the global environments.

ii. General oceanic subsidence increased the total volume of the ocean basins. Seas withdrew from the continents, leaving a regression, an unconformity, a widespread hiatus in stratigraphic successions on the continents.

iii. In a shelf/basin fractionation, the great limestone masses of the shallow seas of Tethys were now eroding, not growing, with the wholesale shift of carbonate from the neritic realm into the oceanic realm, depressing the CCD.

iv. Cooling was a net result of increased continentality and albedo and decreased CO_2.

v. Renewed flooding of continental margins and growth of new neritic limestone masses culminated in MECO.

This transition then was Chill I, the end of Hothouse Earth, and the first step in the palaeoclimatic series culminating in the Oligocene Coolhouse. In our region, lowland forests comparable to the forests of modern New Caledonia were replaced during this cooling by forests more like those in modern New Zealand. The most prominent changes were the diversification and spread of the southern beech, *Nothofagus*, in wetter districts and of *Gymnostoma*, antecedent to *Casuarina* (the she-oaks) which continued to flourish where it was less wet.[1]

The second of the four Cenozoic packages of strata, the Late Palaeogene, is more widespread and more visible than the Early Palaeogene. The region was shaken up in the new spreading regime—for example, Australia's modern inland drainage, centred in Lake Eyre, dates from warping of the Australian crust at this time. The palaeodrainage divide between north-east flowing and south-west or south flowing was shifted to the south, and a series of palaeovalleys either were reversed or initiated at this time (Figure 8.4). The changes were responses to changes in the plate-tectonic regime involving the accelerating separation of Australia from Antarctica, as outlined in Chapter 5.[2]

---

1   Some stratigraphic housekeeping is needed here. The sporomorph and dinocyst zones of palynology are the strongest regional control on our dates and timings through this massive global shift with the passing of EECO. But the sporomorph zone straddling the Lower–Middle Eocene boundary, the *Proteacidites asperopolus* zone, spans 5.5 million years (Partridge, 2006). We have dateable foraminiferal assemblages from the *P. asperopolus* zone no younger than 50–49 Ma (on the last gasp of EECO). The next dateable foraminifera are known from the next sporomorph zone, the Lower *Nothofagidites asperus* zone, which is 6.6 million years long (Partridge, 2006). The foraminifera are dated at 42–41.5 Ma. This is the date for the base of the great Middle Eocene limestones, the Khirthar transgression and the gigantic bryozoan mounds. And there is no evidence from anywhere in the AAG for shelly marine fossils preserved in that time span from 49 Ma to 42 Ma. We called this 7-million-year hiatus in the evidence the Lutetian Gap (McGowran et al., 2004; Frieling et al., 2018). There are sands and muds in this time interval but we know nothing about their ages beyond this vague bracketing. Although we have been aware since the 1970s of the Lutetian Gap and Khirthar Transgression, they have been mostly ignored in seismic stratigraphy, which neglect has masked our ignorance and deformed our tectonic history. See Chapter 5, note 11.

2   Figure 8.4 is modified after Hou et al. (2008): this paper is an especially important contribution to the geohistory of the swathe of territory including the Eucla and Bight basins. The palaeodrainage divide, passing across Yilgarn craton, Musgrave Province and Gawler Craton, the ancient nuclei of the Australian continent, was established in the Early Palaeogene. Growth of the great Cretaceous deltas in the AAG was terminated.

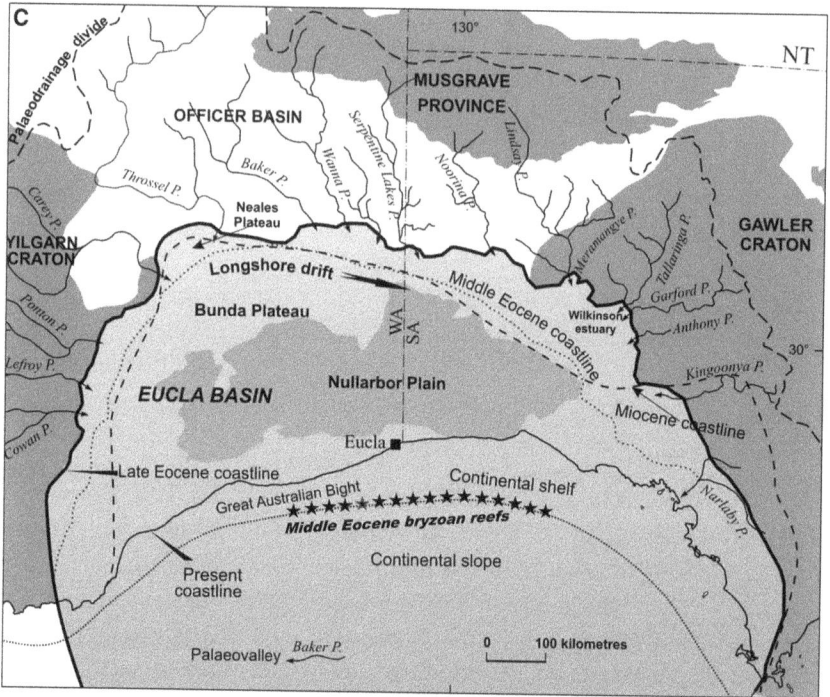

Figure 8.4. Eucla Basin, Eocene and Miocene shallow seas.

The Eucla Basin, outlining the vast shallow seas of Middle and Late Eocene and Middle Miocene times. Their watershed was bounded by the palaeodrainage divide, by then a long way south of its location during Cretaceous times. The palaeovalleys or palaeochannels, now infilled, were carved into the rising landscape under rainforests. The bryozoan mounds extend along more than 500 kilometres of the rim of the continental shelf and 300–500 kilometres distant from the coastline of the Middle Eocene sea.

Source: Modified after Hou et al. (2008).

We see evidence in the rocks of two very large happenings. In the west there accumulated the great carbonate mass called the Wilson Bluff Limestone, marking the Khirthar Transgression around the Indo-Pacific region, which included the western and northern continental margins. In the south-east there accumulated the Traralgon Coal, the largest body of brown coal of Cenozoic age on the planet; and somewhat smaller coals developed around the north shores of the AAG in the Middle and Late Eocene. We might expect major changes in carbon dioxide levels—CO_2 levels rising, as carbonate is precipitated; CO_2 levels falling, as peat is buried and coal accumulated. (When the two trends occur with opposing outcomes …?) Certainly there is extensive evidence in the terrestrial realm for warm and wet episodes; that is, rainforest conditions; and in the neritic realm the photosymbiotic foraminifera expanded to higher latitudes at about 40 million years ago—strong biogeographic evidence for warming. Although this pattern was coming into focus by the late 1970s, the early oxygen-isotopic profiles from the deep oceans showed instead a general cooling, from the Early Eocene high to the drop at the end of the Eocene. There seemed to be a misfit between the terrestrial and neritic realms on the one hand and the oceanic realm on the other. I chafed under this misfit for a quarter-century. The resolution arrived with Steven Bohaty's discovery of MECO, a sharp oxygen-isotopic spike now centred on about 40 Ma and lasting for 5–600,000 years. Oceanic temperatures rose by perhaps 5–6°C. And as we have seen, first the Khirthar Transgression then MECO are natural markers for the onset of the global transformation named (in the oceanic realm) the Auversian Facies Shift.[3]

The Wilson Bluff Limestone accumulated along the north flank of the AAG on seafloor measured in thousands of square kilometres and thicknesses in hundreds of metres. It had several things in common with the massive limestones appearing on two other Australian margins and beyond: they began accumulating at the same time on platforms or ramps in the neritic realm: they were all responses to the onset of the new plate-tectonic regime; they are part of the Khirthar Transgression extending far beyond Australia; they all sit squarely atop MECO; and they would seem to be strongly implicated in the fluctuations in the CCD in the deep, equatorial Pacific Ocean. But the Wilson Bluff at a palaeolatitude of 60°S is distinctive. It is not coral-rich nor is it large-foraminiferal-rich, but bryozoan-rich, and with layers of chert, also biogenic. Indeed, the Wilson Bluff has much more in common with the famous chalks of the shallow Cretaceous seas, the very chalks that TH Huxley compared with the calcareous oozes of the modern deep ocean, than with the limestones of the hothouse and warmhouse Eocene seas with their ecological partitioning.

It has recently been discovered that these diverse communities of bryozoans built huge mounds or reefs extending along the margin of the continental shelf at about 60°S palaeolatitude for more than 500 kilometres, hundreds of kilometres distant from the coastline of the times (Figure 8.5).

---

3   Bohaty and Zachos (2003); Bohaty et al. (2009).

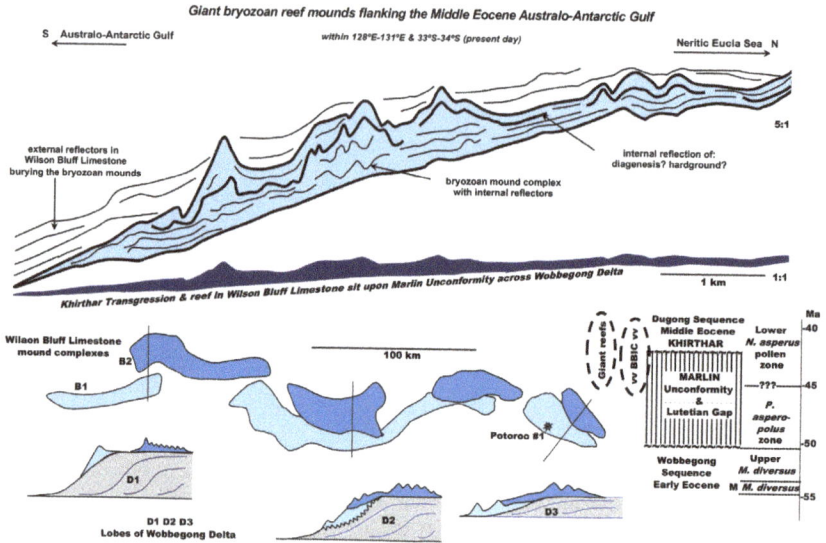

Figure 8.5a. Sharples's discovery of the giant bryozoan reefs.

Above, Alexander Sharples's interpretation of a seismic profile (shown both at 1:1 and 5:1 scales) through a bryozoan mound within the Wilson Bluff Limestone. These biogenic mounds attained hundreds of metres in thickness. Internal reflectors at different strengths of uncertain meaning, and external reflectors in the burying strata could be distinguished. Also distinguishable in seismic surveys (but not shown here) are numerous mounds, somewhat similar to the bryozoan mounds in both seismic form and apparent geological age, which turn out to be volcanoes instead.

Below left, the Wobbegong Supersequence included three deltaic lobes — three piles of siliciclastics — colonised by giant bryozoan reef mounds, one of which was penetrated by the all-important and precious drillhole Potoroo #1. The mounds at the base of (but within) the Wilson Bluff Limestone and the Dugong supersequence seem to sort into two complexes, B1 and B2. Note the scales at 1:1 and 5:1.

Below right, plotted against a time scale, the major unconformity identified by the name 'Marlin' from south-eastern Australia is shown here as a hiatus spanning the *Proteacidites asperopolus* sporomorph zone. Buried within the Lutetian Gap in time are two world-shaking events, namely the Mammerickx pivot and the end of Hothouse Earth. Indeed, this figure could comprise a case for oceanic drilling and coring to meet three lots of ground-truthing for all this seismic imaging — to recover some volcanic rocks, to extend our tangible evidence for unique reefs, and to inject some precision and accuracy into our age determinations. BBIC, Bight Basin Igneous Complex.

Sources: Developed and adapted from Sharples et al. (2014). Dating the Wilson Bluff in Potoroo #1 hole was by Taylor (1975), exploiting the recent discovery by McGowran and Lindsay (1969) of what was dated and named subsequently the Khirthar transgression.

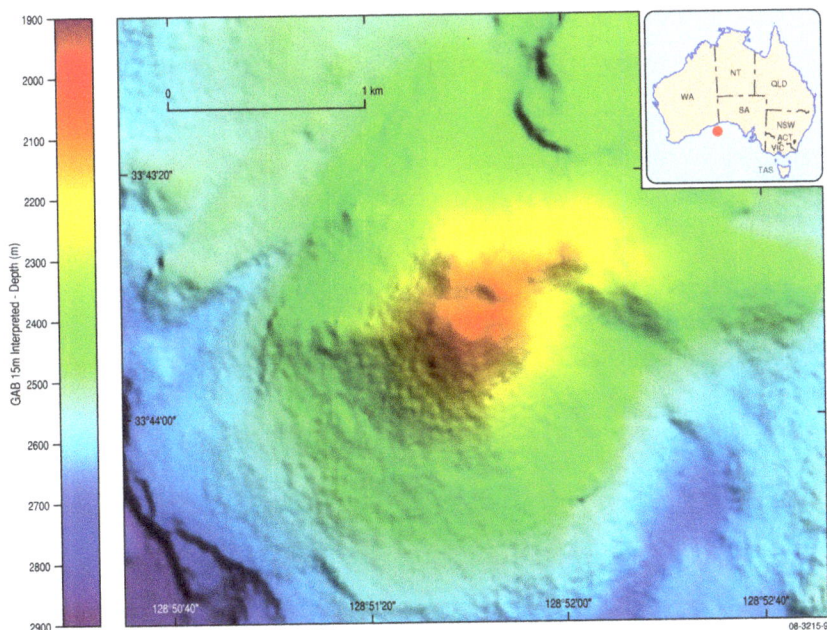

Figure 8.5b. Ceduna volcano.

A 'swathe-bathymetry' image of a volcano in the western Ceduna Sub-basin, part of the Bight Basin Igneous Complex (BBIC). There is a crater, and images suggest lava flows.

Source: Image from Schofield and Totterdell (2008) © Commonwealth of Australia (Geoscience Australia)

The mounds individually attained 60–150 kilometres in length, becoming as wide as 15 kilometres, and growing to 75 metres thick on average but up to 200 metres thick in places. In the world of the bryozoans, living and fossil, these numbers are utterly unique, by an order of magnitude and more. It were as if the Great Barrier Reef were built by bryozoans not corals. The structures were known only through seismic surveys and were thought to be volcanic. To date, there is almost no ground-truthing of the geophysical images. Only one mound is actually known hands-on, from the rock cuttings of but one petroleum-exploration drillhole; that was sufficient to establish the ground truth of its bryozoan nature and its neighbours' too, but the distribution and dimensions of the structures are based entirely on seismic interpretation. And they are not volcanoes.

The conditions, especially copious food supply lines, must have been as special as these colossal edifices themselves. The bryozoan builders were mesotrophic filter feeders, not oligotrophic gardeners of photosynthesising microbes like the corals with their endosymbiont *Symbiodinium*. Recall that in Early Eocene times there was no shortage of nutrient in the AAG but the conditions—low and fluctuating levels of salinity and oxygen—mostly were hostile to organisms with calcareous shells. For the carbonate factory to begin operating, that situation had to change while abundant food remained available—a balancing act. The post-EECO cooling steepened the longitudinal temperature gradients, which invigorated mixing and ventilated the upper waters of the gulf.

A plausible scenario: the reorganising of the oceanfloor spreading systems in the earth's crust in the Early Eocene led to a surge in spreading in the Middle Eocene, spilling shallow seas across continental margins. That was the general, global driving force. Locally, the accelerated spreading within the AAG (well known on good evidence) also warped the continental crust, reorganising the palaeodrainge divide and causing subsidence all around its northern flank, forming new sedimentary basins and rejuvenating old ones (also on solid evidence). Together, these events caused the flooding of thousands of square kilometres to form a warm and shallow sea. Rimming that sea oceanwards, unreached by crises in oxygen and salinity—and with no competition from photosymbiotic corals—the mound-building colonies of bryozoans spread along the continental margin, hundreds of kilometres from the shoreline. The AAG was a gulf (a very narrow and shallow connection with the south-west Pacific at the eastern end notwithstanding) draining richly forested lands, and estuarine in circulation. Its deeper waters, still somewhat stagnant, accumulated more organic material than did the large open oceans. Favourable winds blowing surface waters offshore allowed the fertile deep waters to well up to within reach of the suspension-feeding colonies rimming the platform. Subsidence of the crust for some time encouraged the colonies to keep building upon their predecessors. Eventually, however, this extraordinary episode of construction was terminated, perhaps by drowning under steady crustal subsidence, perhaps the sudden onset of MECO upsetting the balance of ideal conditions. Indeed, it is plausible and even likely that the tectono-eustatic transgression and the enormous neritic carbonate factories at low latitudes (and now, it appears, high latitude) actually brought on the MECO warming spike itself.[4]

---

4   I must keep harping on the importance of correct geological ages. The Khirthar Transgression and the onset of limestone generation on the Bight–Eucla platform happened perhaps as much as 2 million years before the peak time of MECO. Therefore, a suggestion (Cramwinckel et al., 2019) that MECO *caused* the onset of the Khirthar Transgression in the eastern AAG (and by implication the accumulation of these massive limestones) is wrong. Dead wrong, actually, and precisely the other way around, the Khirthar Transgression being the cause not the outcome.

And what about the MECO forests around the AAG? The macrofossil assemblages of plants at Maslin Bay and Golden Grove near Adelaide and Anglesea in south-western Victoria indicated warm and wet conditions resembling today's tropics and subtropics in Australia and New Guinea. On top of that, the broad-leafed angiosperms from Maslin Bay carried beautifully preserved fungi on their leaves with a still-clearer tropical signal. This was in the 1970s, when researchers were having not only to accept a latitudinal displacement of 25 or 30 degrees of their fossil assemblages, but also to contemplate a tropical-type biome in winter darkness. But the evidence for tropical (megathermal) rainforests so far south was inexorable. For example, leaf analyses from the flora at Maslin Bay yielded mean annual temperatures of 23–26°C, perhaps an underestimate.[5]

About the anatomy of our strata

The sea comes in (marine transgression), the sea goes out (marine regression). Repeat, indefinitely, to generate transgressive–regressive cycles wherein marine facies succeed nonmarine facies both laterally and vertically ('spatiotemporally'). The aforementioned Johannes Walther is credited with stating this principle clearly, but the great chemist Lavoisier had discovered it too, and a century earlier.[6] These cycles exemplify the common problem in historical science of a single pattern or signal generated by multiple variables. The basement of the basin may be rising or subsiding, sea level may be falling or rising—and how do we tell the difference? Meanwhile, the plot thickens twice more. First, can we usefully compare local and regional patterns in strata with patterns between continents and globally, or between the great terrestrial, neritic and pelagic realms? Second is the

---

5   For an accessible work, see Hill et al. 'The vegetation history of South Australia' (2018). The Maslin Bay and Golden Grove floras have been known and publicised for several decades as paratropical in winter darkness; even so, Hill points out, they are disappointingly under-researched. The ages of the billabong muds housing the rainforest plants and sporomorphs from Maslin Bay and Golden Grove are somewhat less secure than the ages of the relevant marine shells (Lindsay and Alley, 1995; Greenwood et al., 2003). Me, I am quite sure that (i) the North Maslin Sands of the Adelaide district are part of the new regime ushered into the new St Vincent Basin on the Khirthar/Wilson Bluff Transgression; and if that is so, then (ii) these tropical floras sit squarely atop MECO. It all fits. But the slight possibility of a somewhat older age remains.

6   Lavoisier (the great chemist) did a memoir in 1789, *General observations on the marine horizontal beds and on their significance for the history of the earth*. Using measured sections of strata in the Paris district, he demonstrated a cycle: as sea level rises, the pelagic sediment advances landwards; as sea-level falls, the littoral advances above the pelagic. See Carozzi (1965) and Rudwick (2005). This was decades before the Pleistocene ice ages were discovered; Lavoisier scorned biblical Flood geology and no cogent mechanism for shifting sea level was in sight.

possibility of temporal hierarchy, from the cycles in the earth–moon–sun system (within human experience, such as the seasons recorded in tree rings or glacial lakes) up to periods in the thousands and millions of years? These questions led to interesting insights in biogeohistory, but they became more than academically important when petroleum geology and geophysics progressed beyond the elementary 'find me an anticline, and I will drill it!' to realising how vital was the geohistory of the basin—of strata and cycles and unconformities and all. So arose the discipline known as sequence stratigraphy.[7]

Marine transgressions were interesting and important because they formed a coherent pattern, not merely of lithostratigraphic formations, or biostratigraphic zones or chronostratigraphic stages but morphed instead into *allostratigraphic unconformity-bounded entities*.[8] In the later Eocene in southern Australia there turned out to be six. It also turned out that the six regional entities matched six in the apparently global 'Exxon sequence' (Figure 8.6).

The inevitable question was, what was the mechanism? The significant candidates for causation have been only two in number. One is glaciation, driving glacioeustasy at time scales of 10^4–10^5 years. The other is global tectonics, driving tectonoeustasy at scales of 10^7–10^8 years. These Late Palaeogene units have not been comfortably understood at 10^6 years' scale, the so-called third-order, seemingly too brief for serious tectonic processes to operate and seemingly occurring during times lacking the major glaciation required for glacioeustasy. They are our basic working units, which enhances the discomfort of our ignorance of causes.

Drilling in the Great Australian Bight offered Qianyu Li a possible test. He could distinguish biostratigraphically four 'packages' in the mass of limestone, packages defined by unconformities. He developed a somewhat speculative but evidence-based reconstruction of how this pattern, both regional and seemingly global, might come about. The Traralgon Coals and the oilfield unconformities are in the Gippsland Basin, facing the south-west Pacific Ocean not the AAG. This encourages us to raise our sights—if not just regional but supra-regional, then why not global?

---

7   Cycles and sequences in Cenozoic strata return in the next chapter.
8   Now, there's a mouthful to savour.

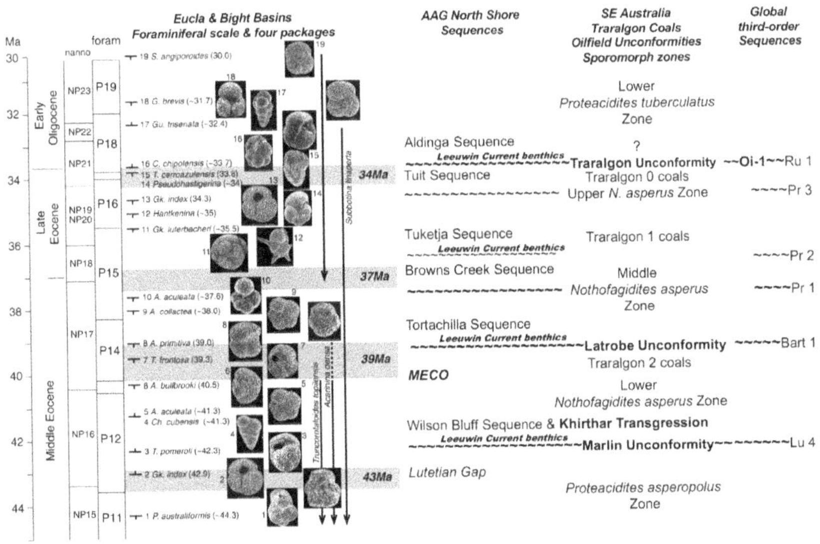

Eucla & Bight Basins
Foraminiferal scale & four packages

Ma		foram	nanno		AAG North Shore Sequences	SE Australia Traralgon Coals Oilfield Unconformities Sporomorph zones	Global third-order Sequences

19 S. angiporoides (30.0)

18 G. brevis (~31.7)

17 Gu. fraseriata (~32.4)

16 C. chipolensis (~33.7)

15 T. cerroazulensis (33.6)
14 Pseudohastigerina (~34)

13 Gk. index (34.3)

12 Hantkenina (~35)

11 Gk. iuterbacheri (~35.5)

10 A. aculeata (~37.6)

9 A. collactea (~38.0)

8 A. primitiva (39.0)

7 T. frontosa (39.3)

6 A. bullbrooki (40.5)

5 A. aculeata (~41.3)
4 Ch. cubensis (~41.3)

3 T. pomeroli (~42.3)

2 Gk. index (42.9)

1 P. australiformis (~44.3)

Early Oligocene — NP23 P19, NP22 P18, NP21
Late Eocene — NP19/NP20 P16, NP18 P15, NP17 P14
Middle Eocene — NP16 P12, NP15 P11

34Ma, 37Ma, 39Ma, 43Ma

Subbotina linaperta
Acarinina

AAG North Shore Sequences:
- Aldinga Sequence — *Leeuwin Current benthics*
- Tuit Sequence
- Tuketja Sequence — *Leeuwin Current benthics*
- Browns Creek Sequence
- Tortachilla Sequence — *Leeuwin Current benthics*
- MECO
- Wilson Bluff Sequence & **Khirthar Transgression** — *Leeuwin Current benthics*
- Lutetian Gap

SE Australia:
Lower *Proteacidites tuberculatus* Zone
?
Traralgon Unconformity
Traralgon 0 coals
Upper *N. asperus* Zone
Traralgon 1 coals
Middle *Nothofagidites asperus* Zone
Latrobe Unconformity
Traralgon 2 coals
Lower *Nothofagidites asperus* Zone
Marlin Unconformity
Proteacidites asperopolus Zone

Global third-order Sequences:
~~Oi-1~~Ru 1
~~~~Pr 3
~~~~Pr 2
~~~~Pr 1
~~~~~Bart 1
~~~~~Lu 4

Figure 8.6. Time chart for 10 million years' worth of Palaeogene strata.

A time chart (correlation chart) for a 10-million-year slice of Late Palaeogene time.

Right side, from right to left, six global sequences defined by unconformities named after the stages (Lutetian, Bartonian, Priabonian, Rupelian). In the coalfields and oilfields of Gippsland, the units T2, T1 and T0 of the huge Traralgon Coal, and the three regional unconformities, the Marlin, Latrobe and Traralgon unconformities. The six packages of mostly marine strata along the neritic north flank of the AAG, named Wilson Bluff to Aldinga, were found to match, one by one, the claimed global sequences. 'Leeuwin Current benthics' are pulses of warmer-water and more oligotrophic species of foraminifera. The species arrive at southern latitudes on the transgressions, but they don't linger if the waters get too cold or too fertile. Some of the taxa are discussed and illustrated in Chapter 9.

Left side, a test of the sequences by Qianyu Li in deep-water strata in the west, showing the controlling foraminiferal–biostratigraphic succession. That the match is quite good (see text) suggests that our sequences are not basically controlled by glacioeustasy. We see too that regional tectonic events can be both geologically near in time and potentially supra-regional.

Source: *Right*, from McGowran et al. (2004); *left*, from Li et al. (2003).

After MECO and into the icehouse

Figure 8.6 is about sequence stratigraphy. Li identified sequences bounded by unconformities in the west, and we are suggesting that there is significant matching to the east in geological correlation and age determination. So, we turn our attention to the limestones and coals of Gippsland, a territory which has had a lot of attention since the 1950s thanks to its fossil fuels, coal, oil and gas, on land and at sea (Figure 8.7). And we find that Cenozoic faulting and folding in southern Australia, discussed in Chapter 5 (Figures 5.14 and 5.15), is quite intimately involved with the allostratigraphic sequences. The unconformities have both time-significance (hiatuses) and structural-significance (the timing of faulting and folding).

In Chapter 3 we saw the problems encountered in sorting our southern limestones, their fossils notwithstanding. It was a double problem. First, ordination: getting the succession right. Second, correlation: accurately using the epochs and ages established in Europe. It was as recently as Walter Parr's and Martin Glaessner's rigorous scrutiny of the foraminifera in the 1940s–1950s that progress is discernible. So imagine how much more difficult it was to impose ordination and correlation upon the brown coals of southern Australia without those microfossils. About all that could be said by 1940 was that the nonmarine series with their coals in southern Australia seemed to be older than the marine series with their limestones.[9] Palynology changed that—more specifically, the palynology that was stimulated and systematised by the economic interest in our Palaeogene strata found its way into the Latrobe Valley and up through the coals and into the Miocene.

The upshot is illustrated most clearly in terms of allostratigraphic units, sequences (Figures 8.8a and 8.8b).

---

9   Singleton (1940); Carter (1964).

COMPOSITE LINE B - B' : SEISMIC

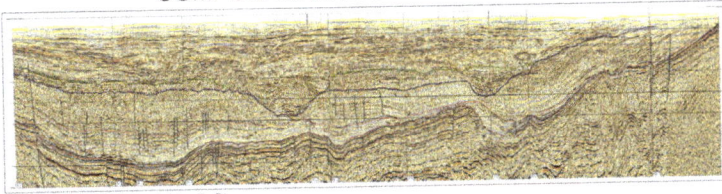

(a) Uninterpreted Seismic Line B- B'.

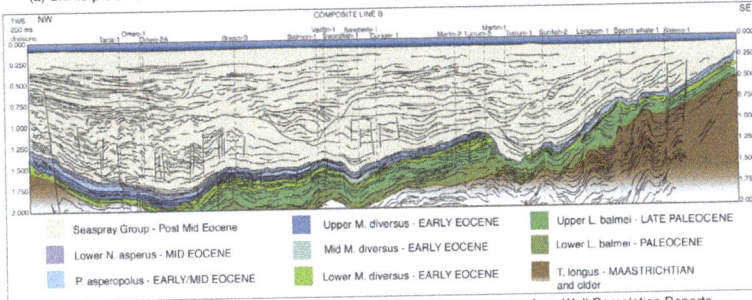

(b) Interpreted Seismic Line B - B' showing Palynology Zones. Palynology zones from Well Completion Reports.

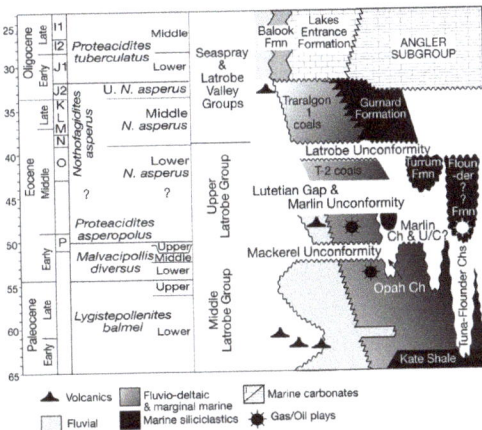

Figure 8.7. Gippsland seismics, the old Anglesea coal mine and Latrobe Unconformity.

Above, a seismic line in offshore Gippsland, pieced together from numerous traverses and displaying complex pattern in the uninterpreted profile. The wells shown could be sampled for sporomorphs which identified the palynological zones, shown in nine

243

colours and exemplifying the power of biostratigraphy in interpreting seismic images. The main discontinuity cuts the Middle Eocene zone and is buried in Upper Eocene to Neogene strata. It is the Latrobe Unconformity, which can be recognised across southern Australia.

Middle, strata and events in Gippsland plotted against time, whereby unconformities become hiatuses and downcuts and infills are clarified.

Below, the southern wall of the old Anglesea coal mine in south-west Victoria, looking south. The composite photograph spans about 150 metres. The Eastern View Coal Measures are dipping about 5° east (lower dashed line). The overlying Boonah sands are flat. Hence an angular unconformity at the top of a thin zone of weathered coal (upper dashed line). This is the only known exposure of the top Latrobe Unconformity in south-eastern Australia.

Source: *Above and below*, from Holdgate et al. (2003). *Middle*, McGowran et al. (2004) modified after Holdgate et al. (2003).

The above-mentioned double problem is seen more clearly as a triple problem—ordination, correlation and facies. Packages of strata onshore could be matched—integrated—with packages offshore by means of seismic geophysics, tracing the reflections marking physical discontinuities that turned out to be unconformities. At the heart of this sequence–stratigraphic synthesis is the limestone–coal couplet. It is not a lithostratigraphic formation, it is not a biostratigraphic zone, it is not a chronostratigraphic stage, and yet this sequence is all of them[10] and is more fundamental than any of them.

We will encounter more sequence stratigraphy in the next chapter. Meanwhile we note that this sequence diagram clearly has superposition and succession as well as lateral changes in facies, but it does not have a linear time scale. We now restore time while looking at those coal forests of south-east Australia, most incisively via palynology and its biostratigraphic succession, driven by the economic imperative of finding and exploiting the oil and gas fields offshore and the coalfields onshore in the Latrobe Valley in Gippsland. There has been much drilling, much digging, much sampling and intensive tallying of many thousands of sporomorphs, sample by sample, biostratigraphically ordered in space and time and rich in information about environments and environmental change (Figure 8.9).[11]

---

10   Holdgate et al. (1995); Holdgate and Gallagher (1997).
11   The most incisive study of the Middle and Late Eocene coals in their context across southern Australia is Holdgate et al. (2017), also the source of Figure 8.10 and Figure 8.15 (right).

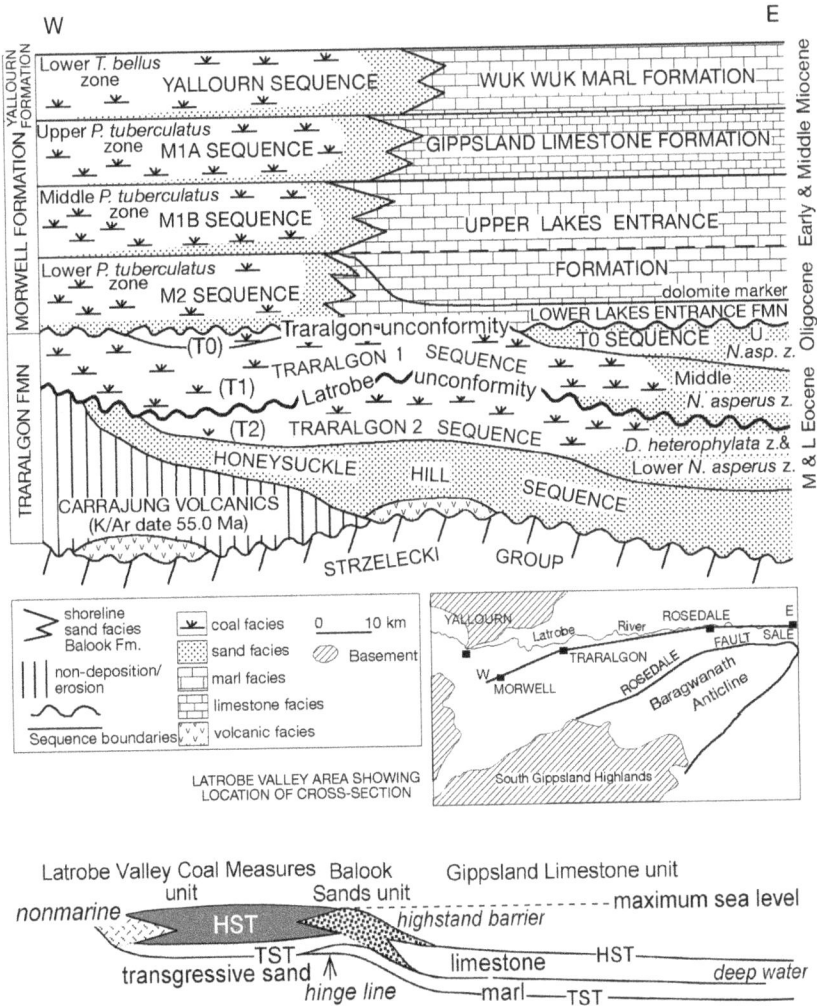

Figure 8.8a. Holdgate's coal–limestone couplets in Gippsland.

Stratigraphic diagram relating the nonmarine Latrobe Valley Coal Measures eastwards to the marine limestones in the Gippsland Basin. There are rock units (formations) and time units (Eocene–Miocene epochs) and sporomorph units (zones); but these sequences are the meaningful entities in terms of understanding and questions arising. The coal–limestone couplet is called a third-order sequence. It has two parts which, in sequence jargon, are the transgressive sequence tract (TST) and the highstand sequence tract (HST). Notice how the shallow marine sands penetrate the coal swamps at the base of each sequence: this is where one finds marine dinocysts intermingling with sporomorphs, signalling the marine ingressions. Notice too that the Oligocene–Miocene Morwell and Yallourn coal swamps face the limestone seas, unlike the huge Traralgon swamps in the Eocene, when there were no carbonates in the district (unlike the huge Wilson Bluff Limestone in the western reaches of the AAG).

Source: Couplets, Holdgate and Gallagher (1997). Gippsland sequences, Holdgate et al. (1995).

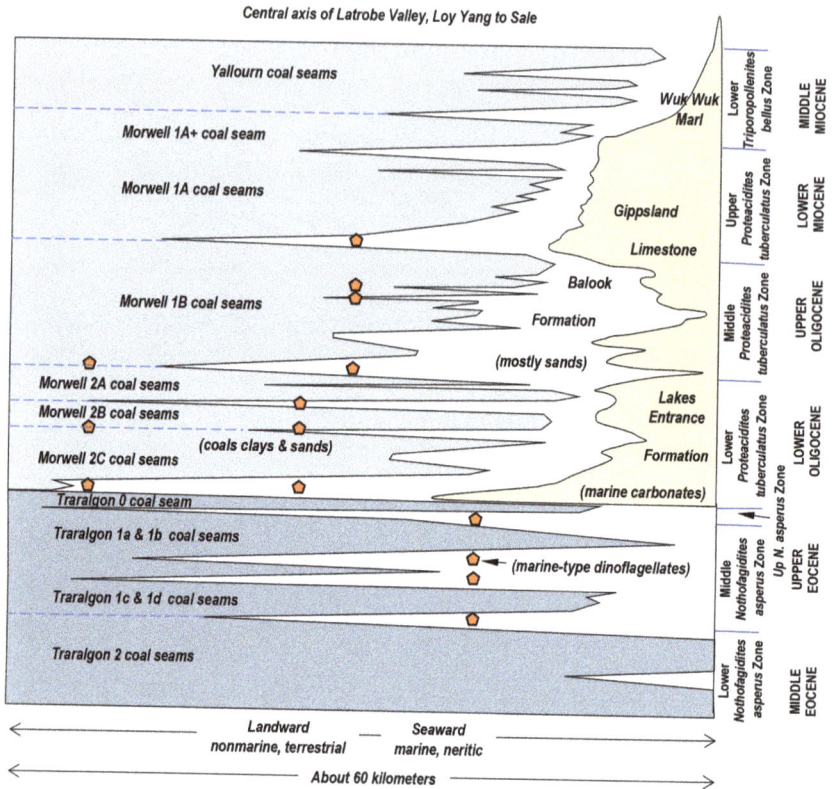

Figure 8.8b. Marine-nonmarine interactions in the Latrobe Valley.

This view pays more attention to the marine penetrations of the terrestrial realm, the coal measures.

Source: Adapted from Holdgate et al. (2021).

The coals are portrayed here at the broadest scale, namely the ratio of gymnosperms to angiosperms in the sporomorphs, expressed as per cent (%) gymnosperms. Although large numbers of samples were employed, the nine coals are distributed through about 35 million years. The gymnosperms include the conifers, the araucarians, pines and podocarps. The angiosperms are predominantly the relatives of either *Casuarina* or *Nothofagus*. In the Eocene Traralgon seams we see an overall shift towards the 'southern Beech', *Nothofagus*, which had diversified and expanded during the Lutetian cooling, its pollen taxon *Nothofagidites* giving its name to the three sporomorph zones Lower, Middle and Upper *N. asperus* zones. But note the gymnosperm peak in the Traralgon Coal T0, hard against the boundary and against the surge in *Nothofagus*.

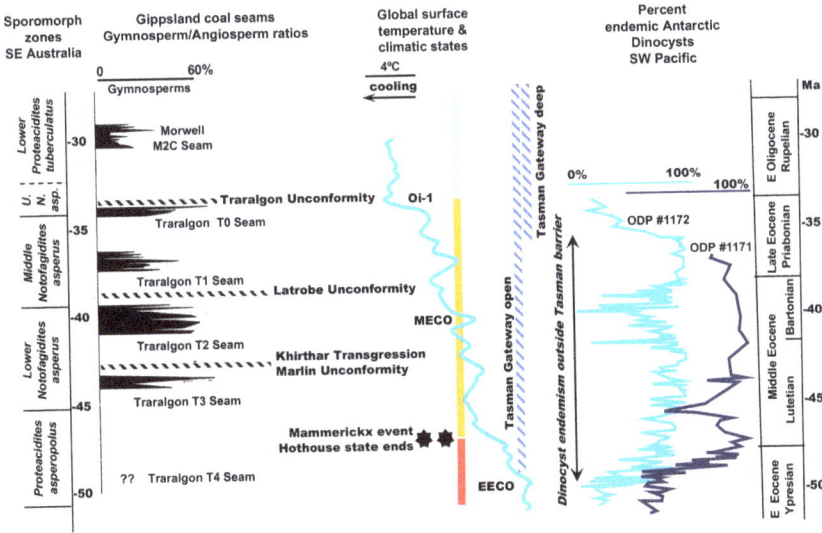

Figure 8.9. Gippsland sporomorphs vis-à-vis south-west Pacific dinocysts.

Latrobe Valley Coals plotted against time and oceanic record. Note the geological time scale of epochs and ages, and the numerical scale in Ma.

Left, on land, sporomorph zones, geological ages in millions of years and coal seams. The major botanical ratios (% gymnosperms) are mostly Podocarps and Araucarians on the one side and *Nothofagus* and the Casuarinids on the other.

Right, at sea, the deep-ocean oxygen proxy for temperature, with the by-now familiar EECO, MECO and Oi-1 (isotopically signalled glaciation) and the global states of hothouse, warmhouse and coolhouse. The curve of endemism gives the time perspective to the biogeographic concept of cosmopolitan versus endemic in the southern dinocysts, as in Figure 7.6. It is the main evidence first for the timing estimated for opening the Tasman Gateway, and then for deepening it as the AAG disappears into the Southern Ocean. In between, note the strong response of MECO in denting the curve, clearly a thrust southwards by the East Australian Current. So the bubble of dinocyst endemism arose when EECO fell and the bubble collapsed when the Southern Ocean emerged.

Source: Correlation of Gippsland coals, Holdgate (pers. com. 2018); plots of per cent gymnosperms are from Holdgate et al. (2009), also in McGowran and Hill (2015).

We can now detect a spatial pattern around the northern–eastern shores of the AAG, where there are numerous Late Eocene deposits of brown coal, close to sea level and not far from the Eocene sea. This broad-brush story is told in pie diagrams of the four classes of sporomorphs (Figure 8.10).

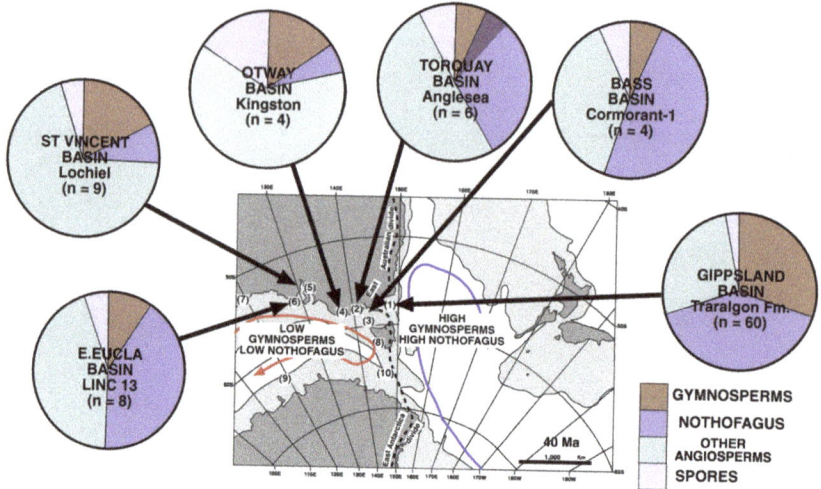

Figure 8.10. Holdgate's pollen pies of the Priabonian.

The dinocysts emphasise the marine–biogeographic contrast that grew in the Eocene between the eastern AAG and the south-west Pacific, a relatively short distance across the Tasmanian barrier. What about the terrestrial situation? This pie diagram of sporomorphs in coal measures displays a similar contrast at the same time, the later part of the Eocene Epoch. Palynology perceives an East Australian Divide (and for good measure continues it via the South Tasman Rise to become the East Antarctica Divide). To the east, facing the Tasman Current, the pollen assemblages are high in *Nothofagus* and high in gymnosperms, especially *Phyllocladidites mawsoni*, the pollens of podocarps like *Lagarostrobus*. To the west, flanking the AAG and the proto–Leeuwin Current, the assemblages are low in *Nothofagus* and low in Gymnosperms; and the family named after the she-oak *Casuarina* is prominent in the 'other angiosperms'.

Source: After Holdgate et al. (2017).

The forests in Gippsland facing the south-west Pacific had high gymnosperms, samples often characterised by high counts of *Phyllocladidites mawsoni*, and high *Nothofagus* numbers; the forests flanking the AAG, in contrast, have low gymnosperms and low *Nothofagus*. The category 'other Angiosperms' included several families of flowering plants and many species of pollens, but persistently prominent are the Casuarinaceae (including *Casuarina*). This contrast parallels the marine contrasts based on the biogeographic distribution of the dinocysts in the AAG and the south-west Pacific that we have already discussed. Perhaps the AAG and its shores were warmer than the south-west Pacific thanks to the proto–Leeuwin Current. But note the added touch of a physical barrier in the form of an East Australian topographic divide, continuing through Tasmania and across to becoming the East Antarctica divide, but now breached by the deepening Tasman Gateway. Perhaps this barrier formed a rain shadow.

Getting seriously cold down here:
Just across the water in Adelaide

So now to the global transition from greenhouse to icehouse—the continent-scale growth of the southern polar icecap. The sudden, oceanic, oxygen-isotopic signal known as Oi-1 records both a deep-ocean temperature effect and an ice-volume effect, both pointing to suddenly very cold. It seems that Antarctica became comprehensively iced over within a couple of million years, making this the most critical interval of time in Cenozoic biogeohistory. What can we say about it from our box seat, perched on the north shore of the AAG, just across the narrow water? Well, there are three horizons or levels punctuating the Eocene to the icehouse. We have encountered the first horizon already, namely the Khirthar Transgression forcing MECO at 40+ million years ago. The second is the disappearance of the shallow-water carbonates at 39–38 million years ago. The third horizon is the big chill itself, Oi-1, 34–33.6 million years ago.

We begin locally, at the section of strata exposed at Maslin and Aldinga Bays south of Adelaide, where the succession spans this critical interval.

(Figures 8.11, 8.12 and 8.13.) It begins with the Maslin billabong flora basking in MECO, preserved in the North Maslin Sand just below the strata shown here. First striking an observer are the colours, browns and yellows (Tortachilla Limestone) giving way to greys, with some almost black and some green (Blanche Point), and returning to browns and yellows (Port Willunga). Modern weathering and erosion in these west-facing cliffs might produce the colours of rusting (as it has the bleaching, hence the name 'Blanche Point'), but there is something deeper and more meaningful here, and the macrofossils display it. The Tortachilla Limestone has a rich and diverse assembly of clams and snails, of sea urchins and bryozoans; and the Port Willunga Formation too is bryozoan-rich. But the grey-green-black Blanche Point Formation has layers crowded with marine snails of *Spirocolpus* known to live in the mud, other layers with intense burrowing by prawn-like animals, clearly prospecting a rich source of food buried in the muds, and still other layers with untold millions of sponge spicules. So the rock is rich in silica (opal) with not much carbonate. These rich macrofossil assemblages were abruptly terminated, truncated, replaced (all these adjectives are used by stratigraphers to emphasise a sudden change going up a section of strata), never to return. They were succeeded by poorly fossiliferous sands and clays in a succession of strata from nonmarine to beach sands and probably mangrove-muds; and then the marine environments returned with their bryozoans, clams and sea urchins. Thus we have a scenario of a warm shallow sea, a normal marine environment ('normal' meaning stable oceanic levels of the two most influential parameters, oxygen and salinity) interrupted by a very strange environment, in some way a stressed environment. The main candidates for stress would be low oxygen levels, raised salinity levels or lowered salinity levels. In all these situations the animals mostly go missing—especially those that grow the calcareous shells familiar to us all and described by Lamarck and his successors. Those opportunists that can cope with the stresses flourish in huge numbers in the absence of the 'normal' competition.

Figure 8.11. Two views of Maslin Bay.

Two views of Maslin Bay, located nowadays within suburban Adelaide. The sea floor exposed at low tide, and resembling what geographers call a wavecut platform, actually is a hardground, a floor of the sea preserved upon the Tortachilla Limestone when its abundant shells of aragonite dissolved and the available bicarbonate could form a cement of calcite. It is not being cut today; it is being exhumed today. The hardground marks a pause in the accumulating of sediment, a hiatus, at Pr-2 in the sequence tabulated in Figure 8.6. Actually surface Pr-1 is within the Tortachilla Limestone, and this formation as currently defined straddles the Middle–Late Eocene subepoch boundary, also the Bartonian–Priabonian stage boundary, also sequence boundary Pr-1, and thereby straddles the major shift in global environments. To the north, the Tortachilla is above today's high tide level; to the south, it is below low tide — explaining the beaches backed by soft sediments and why Blanche Point is exactly where it is.

Source: Author's images.

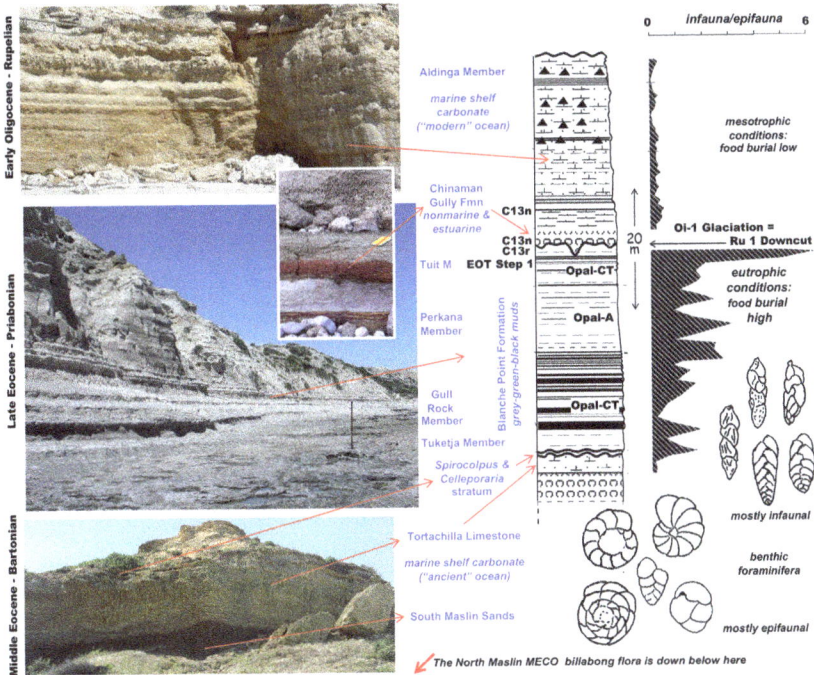

Figure 8.12. Biofacies, Maslin and Aldinga Bays.

The strata in the marine cliffs of Maslin and Aldinga bays boldly display a succession of significant colours — yellow-brown (Middle Eocene), then grey-green, even black (Late Eocene), then yellow-brown (Early Oligocene). The benthic foraminifera explain these contrasts. The spectacular sixfold drop in the infauna/epifauna ratio across the Eocene–Oligocene boundary implies a ventilating, a cleaning-out of more than just this corner of the AAG as oceanic circulation is reinvigorated on the newly Icehouse Earth. The succession of marine strata is interrupted by the Chinaman Gully Formation, a glacioeustatic event marking the sudden growth of the icecap on the other side of the narrow AAG, perhaps getting as close to Adelaide as is Sydney today. The geomagnetic identifications of Chrons C13r and C13n, bracketing the downcut, confirm that the downcut is coeval with the oceanic isotopic event Oi-1, also known as EOT (Eocene–Oligocene transition) Step 2, for it had a precursor, EOT Step 1.

Source: New compilation. Infaunal/epifaunal ratios, Moss and McGowran (2003). C13n, C13r and EOT-1, Haiblen et al. (2019).

Figure 8.13. Snails and events, Maslin and Aldinga bays.

Events preserved in the Maslin Bay – Port Willunga section. In order from older to younger: *5*, a boulder of Tortachilla Limestone spanning the fundamental shift from oligotrophic Bartonian sea to eutrophic Priabonian sea, the latter with abundant snails, the first coming of *Spirocolpus* (though not shown here). *4*, the nautiloid *Cimomia felix* mounted on a slice of mud with *Spirocolpus* in their second coming (Gull Rock). These nautiloid shells usually filled with a watery fine mud which shrank during burial and dehydration, crushing the shell which, however, retained its aragonite, like these snails but unlike the fossils of the same species in the limestones. *1, 2, 3*, Tuit Member and the third coming of the snails. (EOT Step 1 was identified here, but the Step 2 signature was lost in the glacioeustatic downcut.)

Source: New compilation. Model: Stephen Pekar, Queens College, New York.

The rich microfaunas of foraminifera were dominated by a few groups, including the families Cibicididae and Uvigerinidae plotted here. The Cibicidids are mostly epifaunal, living on the sediment surface or perched on rocks, sea grasses and algae, and are very common in shallow seas in the 'normal' marine environment. The Uvigerinids are mostly infaunal, living in the mud. The infaunal/epifaunal ratio gives a strong signal of the fate of organic carbon (C_{org}). When C_{org} is all consumed and/or oxidised on the spot, the ratio is very low, as in the 'normal', well-ventilated environment, because there is not much food in the muds to burrow for. Consumption on the seafloor keeps up with primary production. But a high infaunal/epifaunal ratio suggests some combination of low oxygen in the environment and rapid burial of the sediment with much of its C_{org} intact. Consumption does not keep up with primary production.

All the contrasts between the Blanche Point Formation and the limestones below and above it point in the same direction. The Blanche Point Formation in the St Vincent Basin preserves an episode of poor ventilation in this district of the AAG during Late Eocene time—not in the Middle Eocene, not in the Early Oligocene, as in each of those times it was well ventilated. The Late Eocene was also the age of the brown coals at Lochiel in the St Vincent Basin, and this association is very interesting—coals onshore, grey-green-black sediments in the shallow sea with strange fossil assemblages and abundant silica but not much carbonate. We find a similar situation in Western Australia in the Late Eocene, coals and sediments, some with sponges, others with abundant opaline sponge spicules. But where are the limestones? It seems that the Tortachilla Limestone and its equivalents in the Wilson Bluff in the west represent the last impressive limestones in the AAG, and they are Middle Eocene in age, rarely stretching clearly into the Late Eocene.[12]

Things changed, still more comprehensively, at the third of our three fundamental levels. The Chinaman Gully Formation was the start of a new marine transgression; next were sands and muds and bryozoans and shells. Indeed, this vertical succession from nonmarine to marine depths in tens of metres was remarkably similar to a modern horizontal succession, revealed as one might imagine walking from the beach at Port Willunga along the seafloor from the sands, through the seagrasses and out to the bryozoan meadows in the middle of Gulf St Vincent—very useful for illustrating the basic principle of lateral/vertical facies in stratigraphy. Even more important, the microfauna had a much more modern look than do the Eocene fossils. And the modern-neritic look was reinforced by the infauna/epifauna ratio, so spectacularly different from the nutrient-rich Priabonian sea. The Port Willunga sea was 'normal' and well ventilated, implying much better oceanic circulation and mixing in the AAG!

Why did these changes happen? We identified the base of the Chinaman Gully as the local Eocene–Oligocene boundary. Recently that same local level was determined palaeomagnetically to be at the Chron C13n–C13r boundary—which is the level of Oi-1. This strong correlation allows us to address the question posed above: is Oi-1 an ice-volume effect as well as a deep-ocean temperature effect?

---

12 We have Late Eocene samples of limestone dredged from the slope in the Bight Basin and our coverage is sparse, but the carbonates contracted sharply from their Middle Eocene spreads.

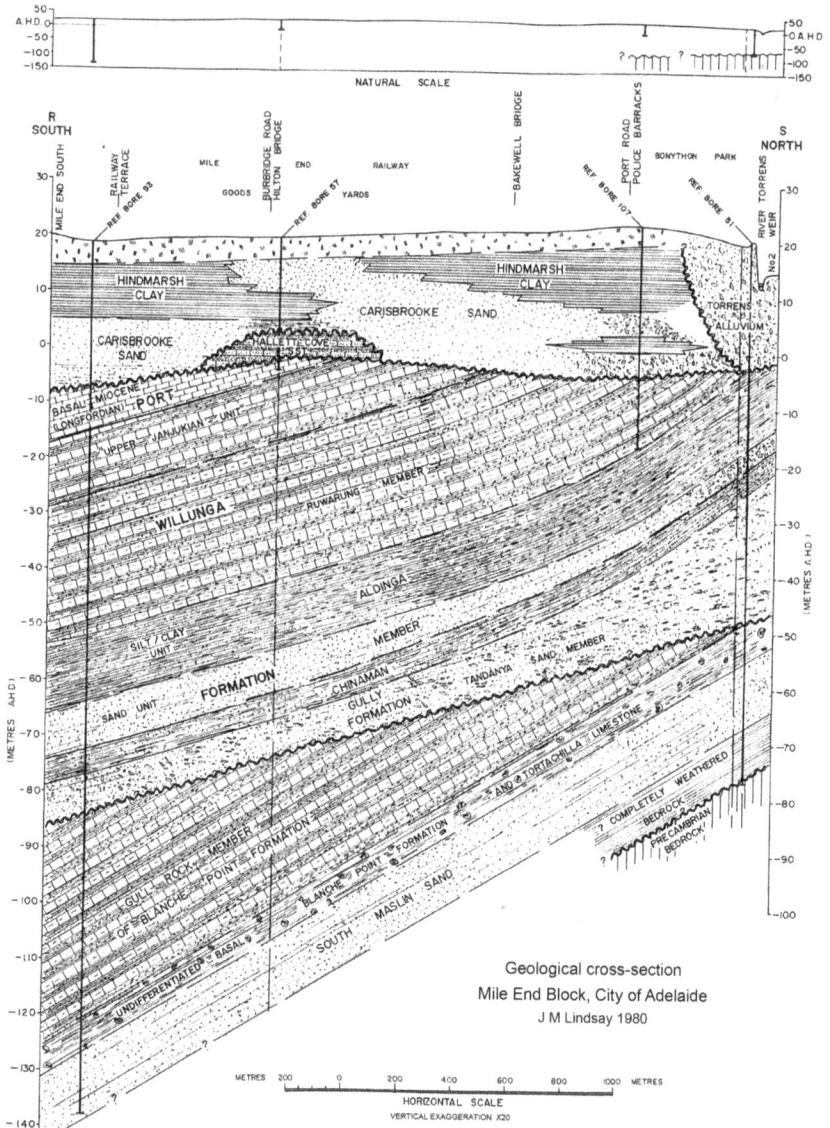

Figure 8.14. Lindsay's section: Eocene–Oligocene glacioeustasy under Adelaide.

Murray Lindsay's reconstructed cross-section under Adelaide is evidence of the glacioeustatic event seen in the oceanic $\delta^{18}O$ signal, Oi-1. Some 35 metres' thickness of Blanche Point Formation preserved in one section (*left*) is missing from another, 2.5 kilometres away (*right*). Overall, at least 50 metres of strata were removed in a glacioeustatic downcutting, then a backfilling of the Chinaman Gully sands and clays and the return of the sea (Aldinga Member of Port Willunga Formation), all within

500,000 years and probably less time. Also note (*right*) the logging of completely weathered bedrock. This was a land surface under the EECO rainforests. (This diagram is a companion to Figure 2.5. Note the 20 x vertical exaggeration.)
Source: Lindsay's (1981) unpublished MSc thesis.

The Chinaman Gully Formation at Port Willunga is very thin, sitting on an unconformity, but thicker under the city of Adelaide, where we have already seen one of Murray Lindsay's painstaking reconstructions (Chapter 1). Here is another (Figure 8.14).

The Chinaman Gully sits on the Blanche Point, as it should, but it can be seen that 35 metres of the latter are missing from Bore #51, compared to Bore #93, 2.5 kilometres away; and this number underestimates the real situation, because more than 10 metres are missing at the latter site too. Overall, at least 50 metres of the Blanche Point Formation were cut out, and the downcut was backfilled with nonmarine sands and pebbles, then marginal marine and then neritic sediments with Early Oligocene foraminifera. The open sea was re-established within the time range of Chron C13n, which was about 500,000 years' duration. This then is the maximum time available for the sea level to fall at least 50 metres, for back-filling to occur and for fully marine conditions to return. (And, indeed, for a brief warming! For the warm-neritic benthic species reappear on a pulse of the Leeuwin Current.[13]) In the Latrobe Valley in Gippsland, the highest Eocene coal seam, Traralgon-0, is cut and backfilled in the same way and at the same time. The backfill contains dinocysts indicating a marine transgression on the southwest Pacific (eastern) side of Tasmania, coeval with the Leeuwin Current pulse that brought the photosymbiotic benthic foraminifer into the AAG in the west. (Figure 8.15.)

Clearly these cuts signal a glacioeustatic event—we are seeing a near-field effect of the initial cycle of growth and decay of the icecap just across the water (Figure 8.16). We can now answer the question. The sharp and strong change in the oxygen-isotopic signal is an ice-volume effect as well as a deep-ocean temperature effect.

---

13   Lindsay (1981); Lindsay and McGowran (1986). The foraminifera of interest include *Halkyardia, Linderina, Maslinella* and *Crespinina*. But the major groups from the tropics, such as the Nummulitids, Orthophragminids or Alveolinids, did not penetrate the AAG. Compared to the Tethyan neritic shelves the neritic realm in the AAG was volatile in the parameters that matter—temperature, salinity, oxygen and nutrient.

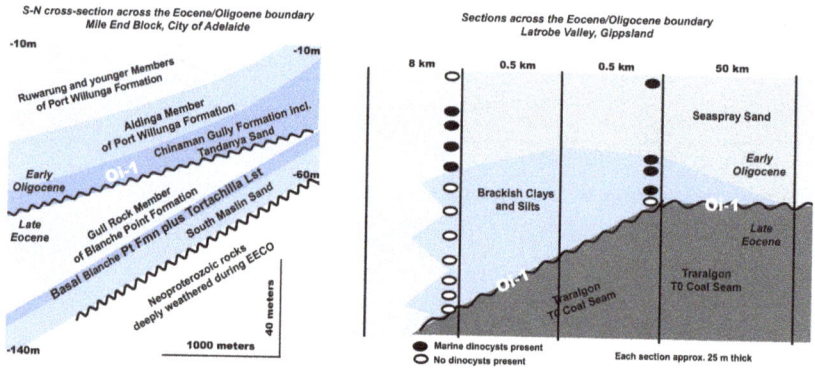

Figure 8.15. Glacioeustatic cartoons, Adelaide and Latrobe Valley.

Left, a cartoon of Lindsay's reconstruction under Adelaide emphasising downcut, backfill and marine transgression within neritic limestones on the north flank of the AAG, all constrained within geomagnetic Chron C13n (about 500,000 years' duration).

Right, a cartoon of four sections in the coalfields in the Latrobe Valley in Gippsland, showing how the T0 coal seam at the top of the Eocene has been cut out and backfilled in earliest Oligocene times by sands and clays containing marine dinocysts. There are no marine shells or geomagnetics to anchor us chronologically here, but otherwise paralleling and surely coeval with the glacioeustatic succession at Maslin–Port Willunga.

Source: Author's depictions based on compilations in Lindsay (1981) and Holdgate et al. (2017).

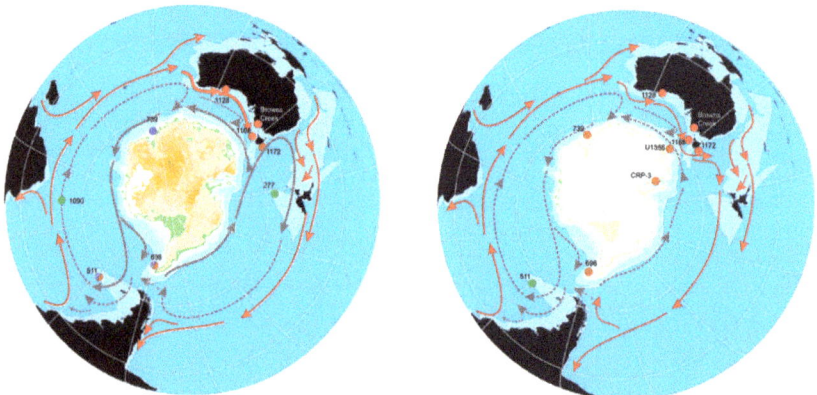

Figure 8.16. Modern-type icecap grows on Antarctica.

Growth of the first modern-type icecap on Antarctica, shown for 34 Ma and 32 Ma. At 34 Ma in the latest Eocene, the ice sheet is small and transient. This would be the EOT, Step 1. And although the full-blown icecap is displayed here for 32 Ma, our local evidence for major glacioeustatic drawdown is that the icecap could grow and decay within only a couple of hundred thousand years at most. The coloured dots denote ecologically significant dinocyst assemblages. We need only note here that the cosmopolitan species have broken through the Tasman Barrier (and indeed are the evidence for that breakthrough). Meanwhile there are now indications of flow westwards off Antarctica.

Source: From Houben, Bijl et al. (2019).

Now compare charts displaying the level of Oi-1 about 34 million years ago (Figures 6.10, 6.11 and this chapter). We see that Oi-1 marks the end of the warm times, notwithstanding significant warmings later in the Cenozoic. We see that it marks the end of the great chemical experiments in the global ocean: it is where the CCD stabilises at depth, acid levels ceasing their rises and falls at the scale of kilometres. We see that it marks a profound change in carbon dioxide levels. We see that profound changes in a small marginal pocket of the AAG are in tune with these global shifts. And we see a backstep in the coal forests by the great Palaeogene gymnosperms, the Araucarians and the Podocarps, in their waltz with the angiosperm southern beeches through the succession of Gippsland coal measures. But the bigger story in Cenozoic biogeohistory is rather like a symphony at a school concert, being played in enthusiastic mode and making progress. It is as if there are minor dramas—false notes here and there, sections getting lost for a bar or two—but they matter not too much while the brass, strings and percussion are holding together pretty well, mostly in tune and in time. As do the narratives from the pelagic, neritic and terrestrial domains of the Eocene world.

I see first the Khirthar Transgression and MECO then the Tasman Gateway's deepening as fundamentally significant in triggering what became the greenhouse–icehouse transition. Eutrophication of the estuarine AAG and expansion of the coals along its northern margin were part of the grander Late Eocene story. By Early Oligocene times, all had changed. The waters, neritic and oceanic, were becoming well ventilated by steepening gradients reinvigorating circulation, exchanging carbon dioxide for oxygen. As it was expiring, the AAG was being cleaned out as it were. In due course the coal swamps return in Gippsland but not around the expiring AAG; the modern-type bryozoan carbonates spread and their foraminiferal faunas, hard-hit by the great change, recover and spread, also looking much more modern than did their Eocene antecedents. Berger's Auversian Facies Shift has run its course; and by Late Oligocene times, about 27 million years ago, really we are in the new world of the Neogene.

9

Hello, Southern Ocean: Into the arid zone and into the Early Neogene

Southern Australia faces the Early Neogene Southern Ocean

The present is the key to the past. The past is the key to the present. Both statements have some truth, although elementary geological education down the decades has favoured the first. We too begin with the present, Australia now isolated among the oceans (Figure 9.1), and we revisit the heterogeneous group of effects known as the Leeuwin Current or proto–Leeuwin Current. The large tropical photosymbiotic foraminifera traced the current back to the Middle Eocene, beyond which the dinocyst story became crucial. The current has had different origins and highly varying influences through the epochs of the Cenozoic; now it's the Neogene's turn.

As we see in the weather reports, the winds in the southern hemisphere spin off the atmospheric high-pressure cells anticlockwise, so that the monsoonal cyclones and the intense lows from the south-west spiral in clockwise. Likewise in the oceans—the big central gyres are anticlockwise, so that, broadly speaking, the eastern seaboards are warmer and wetter at midlatitudes than are the western seaboards of the southern continents. I write 'broadly speaking'—for Australia has an important modification of that pattern in the form of the Leeuwin Current. Equatorial water, known as the Pacific Warm Pool, piles up in the constricted Indonesian region, causing a current to flow, literally down-slope, along the western margin of Western Australia, around the corner at Cape Leeuwin and along the southern continental margin. Its ultimate cause is the oceanic bottlenecking as Australia approached Southeast Asia during later Neogene times, sharply constricting the flow of warm water from the Pacific into the Indian oceans. Its biogeographic effect is to freight tropical-type organisms into extra-tropical places—corals and clams, for example; or individuals of *Nautilus* straying far from their population centre in the north-eastern Indian Ocean to South Australia; or plant-produced bitumens and resins also stranded as lumps on the coasts of southern Australia, containing organic biomarkers of their provenance in modern tropical rainforests in the Indonesian Archipelago.

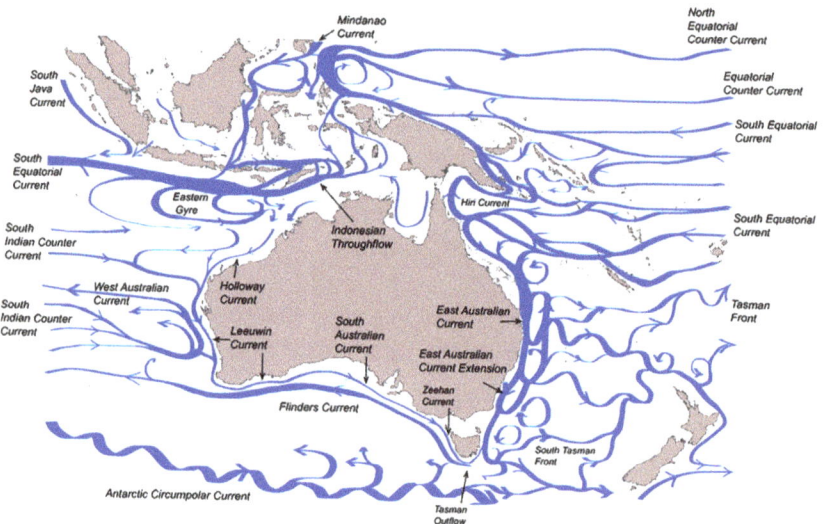

Figure 9.1. Australia's oceanographic situation.

The continent, including the island of New Guinea, approaches Asia to form one of Earth's major tectonic hotspots, the region known biogeographically as Wallacea. In the modern oceans the warm water 'wanting' to flow from the Pacific Ocean to the Indian is bottlenecked (Indonesian throughflow) and piles up in the Pacific Warm Pool. Its deflection along our eastern margin promotes growth of the Great Barrier Reef and enhances the La Niña effect in eastern Australia. Meanwhile, the Antarctic Circumpolar Current flows wild and free to our south. The Leeuwin Current is prominent in this chapter, but note that we use the name in deep time to include what here are distinguished as the Holloway, South Australian and Zeehan currents flowing today (i.e. in shallow time).

Source: From Wijeratne et al. (2018).

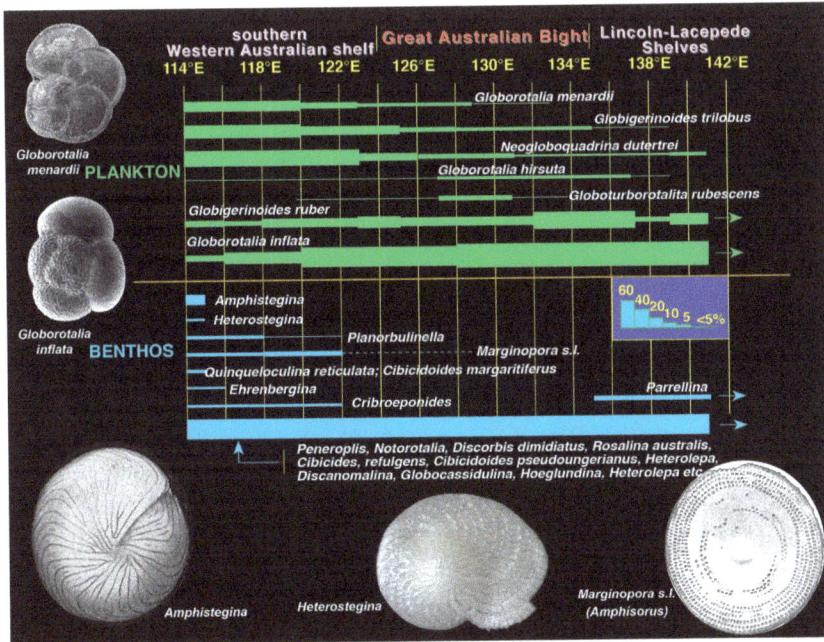

Figure 9.2. Climate and biogeography across 30° longitude in southern Australia, the longest coastline facing the Southern Ocean.

Planktonic and benthic foraminifers show the same pattern, namely that the warmer water comes from the west on the Leeuwin Current. Note the large benthic photosymbiotics, *Amphistegina, Heterostegina* and *'Marginopora'*. The latter was particularly prominent entering the South Australian Gulfs in Pliocene and Pleistocene times.

Source: Adapted from Li and McGowran (1998).

The foraminifera contribute most of our information in detail and precision, as in this survey of a segment, extending 28° longitude, of the uniquely long southern continental shelf lining the Bight and facing the Southern Ocean (Figure 9.2.).

Recall Sir John Murray's sketch (Figure 4.8) of the biogeographic spread of modern planktonic foraminifera, such as *Globorotalia menardii* in tropical waters and the temperate *Globorotalia inflata*. In the neritic waters of the Bight, *G. menardii* and several other warmer-water species fade from west to east; *Globigerinoides ruber* continues robustly through; whereas *Globorotalia inflata* fades from east to west. The large photosymbiotic benthics *Amphistegina, Heterostegina* and *Amphisorus* have migrated down the western margin and managed to sneak around the corner. Cape Leeuwin is right on a major oceanic-biogeographic boundary (between the Subtropical and Transitional provinces), which was forced northwards about 14° to

Northwest Cape in cold Pleistocene times ('glacials'). It could be said that the warmer-water species are interlopers here—that, like the pearly nautilus, they are out of place. It is not difficult to imagine that those species would disappear during a glacial period, when the Subtropical Convergence forced its way right up into the Bight, the Roaring Forties became the Roaring Thirties and the Leeuwin Current was shut down, at least in its southern reaches. Alternatively, in warmer times than the present, the invaders from the north should be more in evidence (Figure 9.3).

That situation was confirmed when Cann and Clarke (1993) showed that *Marginopora* (*Amphisorus*), thriving off Esperance (south-western Western Australia) today, was common much further east in gulfs St Vincent and Spencer (South Australia) at the time of the last interglacial 125,000 years ago. The tropical–subtropical molluscs *Anadara trapezia* and *Pinctada carchariarium* were also common at that time but are living in southern Australian seas no longer.

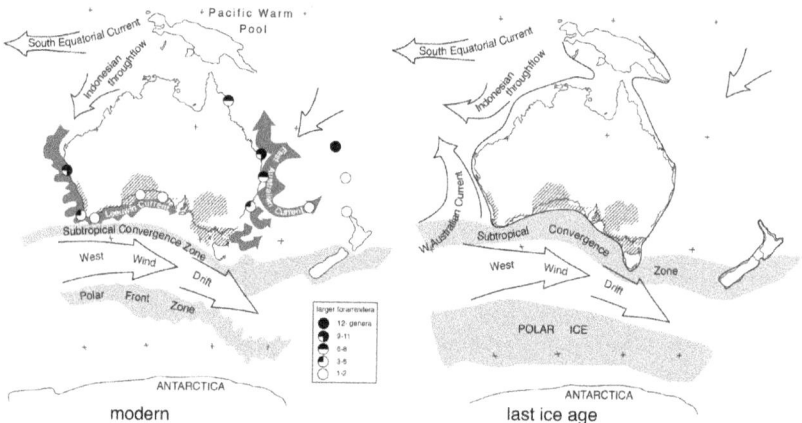

Figure 9.3. Pleistocene climatic alternatives in southern Australia.
The 'last ice age' was tens of thousands of years ago. The effects of oceanographic shifts northwards in the ice times were amplified by shutting down the warm currents near the coasts which had been supporting their large photosymbiotic foraminifers.
Source: McGowran and Li (1998), published before the input of the South Indian Countercurrent was appreciated (Figure 9.1).

MD03-2611 & MUC 3 combined

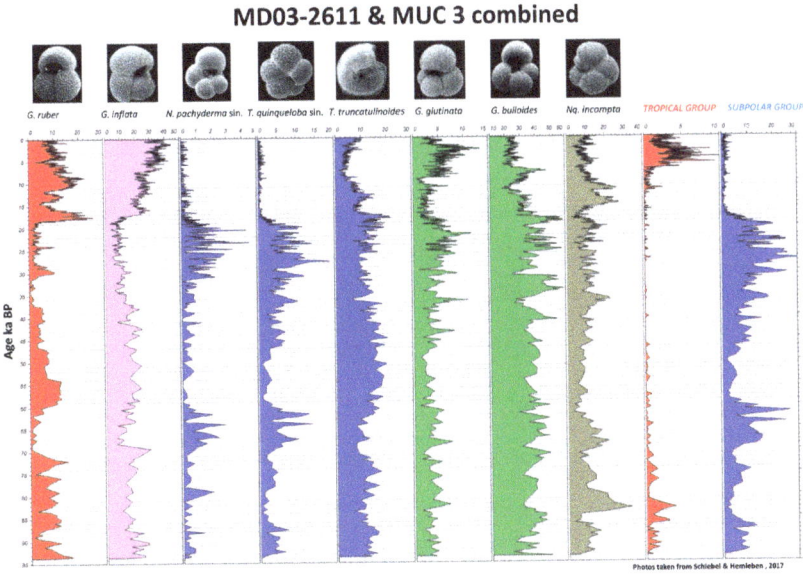

Figure 9.4. Leeuwin Current ice age signal south of Kangaroo Island.

Cores drilled south of Kangaroo Island (see Fig. 9.23) are rich in planktonic foraminifera. The quantified vertical profiles of environmentally sensitive species have now shown the Leeuwin Current arriving off central–southern Australia, marking the end of the last ice age, particularly by the sudden arrival of the tropical planktonic species *Globigerinoides ruber*.

Source: Kindly supplied by Patrick DeDeckker (Perner et al., 2018).

The molluscs and the large foraminifera are from the marine shallows—from seas metres to tens of metres deep. It is too easy for sceptics to argue that a discontinuous pattern is merely a discontinuous or imperfect geological record biased towards the warmer times because shallow-water strata are readily removed by erosion. Although Cann and Clarke could anticipate that argument pretty convincingly, we were relieved to corroborate the switching on/off hypothesis on a core in oceanic sediment in the Bight, using a continuous record of planktonic foraminifera.[1] Drilling south of Kangaroo Island has shown elegantly that the Leeuwin Current was switched on at the end of the last Pleistocene ice age (Figure 9.4).

Tropical species tending to displace subpolar species signalled the arrival of the Leeuwin Current at the far eastern end of the Bight, and that implied that the ice age was ending about 20,000 years ago.

---

1   Cann and Clark (1993); Almond et al. (1993). The history and prehistory of the Leeuwin Current was argued in McGowran, Li, Cann et al. (1997) before the dinocyst narrative clarified the Palaeogene.

So, the Leeuwin Current is of real biogeographic significance in exporting warm biotas into cooler domains. It is a Late Neogene effect caused by the Late Neogene squeezing of the Indonesian archipelago between the Australian continental mass (Sahul) and the Southeast Asian continental mass (Sunda) and through what is known biogeographically as Wallacea. Somewhat inelegantly, I called this the squirting hypothesis. We will see this effect amplified in the Miocene; we have seen something like it already in the Middle Eocene, when Sahul and Sunda were far apart and not squeezing; and we have seen an Early Palaeogene 'proto–Leeuwin Current' as the nominated importer of cosmopolitan marine dinocysts from the far side of the Indian Ocean into the Australo-Antarctic Gulf (AAG).

Repopulating southern Australian neritic seas

By earliest Oligocene times the Tasman gateway had deepened, the biogeographic contrasts between the AAG and the south-west Pacific Ocean had broken down, and the Southern Ocean was in parturition. Palaeogene Australia, quite isolated when still in eastern Gondwanaland, became more isolated still. As eastern Tethys contracted, the vast biogeographic entity known as the Indo-West Pacific region became clearer, with Wallacea and the Pacific Warm Pool at its core. And southern Australia, on the far southern fringes of the region, became known for its endemic biotas on land and at sea. Indeed, the century of frustrations in recognising the Cenozoic epochs in southern Australia strata boiled down to one major reason, endemism—endemism among the molluscs, bryozoans, echinoids, corals and more.[2]

The foraminiferal populations along the north shore, benthos and plankton, came to prominence in questions of age, of facies and environment, and of biogeography. But what of the populations themselves? It turns out that many foraminifera were endemic too (Figure 9.5).

---

2   Endemic is more or less the opposite of cosmopolitan—a species, genus or family is confined to a particular region and is thought to have originated there.

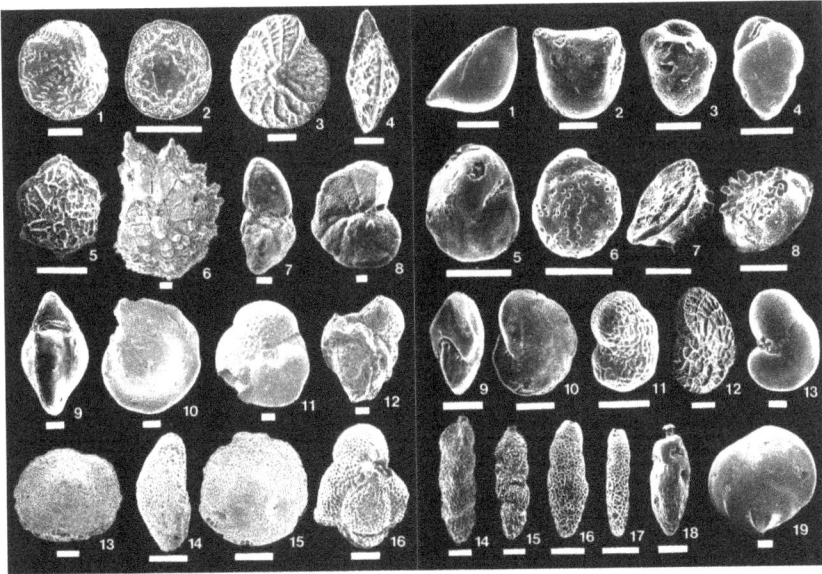

Figure 9.5. Li's plates of (semi)endemic species.

Scale bars are 200 μ (*left*) and 100 μ (*right*). This endemism on the southern fringes of the Indo-West Pacific biogeographic region was particularly strong in the later Eocene after the Khirthar Transgression, and included the species (*left plate*) *Maslinella chapmani* (*7, 8*), *Wadella hamiltonensis* (*12*), *Linderina glaessneri* (*13*) and *Halkyardia bartrumi* (*14, 15*).

Source: Li et al. (1996).

From a 'preliminary' survey of a century's systematics, Qianyu Li made several generalisations beginning with a sorting into three categories: endemic and semi-endemic, migratory, and cosmopolitan. The cosmopolitans included most of the infaunal species and deeper-water species. The migratory taxa are the large photosymbiotics, from *Asterocyclina* and *Operculina* in the Middle Eocene to *Amphistegina* and *Marginopora* at the young end of the scale. The endemics are mostly epifaunal and mostly shallow-water. Semi-endemics broadens to include New Zealand and South America. It may be that endemism is stimulated by the interplay down the ages between warm and cool, between Leeuwin and East Australian waters from the north and West Wind Drift and Circum-Antarctic Current from the south. At any rate, Li perceived a link between benthic endemism and major marine transgression over the continental margins, so as to produce four phases of endemism, in the Middle and Late Eocene (the most pronounced), the Oligocene, the Early Miocene and the Plio–Pleistocene.

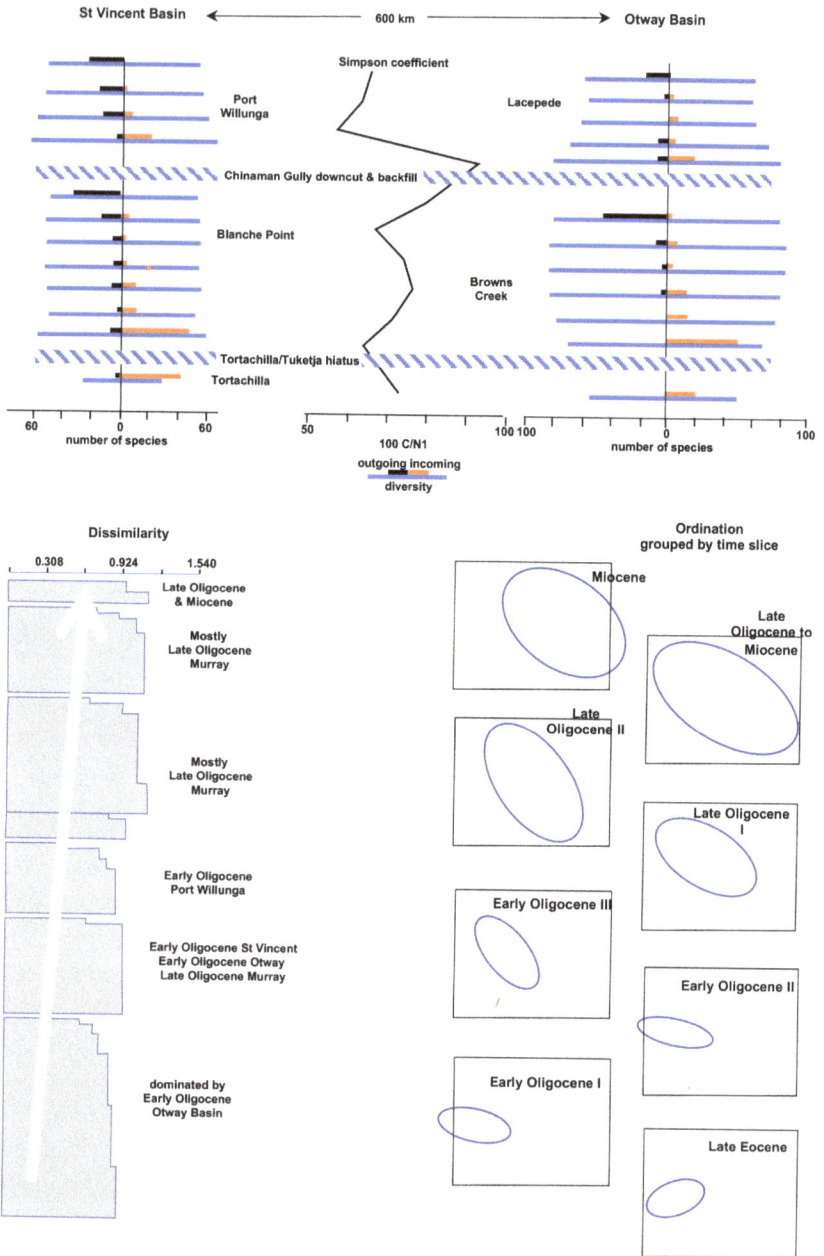

Figure 9.6. Three simple to complex ways of looking at plots of benthic foraminifera in the transitions from Late Palaeogene to Early Neogene.

Above, parallel plots through time in two marginal seas, one restricted in oceanic influence (St Vincent Basin near Adelaide), the other, more open (Otway Basin, south-west of Melbourne). Each bar has total species, species incoming and species outgoing.

The Simpson correlation coefficient measures species in common, C, against N1, the lesser of the two diversities. *Lower left*, clustering of 257 samples (details omitted) shows growing dissimilarity through 10 million years (the ghostly arrow). *Lower right*, ordination through time shows the same growing dissimilarity in the strong ballooning through time, and both detect a jump from Early to Late Oligocene.

Source: The chart labelled Otway Basin (*top right*) needed two sections, Browns Creek in the east (lower half of the chart) and Lacepede in the west (upper half of the chart). Redrawn and adapted from Moss and McGowran (2003).

In the Oligocene phase we wanted to see the impact on southern Australia of the first major ice sheet from just across the water, then the recovery through several million years as the neritic seas returned through the Oligocene. Several kinds of numerical data are packed into one figure—counting, clustering and ordination (Figure 9.6).

First, compare the tallies across the boundary in the restricted-neritic St Vincent Basin (on the evidence of very low numbers of oceanic plankton) and in the open-neritic Otway Basin (with demonstrably more oceanic influence). The assemblages are stable through the Priabonian, with low numbers outgoing and low numbers incoming. The strongest impact—loss of species—is at the boundary. The climatic extreme was influential. At the same time, the contrast between restricted-neritic and open-neritic assemblages reaches its minimum: that is, species-in-common reaches its maximum. But straightaway in this new ocean the contrast snaps back, and even increases. It is as if specialists, those more narrowly adapted, were more vulnerable and the generalists had more options.

Second, inspect the outcome of a cluster analysis of 257 microfossil samples of Oligocene age. The Late Oligocene samples (groups 1–4a) show increasing dissimilarity through time. Third, ordination treats the same data by timeslice, from just below the Oligocene to just above the Oligocene, in examining the distance between samples as a measure of dissimilarity. The successive balloons are increasingly inflated, supporting the clustering in implying an expanding and distancing of the ecological niches in this marginal and shallow sea. Li had found an increase in endemism from the Early to the Late Oligocene, in response to a major transgression and expansion of the neritic sea and global warming. Corroborating the endemism, these balloons suggest that the Early Oligocene sea was restocked with generalists to which specialists were added in the Late Oligocene.

In a nutshell: the Early Oligocene was a time of recovery; the Late Oligocene was a time of expansion. It sounds seductively like good community ecology with its shallow-time notions like pioneering settler communities which mature according to some internal dynamic with the passing of time. But we found the external dynamic to be more persuasive in deep time, especially the onset of glaciation, the marine transgressions and responses to climatic shifts and pulses and the fact that the fossil record of these populous microbes come as meaningful packages.

The Lakes Entrance Oil Shaft and packaging biogeohistory

The Lakes Entrance Oil Shaft was sunk in East Gippsland in 1941–1945 to develop the oil-bearing beds known since 1924. An oil mine, not a drillhole, it encountered the oil-bearing horizon at about 365 metres, but not in the amounts of oil needed in this country in those fraught times. But there was a second objective—to recover a section of Miocene strata longer and more complete than the scenically exposed limestones scattered around the continental margin, a section that might tie together and clarify the succession of fossils and strata in southern Australia, some of the trials and tribulations in which we have met in earlier chapters. The central figure in this objective was Irene Crespin, Commonwealth palaeontologist (who did some of the sampling herself, lowered in a bucket). The shaft is on a narrow platform, situated between hills close by in one direction, the deep sea close by in the other, and the Latrobe Valley with its coal measures not far away. Climatically, then as now, Lakes Entrance is pivotal, what with the Southern Ocean and the Roaring Forties from the south-west and the warm currents and monsoonal impacts from the north. By the early 1990s we were asking what the foraminifera from this densely sampled pile of strata might tell us about Neogene environmental and biohistory and their chronology. Qianyu Li processed 228 samples for 68,000 fossils in 65 planktonic species and 410 benthic species.[3]

With numbers like that, one can count and quantify and spot trends. One can partition benthic foraminifera into cohorts of ecological interest, such as epifaunal, infaunal and photosymbiotic lifestyles. One might look for statistical associations for the deep-time patterns of chronofaunas. These are palaeobiological questions.

---

3   The monograph of the foraminifera (Li and McGowran, 2000) includes an account of the Lakes Entrance Oil Shaft in the context of the times.

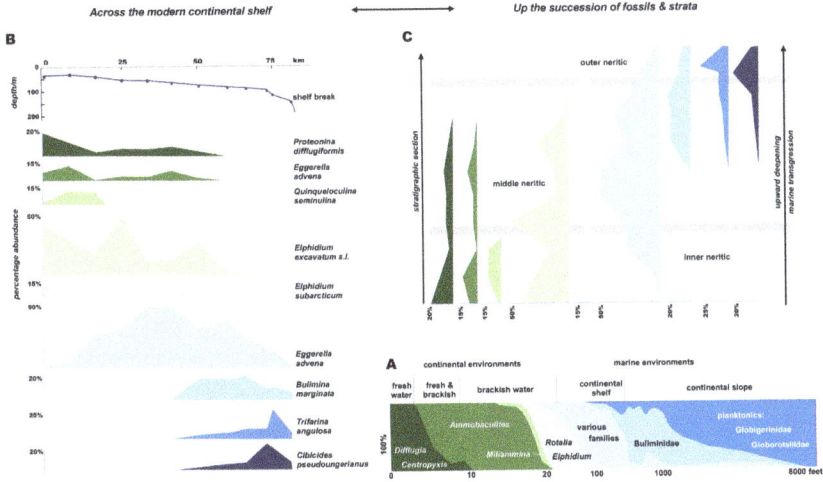

Figure 9.7. Modern neritic biofacies in space and time.

A, percentage abundances of foraminifera delineating modern biofacies in a profile from fresh water to continental slope, in the Mississippi Delta and Gulf of Mexico. Globorotaliidae and Globigerinidae are families of planktonic foraminifera; all others are benthics. B, another profile, this one across the Atlantic coastal margin of New Jersey, of percentages of nine common to abundant benthic species. C, the same profile of the species, upended in a thought experiment to construct a profile through a stratigraphic section and through geological time. The identification of inner, middle and outer neritic facies is real enough but with fuzzy boundaries. This consistent space–time relationship (unrealistically lacking unconformities!) is another example of Walther's Law.

Source: Adapted from Parker (1948), Lagoe et al. (1997) and McGowran (2005b).

But we begin with the palaeogeological or stratigraphic questions of marine transgressions and regressions. Species of living foraminifera are distributed across modern continental shelves and beyond in orderly patterns reflecting adaptation to environmental patterns. Visual inspection suggests a three-part lateral succession from inner, through middle, to outer neritic assemblages of species. As a thought experiment, upend this modern, lateral pattern to generate a vertical pattern (Figure 9.7).

This would indicate an upward-deepening, an advancing sea, a marine transgression, and these fossil patterns are of much interest in petroleum geology and other economic exploration. Add a second stack of hypothetical strata, upended in the other direction, and you have a transgression-regression, a sea advancing then a sea retreating, a stratigraphic cycle.

Although this is too simple to capture the turbulent real world of tectonic disruption, erosion and unconformity, the notion of departure and return, of advance and retreat, manifested most clearly in marine transgression and regression, has underlain geology for two centuries. It is time's cycle and it underlies sequence stratigraphy. We have seen its elegant application to the Eocene coals and limestones in Gippsland and extended across southern Australia and we take it further here, in the Miocene records of the foraminifera.

Li partitioned the hundreds of benthic species and thousands of specimens into two bins, inner neritic and outer neritic habitats, the shifting balance predicted to reflect shallowing or deepening as the case may be. The shifting balance gave a palaeodepth curve. A second binning of the same data was between infaunal and epifaunal habitats, implying more food being buried, and vice versa. The accompanying planktonic species gave the age control within the Miocene (Figure 9.8).

We knew that the earth was in an icehouse state during some of Miocene time, and that the deep-sea oxygen-isotope curves were showing fluctuations implying glacioeustatic global fluctuations. There was a succession of Miocene glaciations, Mi-1 to Mi-6, predicted from the oceanic isotopic proxies. So, should not this pattern be visible on the continental shelves, in the neritic realm? We knew that sedimentary strata come in packages between unconformities indicating gaps and hiatuses; the unconformities in the neritic realm were claimed to be sequence boundaries of worldwide significance and application. The sequence boundaries were named after the stages of the Miocene Series, Aquitanian to Messinian. And to round out the scenario, the glaciations signalled in the deep ocean should coincide with the sequence boundaries in the shallow seas.

The Gippsland curve for palaeodepth, close by the deep ocean, the south-west Pacific, displays an overall shallowing of the sea on the Lakes Entrance platform during the Miocene Epoch, the trend interrupted by a strong reversal towards MICO (Miocene climatic optimum). And the lower part could be matched plausibly with the Waikerie curve, in the Murravian Gulf (Figure 9.9), hundreds of kilometres from the Southern Ocean, where the reversal is more pronounced and a little later.

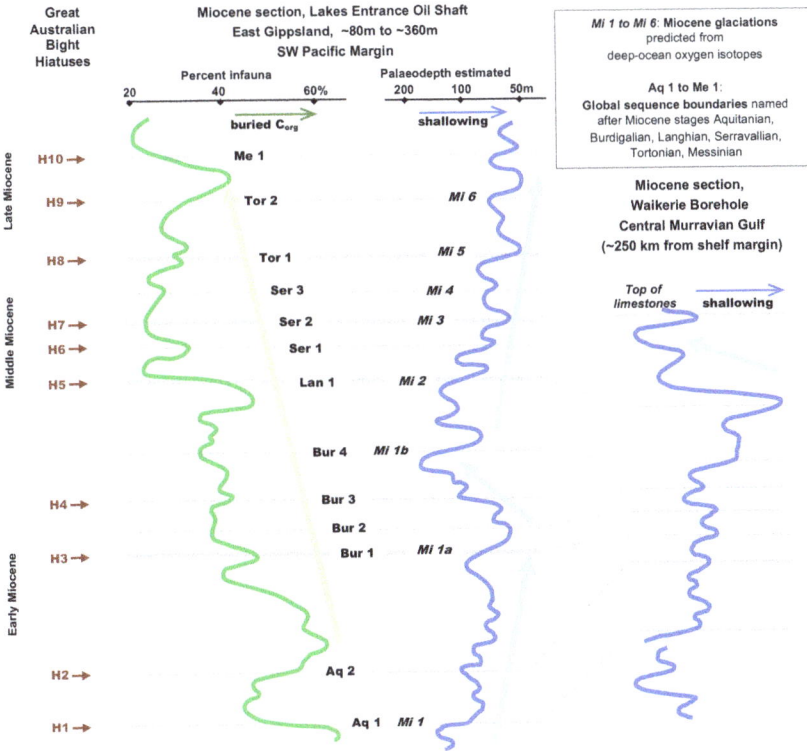

Figure 9.8. Miocene biofacies and sequences, Lakes Entrance and Waikerie.

The section preserving 20 million years of Miocene time at Lakes Entrance is profiled in a two-part binning of the benthic foraminifera into inner- and outer-neritic facies. The resulting curve is a proxy for palaeodepth, controlled by planktonic foraminiferal biostratigraphy, and compared with the 'global' sequence boundaries Aq-1 to Me-1, themselves having been matched with deep-ocean oxygen-isotopic peaks identified as glaciations. And this palaeodepth curve from the very edge of the Miocene south-west Pacific Ocean could be compared with a drillhole at Waikerie, 300 kilometres from the Southern Ocean. Meanwhile, per cent infauna is a proxy for buried carbon (C_{org}), as we saw in the Eocene. The overall trend (green arrow) is decreasing C_{org}, and that may be signalling the general drying out of Australia through the Miocene. However, the overall shallowing (grey arrows, temporary reversal notwithstanding) may be affecting the C_{org} signal. We interpreted the apparent reversals in C_{org} as fertile upwellings from the deep ocean nearby—but might the earlier one not be terrestrial, that is, from the drenched forests nearby? All these fits are perhaps more plausible than compelling, because we do not separate eustasy from isostasy, but surely they demonstrate that the eustatic and climatic rhythms well established in the closely sampled deep oceans were effective in our shallow seas bordering a relentlessly desiccating continent—even as they are aligned, almost one for one, with hiatuses identified in the Great Australian Bight?

Source: Redrawn and modified from McGowran, Li, et al. (2009).

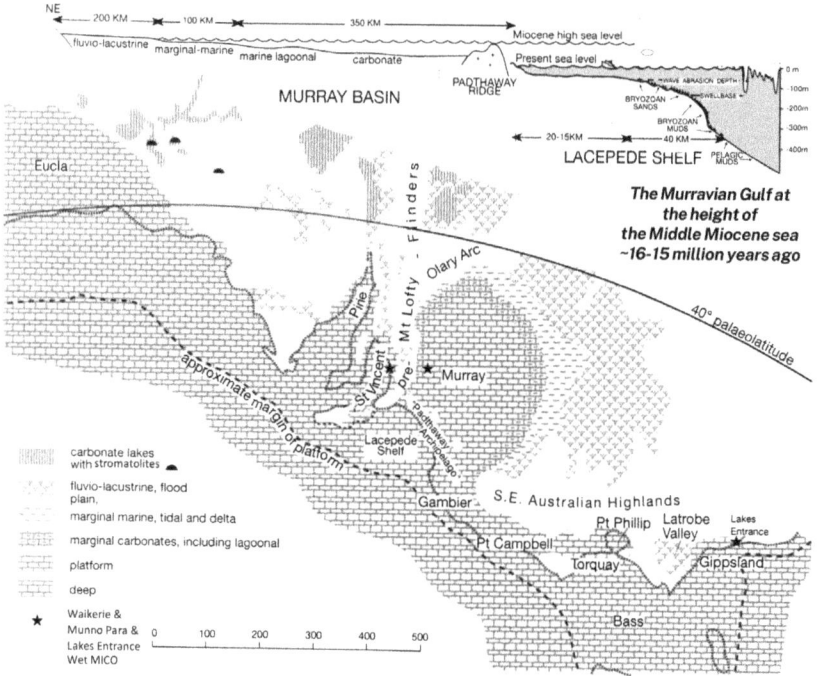

Figure 9.9. Murravian Gulf in the Murray Basin.

The Murravian Gulf in the Murray Basin reached hundreds of kilometres from the Southern Ocean at its peak in the Middle Miocene. The peak marine transgression coincided with MICO. Extreme wetness produced a brackish lid inferred at the three localities starred. The Padthaway Archipelago was a chain of islands formed of ancient granites. We know there were significant tectonic movements in the Mt Lofty–Flinders region in Eocene, Oligocene and Miocene times, but the ranges themselves are younger.

Source: From McGowran, Li, et al. (2009).

At the next level down, the local wiggles show plausible matching with the global pattern, itself seemingly under glacioeustatic control. As for the wiggles in per cent infauna, perhaps we are seeing pulses in the supply of nutrient from the land, although we can only point tentatively to the coal cycles Morwell M1a and M1b and Yallourn. But the main narrative is the overall drying-out of Australia through 20 million years of Miocene time, as displayed off the south-east corner of the continent. Surely the bold arrow spanning the epoch is asserting this trend? The decrease overall in per cent infauna says partly the same thing, in that shallowing implies more nutrient consumed and less buried. The pronounced reversals imply upwellings from the deep water nearby.

Limestone seas expanding and coal forests returning

The still-very-young Southern Ocean spilled across the southern margins of isolated Australia, advancing tentatively at first, in the earlier Oligocene, more boldly during the Late Oligocene global warming, and attaining its maximum in Miocene times, 16–15 million years ago. The limestones in the cliffs facing what are now the Roaring Forties are the most visible rocks, but limey sediments were laid down across the continental shelf and up to 500 kilometres into the continental interior, in the Eucla Basin and the Murray Basin, where this grand Miocene invasion has been called the Murravian Gulf. Vast areas of this sea were only tens of metres deep and well within the photic zone, encouraging the flourishing of sea grasses in its shallower part and 'calcareous algae' (Rhodophytes) in its deeper. In the west, the Middle Miocene sea had dimensions comparable to the Middle Eocene sea (Figure 8.4). In the east, the Middle Miocene sea went well beyond the Eocene dimensions to form the Murravian Gulf (Figures 9.9 and 9.10).

These vast shallow seas are known as platforms where the bottoms are flat, and as ramps where gently sloping—very gently, actually, because they remain almost all within the photic zone with a sunlit floor across hundreds of thousands of square kilometres. With an enormous ratio of surface area to volume in these seas, one might expect the evaporating-dish effect to operate, as in some places in the tropics (recall the halothermal mode of circulation). That serious hypersalinity seems not to have happened indicates lands nearby under high rainfall.[4] And located at 40°S palaeolatitudes and further south they are unique in the Neogene and have no helpful modern analogue.[5]

---

4   The outcropping rocks of the Murray Basin and the Murravian Gulf are known in the cliffs along the east–west then north–south course of the River Murray and to biostratigraphic (foraminiferal and sporomorph) access to the subsurface in drilling for groundwater. For the South Australian part of the Murray Basin, see *Natural History of the Riverland and Murraylands* (Jennings, 2009). All of this is much sparser for the Eucla and Bight basins; even so, see competent accounts in Lowry (1970) and Alley and Lindsay (1995). O'Connell et al. (2012) described the outcropping Nullarbor Limestone.

5   These vast shallow seas at mid-palaeolatitudes have no modern analogue. Certain modern tropical platforms with limey muds have reefs at their margins which give some protection against stormy open oceans, rather like the granites of the Padthaway archipelago across the mouth of the Murravian Gulf but constructed by corals, bryozoans or algae. On the basis of seismic images Feary and James (1998) postulated a 'Miocene Little Barrier Reef' at the edge of the continent. However, when O'Connell et al. sketched its expected position along about 1,000 kilometres of the continental margin (2012, Fig. 8), the reef lurked deep in the shadows of postulation. It still does. Meanwhile, as we have seen (Figure 8.5a), a reef of formidable dimensions is recognised in much the same place in the Eucla sea of Middle Eocene age—but that Eocene reef possesses, in a solitary drillhole, that one precious grain of ground truth missing from the Miocene.

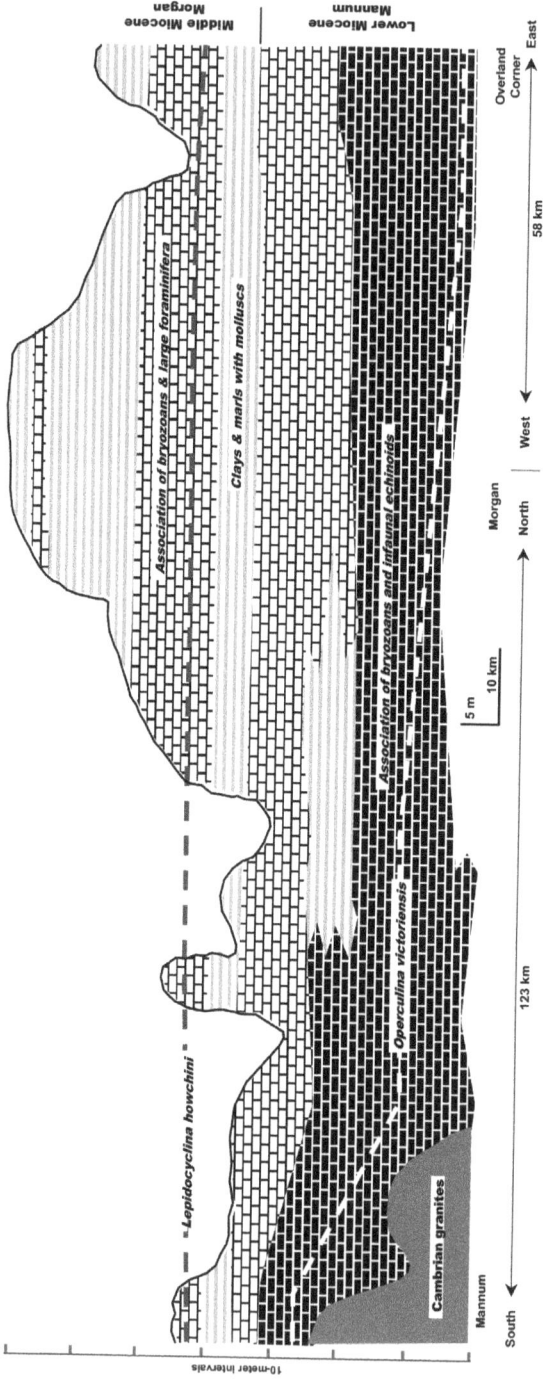

Figure 9.10. Lukasik's Miocene stratigraphy, River Murray.

Long known by the First Nations peoples and documented by Charles Sturt's expedition, the River Murray flows west then sharply south. Its destination was not to be an elusive sea. The change in direction was forced neotectonically. The cliffs display Miocene limestones overlain by Pliocene oyster banks and younger strata (not shown here). Black lines indicate the accessible outcrop at each locality. The diversely fossiliferous strata have been sorted and grouped into four 'facies associations', which are more informative biogeohistorically than the formal names to the right. Note the two 'ecostratigraphic markers', useful controls in this compilation, marking warm southwards incursions by these tropical-type photosymbiotic foraminifers. The Cambrian granites were a partial and protective barrier between this ramp or platform in the Murravian Gulf and the Southern Ocean. And note too the scales and the enormous vertical exaggeration needed for intelligibility. Omitted here are the beach ridges of the Murravian Gulf in the Late Neogene (Figure 9.21).

Source: Redrawn and modified after Lukasik et al. (2000).

274

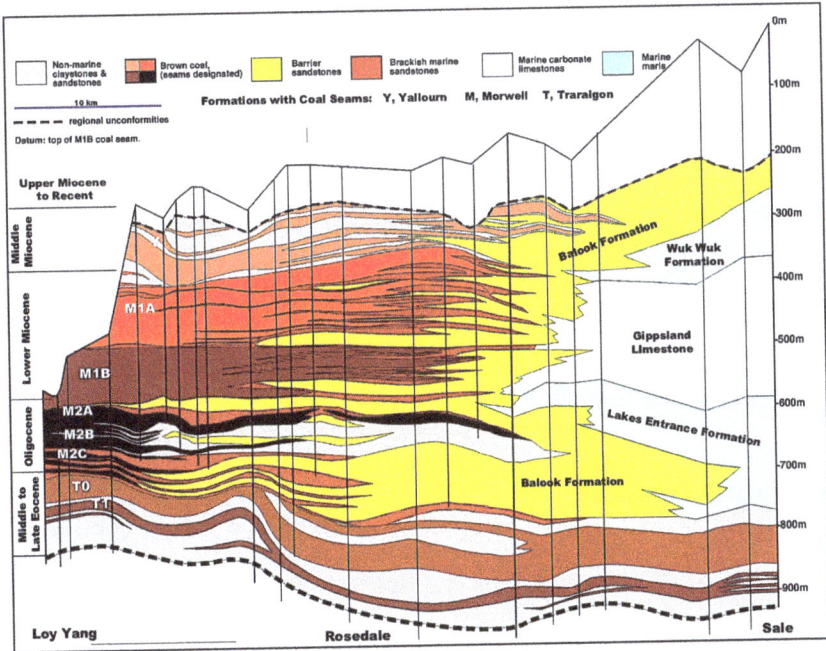

Figure 9.11. Holdgate's recent reconstruction, Eocene–Miocene Gippsland.

Another MECO (Middle Eocene climatic optimum) to MICO portrait of southern Australia, this one possible in such detail because it spans reservoirs of the fossil fuels coal, oil and gas. (Otherwise, most or all the 15 drillholes shown here would not have happened.) Foraminiferal biostratigraphy managed to sort out the ages of the marine strata in southern Australia, as in the right side of this reconstructed section of about 40 kilometres in Gippsland. (Location is shown on Figure 8.8a. The Lakes Entrance oil shaft is 75 kilometres to the north-east of Sale.) Palynological biostratigraphy showed in due course that the Morwell to Yallourn coal measures (*left*) were coeval with the marine sediments (see how the extensive Traralgon Coals underlie the entire panorama). The coal measures are packaged in sequences named after the main coalfields, and the marine packages are identified with the global sequences T1 to TB2.4, in the nomenclature in use at the time. The boundaries, the broken lines, are unconformities.

Source: Holdgate and Sluiter (2021).

And, indeed, returning along with the limey marine muds were the coal forests, now shrunken at the continental scale compared to the Eocene forests—they now were restricted to the Latrobe Valley in the far south-east and were more confined within that district than were their Eocene forebears. (Coal forests, that is; not all forests.) For many years up to 1964 it was thought that all the coals were older than all the limestones, but the picture changed dramatically under the economic imperatives of the exploration and development of the coals, onshore, and of oil and gas, offshore. As foraminiferal micropalaeontology revealed the ages of the limestones, and palynology likewise of the coals, an elegant pattern of the strata could be constructed (Figure 9.11).[6]

Across about 40 kilometres of reconstructed section we can see a threefold pattern—marine strata, brackish sands and clays, and nonmarine coals, sands and clays. As the environments fluctuate, the sea advances and retreats, there is erosion then deposition, and the resulting unconformities can be traced right across the spectrum of environments to define packages of strata, namely sequences. Recall the coal–limestone couplet in Figure 8.8A. It integrates the nonmarine coal-bearing local sequences with the marine limestone sequences, themselves identified as the local representatives of the putatively global sequences. More on this matter of sequences below. Meanwhile, there is no time scale in this reconstruction, and so it should be enlightening to include time.[7]

Miocene climatic optimum

We became interested long ago in the distribution of the large, photosymbiotic foraminifera in space and time (spatiotemporal) in southern Australia. Our foraminiferal predecessors back to Walter Howchin had that interest too, but without the benefit of continental drift, or palaeotemperature curves from the deep ocean, or the notion of a Leeuwin Current, or a more-or-less correct dating of the various fossil occurrences. We have those insights now, and it turned out that the pattern assembled in the 1960–1970s held together (Figures 9.12 and 9.13).

---

6   Figure 9.11 has forerunners (Holdgate and Gallagher, 1997, 2003), republished in McGowran et al. (2004).

7   It is necessary to repeat what we have known for some time, namely that the coals of the Yallourn rainforests are coeval with the Wuk Wuk Marl in Gippsland and with the youngest limestones of the Murravian Gulf and the Eucla Sea. They all sit squarely atop MICO and all predate the onset of floras with common to dominant *Eucalyptus* and *Acacia* and the general onset of cooling and aridity in southern Australia. But Lukasik et al. (2000) and especially Pufahl et al. (2006) presented a different scenario. They inferred a 'progressive' shift, preserved through the Murray Group carbonates, from cool and wet conditions and abundant nutrients delivered from the land (Mannum Formation) to a seasonal and arid climate with a reduced supply of nutrients from the land (Morgan Group). They invoked in support of this a climatic shift from perpetually wet to seasonally dry conditions as revealed by the terrestrial floras, the shift being from a mixed rainforest assemblage with Myrtaceae and *Nothofagus* (Late Oligocene and Early Miocene) to a hot, semiarid ecosystem dominated by *Eucalyptus*. Correlations and age determinations do not sustain this scenario; those fundamental changes in terrestrial vegetation happened 2–3 million years later (McGowran, Li, et al., 2009).

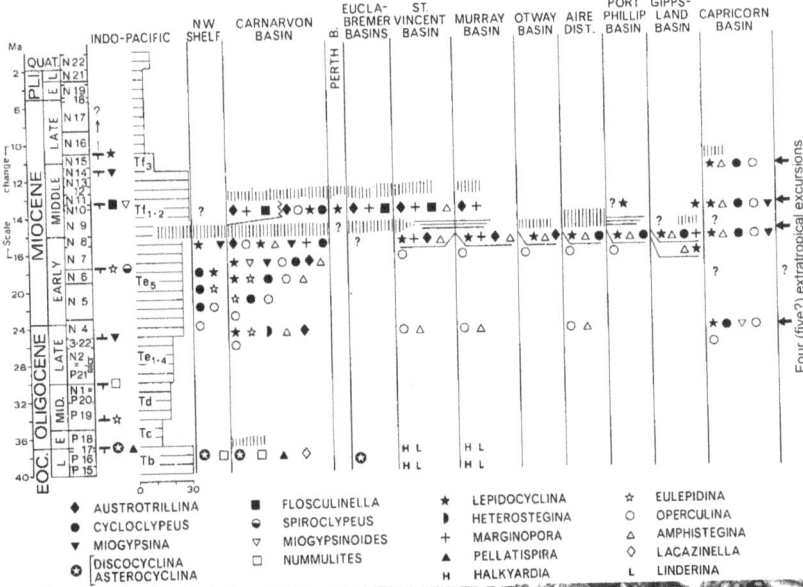

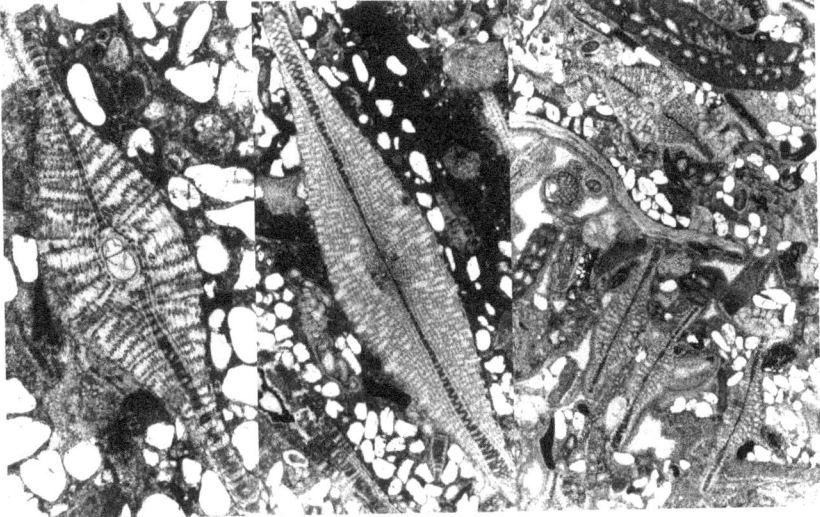

Figure 9.12. Miocene large-foram migrations southwards and Lindsay's *Lepidocyclina*.

Above, the circum-Australian distribution of large, benthic, photosymbiotic foraminifera, as seen in the 1970s as extratropical excursions out of their natural habitat in the tropical Indo-Pacific region. Tb to Tf_3 are the Indo-Pacific zones, and the histogram of genera shows the high diversities in the Early Neogene tropics. The excursions implied episodes of warming and expanding neritic seas, and were congruent with the early signals of deep-oceanic $\delta^{18}O$ warmings. MICO across the Early–Middle Miocene boundary is quite apparent. Vertical lines indicate widespread hiatuses. Horizontal lines include the Munno Para Clay in the St Vincent Basin and the Cadell Marl/clay in the Murray Basin, coeval and

277

critical to the narrative of a very wet MICO (located at the stars in Figure 9.9). The most prominent gap in this 1979 chart was the absence of *Lepidocyclina* from the Eucla Basin (see the query in the Eucla-Bremer Basin column).

Below, in due course, Murray Lindsay removed that query by finding *Lepidocyclina* where it 'should' be in space and time in the Eucla Basin. Cross-sections of individuals in rock thin sections of, from *left*, a megalospheric (large embryo) specimen (4.15 mm length), a microspheric (small embryo) specimen (8.9 mm), and a swarm of specimens (average 3 mm) among bryozoan, echinoidal and sand grains.

Source: McGowran (1979a) and McGowran and Hill (2015).

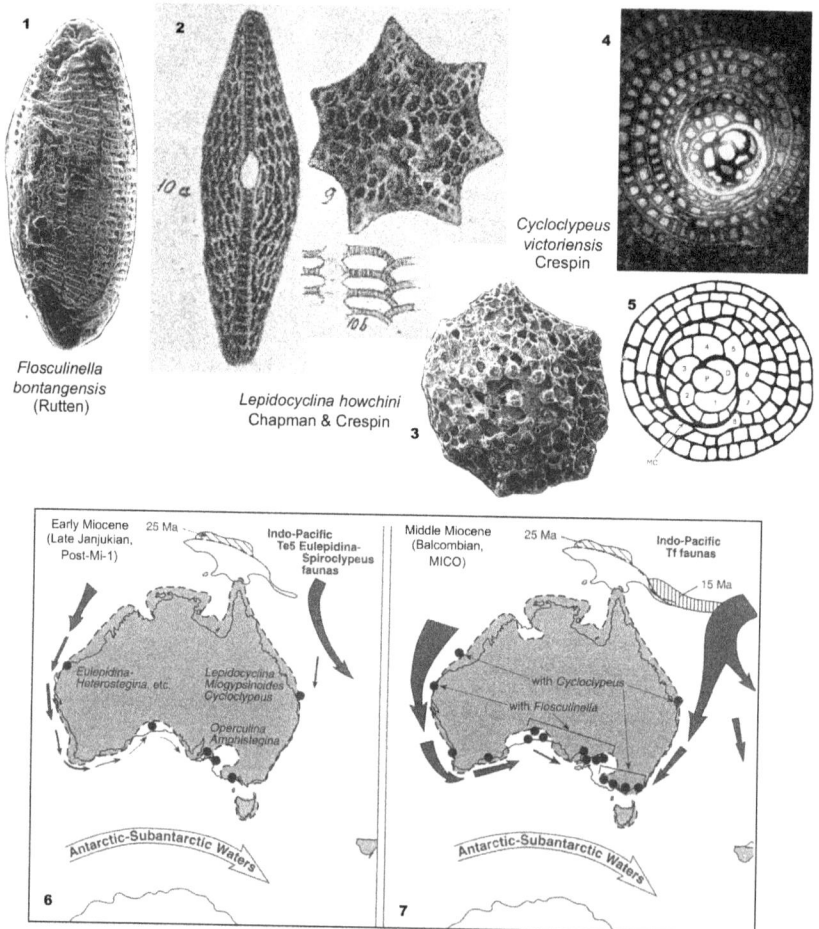

Figure 9.13. Miocene large-foram biogeographic indicators, southern Australia.

Tropical-type large photosymbiotic foraminifera came south episodically on the Leeuwin and East Australian currents.

1, the spindle-shaped *Flosculinella bontangensis* (Rutten) is from the Middle Miocene, St Vincent Basin (axial length ~1.1 mm). The thin outer wall is abraded exposing numerous chamberlets.

2, 3, Howchin (1889) described the discoidal form later named *Lepidocyclina howchini* from the Middle Miocene at Muddy Creek (labelled 9, 10a, 10b). 10a (~2.7 mm diam.) demonstrating in axial section the numerous equatorial and lateral chamberlets, the latter (10b) with thin walls (and now inferred as packed with photosymbionts) and solid clear calcite pillars (not only for structural support but also acting as light shafts). Howchin's 9 and our 3 are whole specimens showing erupting pillar ends like skylights; 3 is from the Middle Miocene, St Vincent Basin (Daily et al., 1976; max diam. ~1.8 mm).

4, 5, a thin section of the juvenile stages of *Cycloclypeus victoriensis* Crespin, photographed and drawn and labelled (diameter ~8 mm). The large (megalospheric) initial chambers seen here (hidden in *Flosculinella*) and in Figure 8.12 are the receptacles for asexually bequeathing (transmitting 'vertically') the photosymbiont protistan stock to the offspring, which otherwise needed themselves to restock from others or from the environment ('horizontal' transmission).

6, 7, two snapshots of the extratropical excursions by Indo-Pacific large benthics to southern Australia: eastern, with the gyre; western, against the gyre on the Leeuwin Current. The Late Janjukian event, on the rebound from glaciation Mi-1 (LOWE in Figure 4.18) is the weaker, but still clear. The Balcombian event is actually a cluster of events sitting atop MICO, the Miocene and Neogene climatic optimum. The extratropical neritic excursions matched chronologically with warm peaks derived from deep-oceanic (benthic) oxygen isotopes. That powerful consilience became apparent in the late 1970s. Note that in the event labelled 'Balcombian', *Flosculinella bontangensis* and *Cycloclypeus victoriensis* do not overlap geographically in the south. A discontinuous record could be (and was) interpreted as collection failure, but episodic excursions into the outer southern reaches of the Indo-Pacific biogeographic province are well established.

Source: McGowran and Hill (2015). Howchin's illustrations are seen also in Fig. 4.11.

The tropical-type benthics came south, down the east and west coasts of the continent. From the west on the Leeuwin Current, the genus *Flosculinella* advanced as far eastwards as the St Vincent Basin. From the east, the genus *Cycloclypeus* advanced as far west as the Aire District (on the Otway coast in western Victoria). The genus *Lepidocyclina* had been missing rather blatantly from our records of the vast and warm Eucla sea; it was hunted and found in due course. My old diagram shouts that the peak migration out of the tropics signalled the peak warming in the Miocene Epoch, actually a double peak roundabout the Early–Middle Miocene boundary. This interval of less than 2 million years also encompasses the Yallourn coal measures in the Latrobe Valley; and the resurgence of the gymnosperms in the Yallourn is coeval with the maximum extent of the Murravian Gulf and of the Miocene sea in the Eucla and St Vincent Basins; and coeval with the influx of tropical foraminifera.

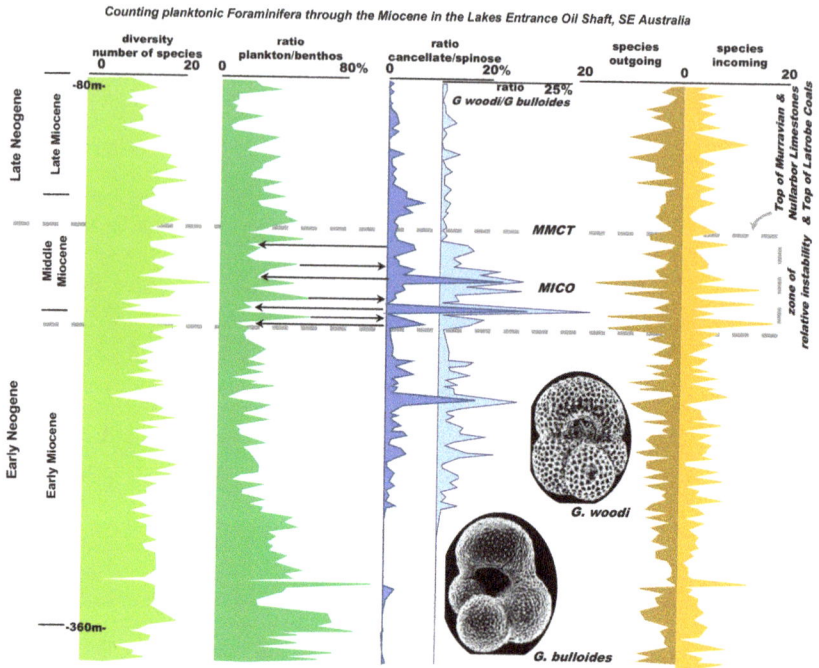

Figure 9.14. Miocene planktonic foram logs at Lakes Entrance.

A log of planktonic foraminifera through the Miocene section at Lakes Entrance. Several metrics increased markedly in amplitude in the Middle Miocene, rising and falling in concert with MICO and its falling away in the later Middle Miocene climatic transition (MMCT). Does warming and maximum transgression imply ecological and evolutionary instability? Some evolutionary-palaeoecological theory suggests that it does. The ratio of *Globoturborotalita woodi* to *Globigerina bulloides* is a subset of the cancellate/spinose ratio (some groups have spines, some don't); its increase implies warming. But the arrows point to anomalies in that the highs in warming match lows in plankton diversity, and vice versa, and this very likely may be a brackish effect consistent with extremely high runoff from the gymnosperm-rich Yallourn forests nearby.

Source: Redrawn and adapted from McGowran and Li (1997).

And now to the planktonic foraminifera at Lakes Entrance in Gippsland. Whichever way we look at the planktonic foraminifera in these plots (Figure 9.14), they fluctuate. The number of species varies from 5 to 25; the plankton/benthos ratio ranges from less than 10 per cent to more than 80 per cent; numbers of species are dropping in and dropping out throughout the Miocene Epoch; most spectacular are the shifts in the balance between two groups represented by *Globoturborotalita woodi* and *Globigerina bulloides*. We should think of organisms and communities as patchy, or clumping, not distributed as uniform sheets across the landscape or seascape, and there will be short-term variation due to fleeting environmental shifts and accidents of sampling. But coming through the variation is a trend. We see the fluctuations increase in amplitude in a well-defined interval from latest Early Miocene into the early part of the Middle Miocene. This is most apparent in the *woodi*/*bulloides* ratio, which is a reliable indication of oceanic warming. Unsurprisingly, not only (i) is this at MICO, right where it should be; but (ii) it fades just where one would look for the climatic transition; and (iii) the increased amplitude in the various fluctuations are reflecting an increasingly unstable, more volatile, warmer world.

Looking closer, we see a pattern within the MICO interval. The arrows indicate that peaks in the planktonic warming indicators match lows in the plankton/benthos ratio, and respective lows match highs. This is back to front. One would expect warming to go with higher sea levels and increased plankton numbers. But MICO is the time of the last hurrah of the gymnosperms such as the Huon 'pines' (*Lagerostrobus*) in the Yallourn coal forests in Gippsland. We suggested that the warmer the climate, the wetter; and the wetter it is, the more likely that runoff and estuarine outflow expanded across the nearby ocean as a less dense, brackish-water lid; and the slightly lowered salinity discouraged the oceanic plankton. This suggestion supports the palaeobotanical estimates of annual rainfall in the Gippsland coal forests of at least 1,500 millimetres and more likely 2,000–2,200 millimetres.

This was the less-dense, less-saline, brackish-lid effect inferred so strongly in the Palaeogene AAG. The outflow of surface water interferes with the air/water exchange of oxygen and carbon dioxide, with implications for ventilation and the supply of nutrient. The Lakes Entrance platform is squeezed between the coal forests and the deep sea. What about the comings and goings of the sea sweeping across the broad and shallow stretches of the Murravian Gulf and the Nullarbor Sea? As mentioned above, the absence of evaporites from such seas implies excessive runoff. And now, note where the Munno Para Clay and Cadell Marl fit in the picture, two fine-grained, dark, organic-rich strata sandwiched between limestones and characterised by high C_{org} levels, low oxygen (dysaerobia), a benthic snail-rich infauna exploiting high levels of nutrients, and a sparse planktonic foram fauna dominated by *Cassigerinella chipolensis*, the opportunistic, dysaerobia-tolerant species that we have met already, back in the Eocene—the snail-rich horizons of the Priabonian (Figure 8.13). The Munno Para was in the proto–Gulf St Vincent, immediately west of the rainforest-covered hills; the Cadell over the hills to the east, in the Murravian Gulf; both were hundreds of kilometres from the continental edge and thousands of kilometres from the Yallourn downpours; and experiencing their own downpours about 15.5 million years ago (Figures 0.4 and 9.9). Marine circulation was surface-water-outflow, or estuarine, in stark contrast to the modern gulfs where it is deep-water-outflow, anti-estuarine or lagoonal. This horizon in three basins marks a brief return in MICO to the kind of ocean seen in Palaeogene times.

But how wet is wet? Or, how closely did the peak conditions of MICO in the southern Australian outback approach the widespread rainforests of the Early Eocene climatic optimum (EECO) in the Early Palaeogene? Our maps indicate that the answer was: not very (Figure 9.15). The reconstruction of the soils at Lake Palankarinna confirmed the warm-wet conditions peaking at MICO on the evidence of the plants and animals, but the soil types underlying true rainforests were not found (Figure 9.16).

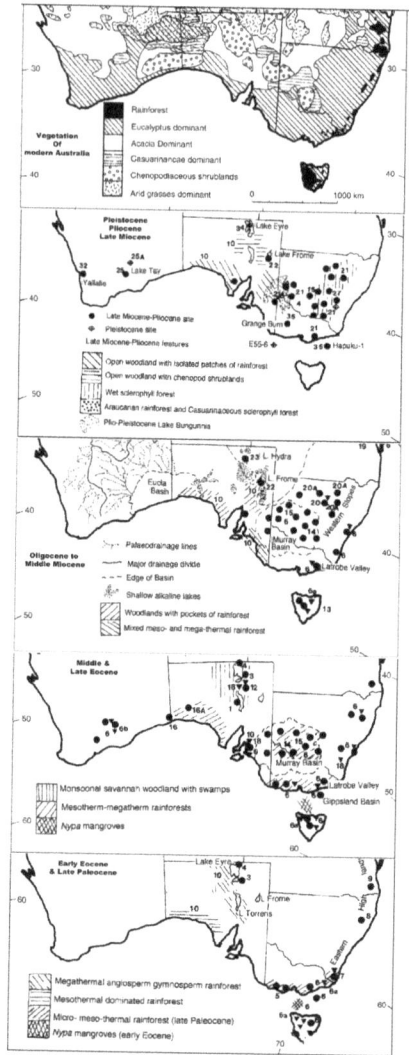

Figure 9.15. Martin's floral development, southern Australia.

These snapshots through the Cenozoic Era illustrated Helene Martin's review of Australia's episodic progress to today's aridity. There were repeated episodes of wetness in southern Australia, wetter than today, but not attaining the extreme levels of the Early Palaeogene. Chenopod shrublands are saltbush and bluebush, and similar plants.

Source: Combines the lower halves of five maps in Martin (2006).

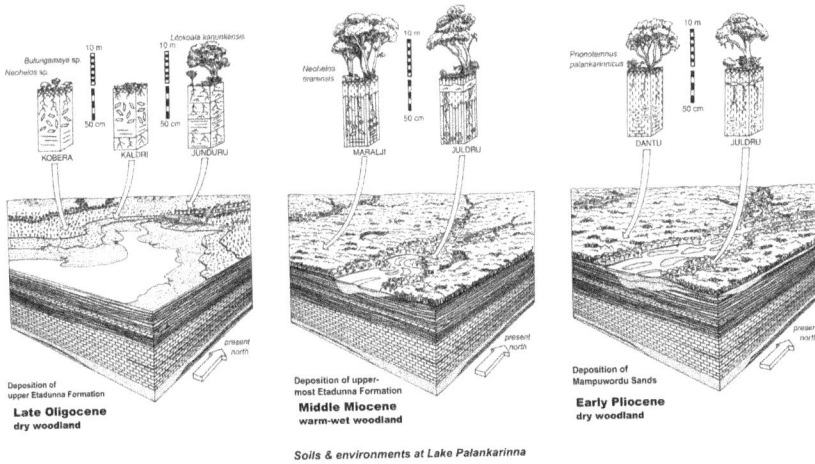

Soils & environments at Lake Palankarinna

Figure 9.16. Retallack's palaeosols and environments, Lake Palankarinna.

Reconstructions of soils (palaeosols) and environments at Lake Palankarinna, east of Lake Eyre in the Tirari desert. The Middle Miocene soils confirmed that times were warmer and wetter than before or after, and soils characteristic of rainforests were not found.

Source: Metzger and Retallack (2010).

There has been argument around the question: more arid or more humid outback in the Neogene?[8] The diverse fossil faunas of Riversleigh in northern Queensland have long been interpreted on abundant evidence as having lived in rainforests. The fossil faunas of the Lake Eyre region in South Australia have taxa in common with the Riversleigh faunas—enough to justify their attribution to rainforests too. And if the Lake Eyre region was not under monsoonal rainforest, instead being woodlands perhaps with pockets of rainforest, then perhaps Riversleigh was something like that, too. Or so goes the challenge to the Riverleigh rainforests. But the evidence keeps on accumulating of rainforest biomes in the Miocene at Riversleigh.

But where are the animals? There were major roles for biogeography in two big questions, Darwin's organic evolution and continental drift; and our large Pleistocene animals, discovered early on, were prominent in both. Richard Owen's diligence in the museum and a century's digging revealed our

---

8   Figure 9.16 is from Metzger and Retallack (2010) who, failing to find rainforest soils in the Lake Eyre Basin implied that similar Riversleigh faunas might not have been living in rainforests after all. Herold et al. (2011) is sceptical about Middle Miocene monsoons but is answered by Travouillon et al. (2012). But all argument about the Miocene in the Australian outback is handicapped by fragile age determinations. The Middle Miocene scene in this figure intended to represent MICO may be several million years older. Another example is the silcrete floras (Lange, 1978), wonderfully preserved as if freeze-dried in the siliceous duricrusts mainly from the Lake Billa Kalina district, and variously dated as Eocene, Oligocene, Miocene and Pliocene. See the comprehensive review by Roger Callen (2020) of this difficult and frustrating subject.

Pleistocene megafauna, but revealed too was our profound ignorance of what came before, of their Pliocene and Miocene forerunners. In 1968 George Gaylord Simpson told us in Adelaide that this lack of discovery was the most pressing problem in vertebrate palaeontology. Progress was underway by then, but Figure 9.17 will show that remarkable advances in knowledge have occurred virtually all in the Neogene and the Palaeogene is almost a blank.

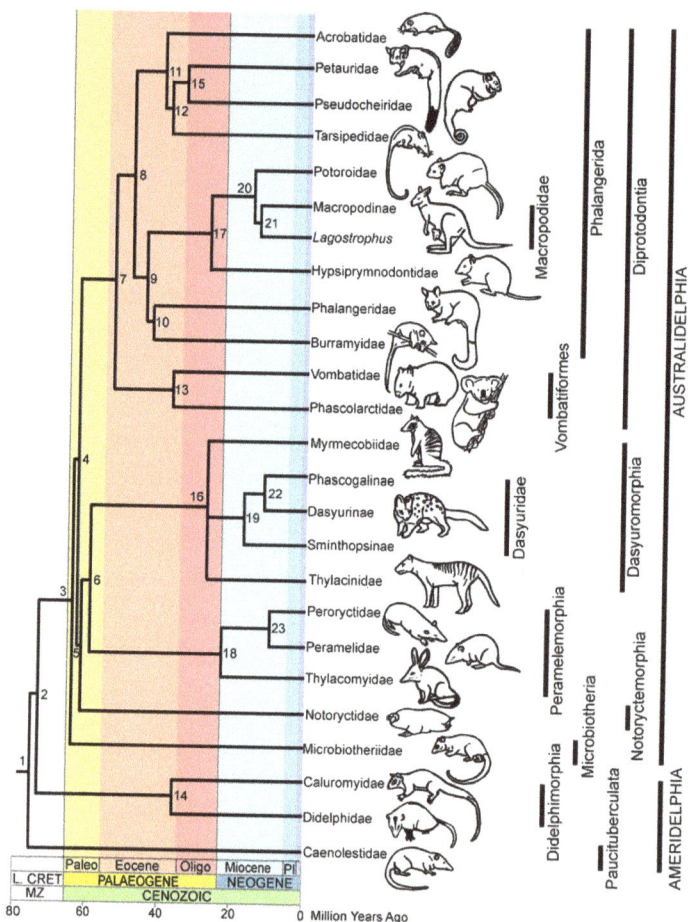

Figure 9.17. Phylogeny of modern marsupials.

A display of modern marsupials as an evolutionary diagram based on nuclear gene sequence data. Except for Microbiotheriids in the Early Eocene, none of the rich array of Australian marsupials (collectively the Australidelphia) are known as fossils before the Late Oligocene. So these divergences displayed between major groups deep in the Palaeogene are based on molecular divergences, dated by assumptions about the molecular clocks.

Source: From Black et al. (2012): the numbers in the figure refer to the nodes of divergence and are not discussed here.

Nothing is known from the Late Cretaceous to the Late Oligocene except for an Early Eocene fauna of Microbiotherians at Tingamarra in south-east Queensland. But the route from Antarctica was about to be closed at the head of the AAG in the Eocene, and so the pioneering stocks of the richest marsupial biota ever to evolve must have been here already, in Australia.

Meanwhile, back in the ocean: The Circum-Antarctic Current

In the greenhouse times of the AAG, we have seen a lot in common between the AAG's northern precincts and their opposite counterparts in the south. What about during the icehouse times, across the ever-widening Southern Ocean and perhaps on opposite sides of the Antarctic Circumpolar Current? Addressing these questions brings us back to drillhole Site U1356 off Wilkes Land, for it encountered Oligocene and Miocene strata (not deposited on the continental shelf, like the Eocene strata, but in the deep ocean due to prolonged subsidence), and there were plenty of information-rich dinocysts. And these dinocysts can be compared confidently with the modern biogeographic configuration in the Southern Ocean, where they characterise the water masses defined by the oceanic fronts (Figure 9.18).

Selenopemphix antarctica is found in the zone of sea ice; its group (the protoperidinioids) indicates seriously cold water; the two major groups (the mixed protoperidinioids and gonyaulacoids with prominent *Nematosphaeropsis labyrinthus*) dominate the zone between the Subantarctic and Subtropical fronts; and beyond the Subtropical Front the mixed gonyaulacoid dinocysts feature the genera *Impagidinium* and *Operculodinium*. These dinocysts as indicators of ancient and modern water masses were and are responding strongly to water temperature and food supply. They don't adapt to changing environments by themselves changing: they stay adapted by following the shifting water masses. We have seen already, for example, that the distribution of *Operculodinium* can be employed confidently as a warm-water indicator deep within Palaeogene times.

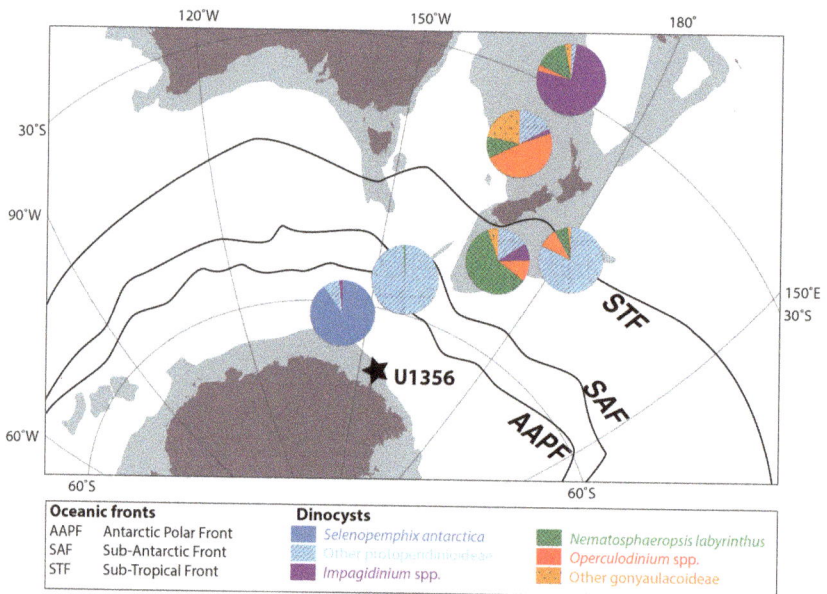

Figure 9.18. Dinoflagellate signatures in southern water masses.

Three 'Fronts' define oceanic water masses at high southern latitudes, and the water masses are characterised by key species and assemblages of dinoflagellates. These dinocyst pie charts show average assemblages in the surface sediments underneath oceanic frontal zones in the Southern Ocean. *Selenopemphix antarctica* is the indicator of sea ice, and twice in the past has been that: for the first 1.5 million years of the Oligocene, and during the MMCT. At other times there was stronger influence of oligotrophic, low-latitude surface waters over Site U1356 off Wilkes Land. Especially noteworthy is the presence of *Impagidinium* and *Operculodinium* at U1356, signalling MICO just before the return of *Selenopemphix*.

Source: From Bijl et al. (2017).

In the Early Oligocene for about 1.5 million years after Oi-1 (Oligocene glaciation 1), *Selenopemphix antarctica* gives a strong sea ice signal off Wilkes Land (weaker than today's signal). This was the expanding ice sheet that caused the Chinaman Gully downcut. After that, conditions were cool during the Oligocene but indications of sea ice were not detected, and there was upwelling of nutrient-rich deep water, but not strongly, and sporadically. Above a hiatus of 8 million years, MICO is signalled off Wilkes Land as clearly as could be by peaks in the occurrences of the dinocysts *Impagidinium* and *Operculodinium*. It was warm then, and warm water was close to Wilkes Land. This is a strong reason for inferring that the Antarctic Circumpolar Current, the engine room of today's global ocean, was not active, and that today's oceanic boundaries would not have been discernible. And then *Selenopemphix antarctica* gives another sea ice signal after the

MMCT and the near-disappearance of *Impagidinium* and *Operculodinium*. By 14 million years ago the parameters on the modern Southern Ocean are beginning to appear.

And then it all changed

And then it ceased—or at least the record of the strata was terminated abruptly. The top of the Nullarbor Limestone is the modern surface, the Nullarbor Plain. It is the top of the limestones in the St Vincent Basin and in the Murravian Gulf. About 14 million years ago the shallow seas drained, just as Australia's desiccation seriously set in. In the sea on the south-east margin, the pronounced swings in the planktonic foraminiferal record at Lakes Entrance became markedly less pronounced. In the wet continental fringes, the Yallourn coal forests disappear at the same time, so soon after the resurgence of their gymnosperms. The biogeographic happenings off Wilkes Land coincided with the contraction of southern Australian limestones and with the disappearance of Gippsland's coal forests. So the peak is succeeded by a crash, on land and at sea. It became apparent in the late 1970s that the regional pattern in the terrestrial and neritic realms would fit, would be congruent with, the oxygen isotopes in the oceanic realm. The carbon dioxide–based contribution to this global pattern arrived in the form of the Monterey hypothesis. As the deep ocean was probed and drilled, the calcite compensation depth was found to form a pronounced and isolated peak—shoaling then deepening—surely having something to do with the temporal pattern of a peaking then disappearing of the great neritic carbonates? This is an exquisitely natural boundary between an 'Early' Neogene and a 'Late' Neogene (informal names, hence the quotes).

The above paragraph is all exogenic. It is about environmental change, icecaps growing and sea levels falling, the biosphere responding, not passively but as an active participant in the story of carbon dioxide. The increasingly unstable, more volatile, warmer world was becoming a less unstable, less volatile and cooler world. The Monterey hypothesis that I've exemplified as a beautiful piece of modern science proposed that burying organic carbon drew down carbon dioxide levels (evidence: the Monterey positive carbon excursion), in turn drawing down temperatures as a reversed greenhouse effect (evidence: the subsequent positive oxygen excursion), during, all up, about 2.5 million years, 17–14.5 Ma. The signals and their chronology have been refined and debated, the warming and cooling have been delineated by multiple methods including astrochronology, and the Monterey carbon excursion enveloping MICO is the most prominent event in Miocene biogeohistory.

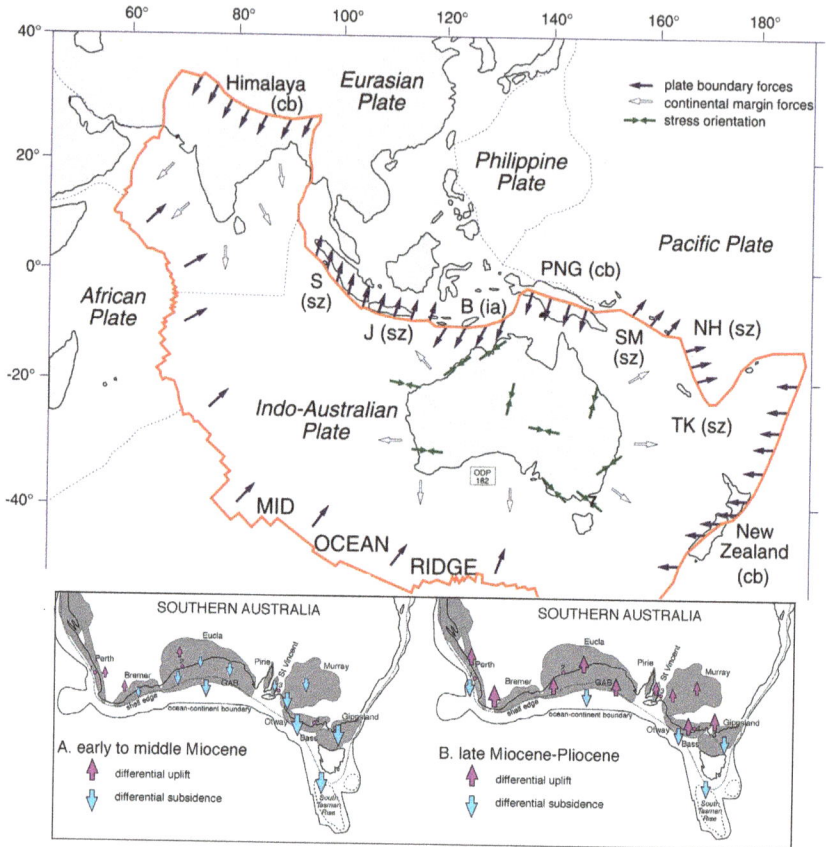

Figure 9.19. Stresses and forces, Indo-Australian Plate.

Above, the Indo-Australian Plate showing plate boundary and continental margin forces and Australia's major stress orientations. Plate boundary types: cb, collision boundary; sz, subduction zone; ia, island arc; S, Sumatra Trench; J, Java Trench; B, Banda Arc; PNG, Papua New Guinea; SM, Solomon Trench; TK, Tonga-Kermadec Trench. Shown in the Bight is the location (ODP182) of the oceanic hiatuses shown in Figures 8.6 and 9.8

Below, differential uplift/subsidence along the southern Australian margin to Australia–Asia collision during (*A*) the Early and Middle Miocene and (*B*) Late Miocene–Pliocene. Relative subsidence mostly occurred onshore and in areas under which the syn-rift faults exist and the thermal regime was declining.

Source: From Li et al. (2004). The map is based upon Coblentz et al. (1995) and Hillis and Reynolds (2003), two papers that changed our perceptions of 'drifting' Australia.

Is there an endogenic factor? To be sure, the warming effects of volcanism became part of the Monterey narrative; the eruption rate of the Columbia River Basalt has been estimated to rise and fall within about 2 million years centred on 15.5 Ma. But recall the problem of stratigraphic sequences from Chapter 8 (Figure 8.6) where hiatuses that might be candidates for glacioeustasy could be traced to water depths beyond the reach of glacioeustasy; and then inspect the Miocene succession of hiatuses in the Great Australian Bight. These alignments of local sequence boundaries with so-called global sequences on the one hand and with hiatuses on the far side of southern Australia—these alignments are plausible matchings, not rigorous demonstrations of coevality. No matter! The important advance is to accept the reality of continent-wide, geologically brief crustal or endogenic disturbance. If these plausible matchings survive improved dates and correlations, then our problem is sharply focused: why should there be a connection between global exogenic shift and regional endogenic disturbance?

Australian continental crust is on a crustal plate subjected to various forces in space and time (Figure 9.19). Australia's migration northwards is not smooth, like a well-maintained machine. It is a bumpy ride over an uneven substrate and with collisions and deflections along the way. Perhaps a brief sharp shuddering in one place can effect a tilt or slumps and hiatus at another, far distant place.

Two of these alignments are quite fundamental—respectively at the horizons of Oi-1 and Mi-3.

Close to the Eocene–Oligocene boundary, we have the isotopically and geologically identified glaciation and glacioeustasy at Oi-1 together with the Traralgon Unconformity. Cretaceous to Eocene structures are dominantly extensional (pull) but Oligocene and Neogene structures are dominantly compressional (push). This is a regional change. The onset of the new regime is concentrated quite strongly within the Upper *Nothofagidites asperus* pollen zone. 'The' point of descent into the global icehouse is now perceived as also the fulcrum in the structural evolution of the Gippsland Basin.

Glaciation Mi-3 at about 14.5 Ma is the turning point marked by palaeoceanographers as the end of MICO and the beginning of the MMCT. It is the surface of the Miocene limestones of the Eucla Sea and Murravian Gulf, and the end of the Yallourn coal forests. Global and regional, oceanic and neritic and terrestrial, all are coeval. The notion of MICO truncated by MMCT could hardly be more robust. And yet, there is a continent-wide tilting of these strata, a tilting suggested as the cause of the Eucla Sea draining dry. The most prominent and unifying geophysical event in the intensely investigated Gippsland Basin is known as the Blue Reflector, which we matched with Mi-3, and this turning point in the south-east corner of the continent can be matched with a comparable signal on the far side, on the Northwest Shelf.

Modern Australia: A desert with damp fringes

The reverse of instability is stability, and stability is one of the Four Horsemen of the Apocalypse long threatening Australia, namely, stability, desiccation, nutrient and fire.

'Stability' in a continent may sound quaint. There was a time when geologists, wrestling with the eternal question of land rising or sea falling? eustasy or isostasy? looked to Australia, the 'old continent', as a promisingly stable land mass where they might discount crustal mobility and focus on the single variable of sea level. But there never was any such stable place. There is no such thing as 'the' global sea level curve to be discovered in the geology of continental margins. But Australia does look relatively passive. It does not sit across any major plate-tectonic boundary; consequently, it lacks mountains of Rockies or Himalaya scale to uplift the land, to cut deeply down into it, exposing high-temperature minerals to weathering and releasing nutrients, to attract rain clouds triggering big and reliable river systems, to shed sediment and volcanic materials—in short, to provide the eternal renewal sought by an Aristotle or a James Hutton. (To be sure, there is a Cenozoic belt of mountains and volcanoes, of strong uplift and erosion and very thick stacks of sediments, but the belt is wrapped around us outside, being on another continent, Zealandia, and in the countries to our north sharing this continent.)

Next is climate and desiccation. Advances by ice sheets thoroughly scouring the regolith during Pleistocene times would have done wonders in exposing high-temperature minerals to rapid breakdown in soil-boosting weathering, as has happened out in northern regions—but Australia was in the wrong place at the wrong time and mostly missed that renewing agency, too.

And still griping about our climate: Australia is perfectly placed for maximum meteorological impact in a perfect storm. From the north-east we have the enhanced El Niño and La Niña climatic phenomena; from the north-west, the Indian Ocean Dipole; from the south, the Southern Annular Mode. In conjunction, they deliver climatic extremes on a relentlessly warming planet. Droughts and flooding rains, indeed.

Third is nutrient—rather, its lack. As one upshot of Australia's deep history, chemical weathering climaxing in EECO could penetrate hundreds of metres, generating leached and impoverished weathering profiles which just squat there, over vast areas of the continent, frequently sheltering under the famous Australian duricrust known variously after the dominant minerals as ferricrete, silcrete, calcrete or gypcrete.

The fourth horseman is fire. Charcoal was preserved in Australian strata as far back as Cretaceous times. Charcoal is well represented in the Oligo–Miocene coal measures in Gippsland—indeed, the most fire-prone vegetation in the peat-forming environments was not the forests of *Lagarostrobus*, the Araucarian Huon 'pine' and or the southern beech *Nothofagus*, but the reeds and rushes in the swamps and the sedge meadows, at the wet end of the spectrum (Figure 9.20).

And as desiccation of the continent proceeded, the aseasonal wet forests contracted and the various dry-adapted plants diversified and expanded. We should finesse this statement thus: the Australian vegetation in the Late Neogene saw an expansion of arid-adapted, low-nutrient-adapted and fire-adapted biomes. Prominent among the plants are the many species of the Eucalypts, which have their roots deep in the Early Palaeogene on such evidence as the pollen *Myrtaceidites eucalyptoides*; and their well-known adaptations to fire also go way back into deep time. Eucalyptus oil contributes to the blue haze in the 'Australian light' celebrated by the Heidelberg painters, especially Arthur Streeton, and their successors such as Hans Heysen. It was suggested during the popularity of the Gaia hypothesis that the Eucalypts actually encouraged conflagration with their oil, conflagration from which they were pre-eminently adapted to emerge, to survive and to prosper. And their present distribution may be more recent than we have thought, under human intervention with firestick farming.

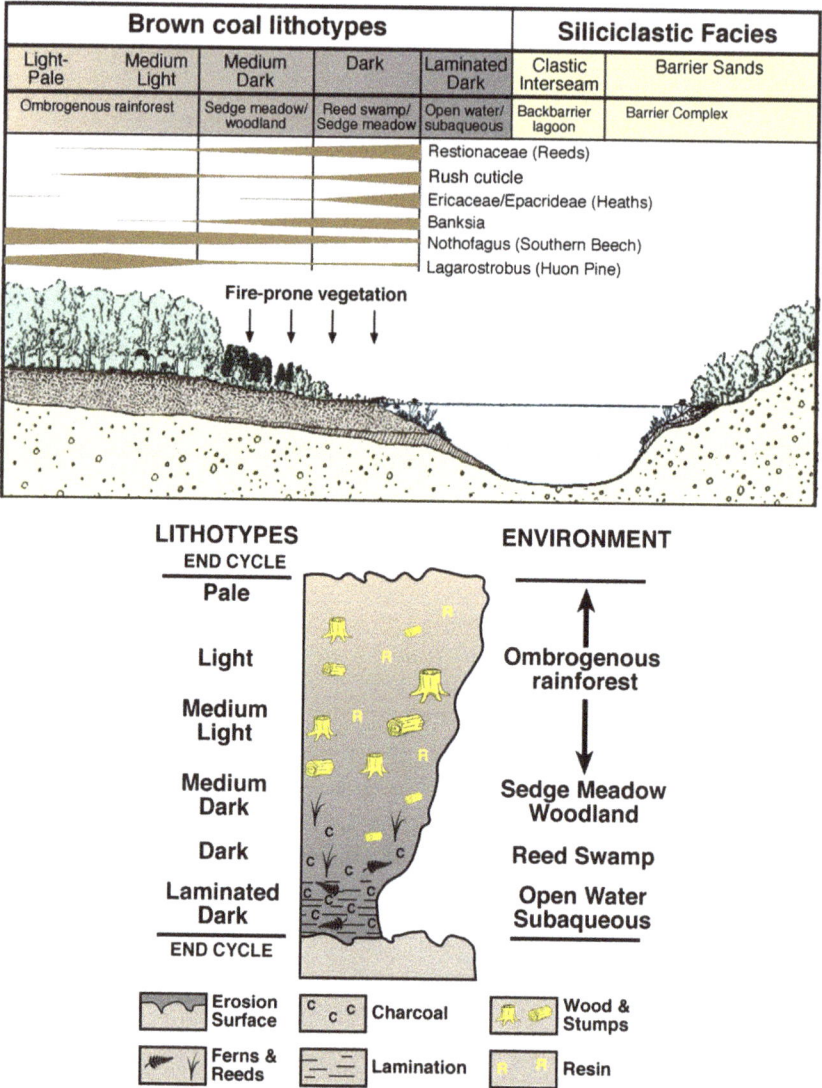

Figure 9.20. Holdgate's fire and rain in Latrobe coals.

The brown coal cycles in the Latrobe Valley show a correlation between coal colour and the plant communities generating the coals. Counterintuitively, perhaps, it is clear from the distribution of charcoals that the most fire-prone community is the wettest. It was within this wetness that the some of the well-known fire adaptedness of Neogene Australia's dry-land vegetation was evolved, for example, in *Banksia*.

Source: From Holdgate et al. (2014).

Descent into the icehouse

We find as we approach modern times that the twin phenomena tectonoeustasy and glacioeustasy accompany us. Geologists realised decades ago that humans or hominins witnessed the growth of the great mountain ranges, the Alps in Europe and the Himalayas in Asia. Imagine a fault, mildly restless, shifting a block of earth's crust by 1 metre once a century on average. One metre (10^0) per 10^2 years becomes 10^3 metres per 10^5 years— a mountain growing by a kilometre in less than a million years. Neotectonics on this scale is scenically familiar to our north but not in southern Australia.

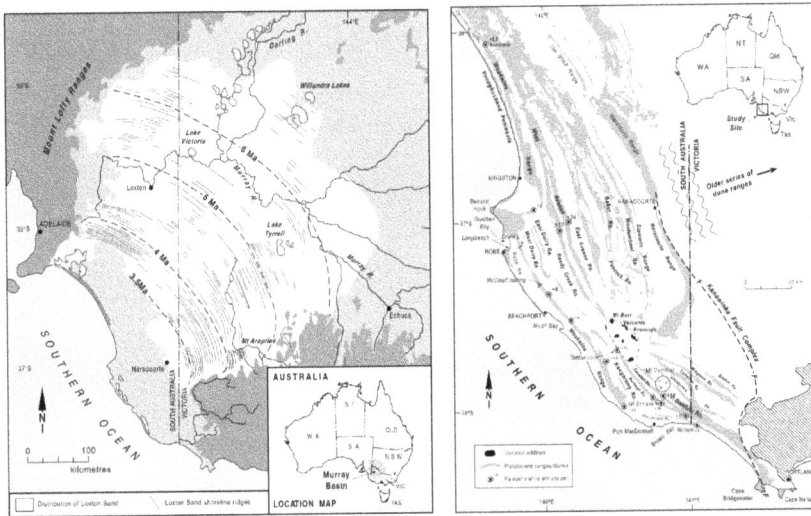

Figure 9.21. Two maps of the Murray Basin and Coorong coastal plain.

Left, distribution of the Loxton Sand (aka Loxton/Parilla Sand) in the Murray Basin as an arcuate series of shoreline ridges tracking retreat of the sea over hundreds of kilometres through more than 3 million years.

Right, distribution of the limestone beach barriers known as 'aeolianites' built across the Coorong Coastal Plain through the past million years.

Source: Both maps from Murray-Wallace (2002).

The Murravian Gulf and the Eucla Sea withdrew comprehensively, several million years after their apogee. In the Murray Basin and coastal plain the record of the Late Neogene is spread out, displaying sea levels and neotectonics in gentle interplay.[9]

More than 600 'ranges' each of about 20 metres relief extend up to 500 kilometres inland from the modern coastline and up to 60 metres above present sea level (local exceptions are higher). In the Murray Basin and older, they are the quartzose Loxton-Parilla Sands. In the coastal plain and younger, the ranges are the carbonates of the Bridgewater Formation.

These arcuate ranges have been known since Julian Tenison Woods who, in the days before the paradigm of global ice ages and glacioeustasy, recognised them as shoreline ridges and postulated their origin in tectonic movements. Thus the coastal plain had been uplifted geologically very recently but sporadically, the ridges accumulating in quiet between-times. In due course multiple ice ages became accepted and geologically recent global changes in sea level became recognised. But sorting the story by correlation and age determination was a problem: too young and too quick for fossils to help; geochemical and geophysical techniques still in the future. However, Alpine glacial geology delivered a succession of glaciations which became a time scale by default: the fourfold Günz, Mindel, Riss and Würm events. The notions took hold of multiple glaciations and changes in global sea level through Pleistocene. Here is Reg Sprigg reminiscing on understanding the aeolianites of southern Australia:

> Practically all contemporary workers of my day, and led by Dr Norman Tindale [in the 1930s and 40s], agreed that these great beaches were products of a massive and continuous, if hesitating, global sea level decline. None seriously considered the alternative of landward uplift, literally pushing the sea out. Each aeolianite beach is composed extensively of coarse, now consolidated, shell sand, but with an increasing proportion of riverine quartz sand northwards towards the previous outlets of the Murray River. The oldest and most spectacular of such dune structures, the Naracoorte Beach now lies stranded inland by 80 kilometres and elevated by seventy metres above modern sea level. The most seaward originally then known

---

9   For the siliceous beach ridges of the Loxton-Parilla Sands and much more about the dying or transforming Murravian Gulf in southern Australia, see especially Bowler et al. (2006) and McLaren et al. (2011). The history of research into limestone beach ridges of the Bridgewater Formation in Pleistocene context is written up in Colin Murray-Wallace's book, *Quaternary history of the Coorong coastal plain, southern Australia* (2018).

example on the other hand, the 'Robe Dune', now forms the present day coast and has its seaward toe submerged to twenty metres below sea level. No simple decline in sea level could fit this scenario.[10]

Sprigg accepted the general view that these beaches were a punctuated succession of high sea-level beaches, to which he added his own arguments that they were preserved by the rise of the (volcanic) Mount Gambier district, not stranded by a straightforward decline in sea level. That there were at least 16 high sea levels related to global fluctuations in Pleistocene glaciation seems to have become generally accepted. But how to date them in a meaningful chronology?

Milutin Milankovitch had reconstructed his series of fluctuations in incoming solar radiation in works culminating in the early 1940s. Milankovitch's central tenet was that the ice-age fluctuations were driven by the amount of solar radiation received at high northern latitudes during the northern summer. Although Milankovitch's reconstructions were well respected at the time of Sprigg's studies, geologists were at a loss as to how they could inform the known geological-palaeoclimatic record in a rigorous chronological framework—beyond a tentative identification of the four Alpine glaciations. Still in the future was access to the long cores from the deep oceans, radiocarbon chronology of the youngest levels and geomagnetic correlations, stable-isotopic chemostratigraphy and palaeotemperatures, and the ecostratigraphy of planktonic foraminiferal assemblages tracking water mass shifts back and forth across latitudes through time.[11] Having access to none of this and working quite outside the paradigm of Pleistocene or Quaternary studies, Sprigg made the intuitive leap of relating the sequence of aeolianites, marking high sea levels, to the reconstructed peaks of summer radiation (Figure 9.22), and his 1952 *Bulletin* 29 of the Geological Survey of South Australia, *The geology of the South-East Province, South Australia, with special reference to Quaternary coast-line migration and modem beach development*, has become a prescient classic.

---

10   The quote is from Sprigg's popular 1989 book *Geology is fun (recollections) or, The anatomy and confessions of a geological addict*. Reg Sprigg (1919–1974) was an exuberantly larger-than-life and highly innovative South Australian geologist (he saw himself as legitimately a zoologist, too). Sprigg the person is portrayed in Weidenbach's *Rock star: The story of Reg Sprigg—an outback legend* (2008), and Sprigg the scientist is in McGowran's *Scientific accomplishments of Reginald Claude Sprigg* (2013).

11   See for example Imbrie and Imbrie (1979) and Berger (2009).

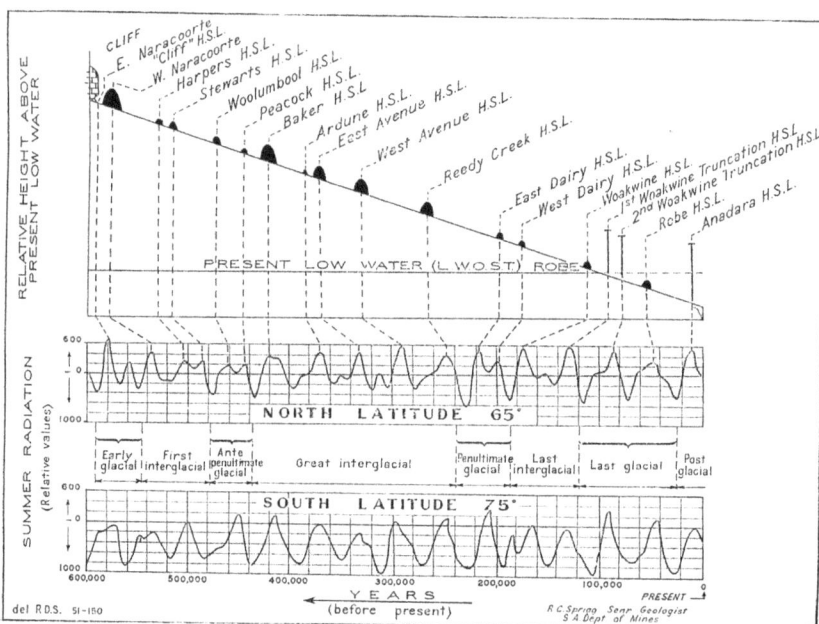

Figure 9.22. Sprigg's Milankovitch theory.

The succession of beach dunes on the Coorong coastal plain which Sprigg correlated with Milankovitch curves of summer radiation at northern and southern latitudes, and with the reconstruction of four glaciations in the Alpine–Mediterranean region (Sprigg, 1952). HSL, high sea level. The HSL named for the clam *Anadara* should be at the peak of the last interglacial, now dated at about 125,000 years before the present. The West Naracoorte HSL now dated at about 800,000 years ago is around the onset of the modern high-amplitude glacial cycles.

Source: Sprigg (1952) (Geological Survey of South Australia, Bulletin 29), also McGowran (2013a).

Sprigg reviewed relative chronology and global correlations, relying mainly on Zeuner (1945, 1946). Much more satisfying to him was his review of 'absolute' chronology (meaning numerical). The summer radiation curves in Figure 9.22 were taken from Zeuner for the last 600,000 years, the date Zeuner assigned to the onset of the first major Pleistocene glaciation. The sloping line represents an assumed overall decline in global sea level to the present. Sprigg saw two options for calibrating his succession. One was to correlate the last high sea level with the latest radiation maximum and then work backwards. The other was to take the first Pleistocene high sea level beach as having formed following the early glaciation, inferred to be at 590,000 years before the present and working forwards. He found that both strategies matched the field evidence to the theoretical inferences remarkably closely—almost 'suspiciously' good. For the second strategy,

Sprigg shifted the two Woakwine truncations and the *Anadara* high sea levels to the right 'from their true altitudinal positions to protect the correct relative age relations' (1952, p. 108)—for he had concluded that the *Anadara* sea floor flooded the interdune corridors after the Robe dune had formed (1952, Fig. 18), still implied in 1979 though shifted back in time. However, subsequent ordination and numerical age determination show that the *Anadara* assemblage is within the Woakwine complex.[12]

Leeuwin Current in southern Australia — one last time

The Leeuwin Current or proto–Leeuwin Current has left us an episodic fossil record in southern Australia, a biogeographic record of pulsating or fluctuating climatic change. We have thought for some time that the current shuts down during ice ages (even though in warmer times it is stronger in winter). Coming to the Late Pleistocene and Holocene, emerging from the most recent ice age (deglaciation), we are discovering that the current is a major player in our ongoing *terrestrial* question. This question is, simplistically: was the extinction of the megafauna due to climatic change or due to human interference?

Humans arrived at ice age Australia before 50,000 years ago, perhaps by 65,000 years ago. They spread along the eastern and western continental margins, were in Tasmania by 39,000 years ago and in arid central Australia by 35,000 years ago. These bracketing numbers also invite comparison with the somewhat elusive, final extinction dates for the Australia megafauna— members of which had survived several glaciation–deglaciation cycles before human advent. Dating the last megafaunal extinction has attained new levels of precision in evidence from palynology. *Sporormiella* is a fungus living and flourishing on the dung of large herbivores. The pattern through time of its spores serves as a proxy for the biomass of herbivores; the sudden decrease in its abundance points to megafaunal collapse in southern Australia just after 45,000 years ago.

---

12 See for example Belperio (1995); Murray-Wallace and Cann (2007).

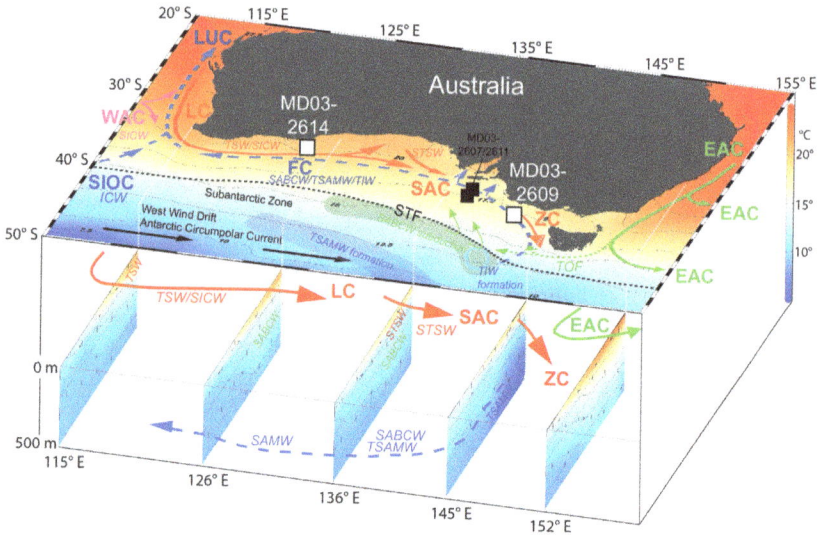

Figure 9.23. Modern oceanographic situation off southern Australia in three dimensions.

The colours show the water temperature in the southern summer (*at right*) in the map and in the five north–south profiles across the top 500 meters of the Southern Ocean. Surface currents are in red and green: LC, Leeuwin Current; WAC, West Australian Current; SIOC, South Indian Ocean Current; SAC, South Australian Current; ZC, Zeehan Current; EAC, East Australian Current. Subsurface currents are in blue: FC, Flinders Current; LUC, Leeuwin Undercurrent. STF, Subtropical Front (also on Figure 9.18). Other initials refer to the water masses being moved around by the currents. The two white squares locate cores of sediment together, capturing the behaviour of the Leeuwin Current along southern Australia during deglaciation. The two black squares south of Kangaroo Island also contribute (one of them, MD03-2611, is displayed in Figure 9.4).

The profiles demonstrate how thin-skinned is the ocean's warm surface, and how easily slight shifts in global temperature will affect the sea and its bordering lands, in this case via the Leeuwin and East Australian currents.

Source: Nürnberg et al. (2022).

The answer to the rhetorical question seems to be: both. The vegetation on southern Australia, fluctuations in its response to temperature and rainfall, its susceptibility to fire, and the dynamics of its animal populations—the variability in all these we now see as intertwined with the variability in strength and reach of the Leeuwin Current.[13]

---

13   Nürnberg, Kayode, Meier and Karas (2022) give a superb analysis of our region over the last 60,000 years, physical, biological and biogeohistorical. The significance of the dung fungus *Sporormiella* is discussed by van der Kaars et al. (2017).

10

Contingency, consilience and historicity are the guts of biogeohistory

The sweep of biogeohistory in portraits: The labours of the conchologists

Our account of fossils and strata has traversed a quarter-millennium of names, dates and concepts. I have summarised and recapitulated the progress of knowledge in several charts, because it has helped me to visualise things on a time axis. Most of this parade of dead white males is mentioned in the earlier chapters. The terse sentences (memes?) track the progress of knowledge up the chart. Portraits and sentences are placed deliberately if approximately and time runs up the page.

The first and most directly biostratigraphic of the charts (Figure 10.1) is about the discovery of biogeohistory in the late eighteenth and early nineteenth centuries. Rudwick (2005) called the discovery the Cuviero-Lyellian Revolution and front and centre are the two central pairs, Cuvier and Brongniart then Deshayes and Lyell. Listed in the earlier days of Arduino, Lamarck and Lavoisier are what, for me, are the four foundation statements of biogeohistory. Arduino demonstrated that a great, orderly stack of strata in the Alps not only were incompatible with a vastly inadequate biblical time scale but also were orderly in time. Lavoisier glimpsed ancient environmental change in the rocks. It was Lamarck's 1802–1809 monographing of the molluscan assemblages in the limestone of the Parisian region that became the touchstone for biostratigraphic progress. Lamarck discovered evolution, too, becoming the only evolutionist in this display.

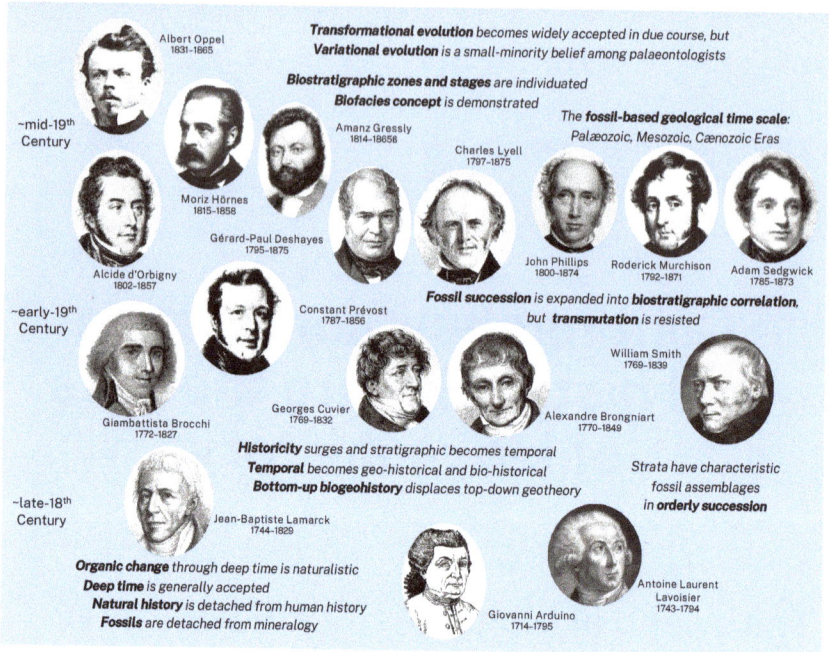

Albert Oppel
1831-1865

Transformational evolution becomes widely accepted in due course, but
Variational evolution is a small-minority belief among palaeontologists

Biostratigraphic zones and stages are individuated
Biofacies concept is demonstrated

~mid-19th
Century

Amanz Gressly
1814-18656

The *fossil-based geological time scale:*
Palæozoic, Mesozoic, Cænozoic Eras

Moriz Hörnes
1815-1858

Charles Lyell
1797-1875

Gérard-Paul Deshayes
1795-1875

Alcide d'Orbigny
1802-1857

John Phillips
1800-1874

Roderick Murchison
1792-1871

Adam Sedgwick
1785-1873

~early-19th
Century

Constant Prévost
1787-1856

Fossil succession is expanded into *biostratigraphic correlation,*
but *transmutation* is resisted

William Smith
1769-1839

Giambattista Brocchi
1772-1827

Georges Cuvier
1769-1832

Alexandre Brongniart
1770-1849

Historicity surges and stratigraphic becomes temporal
Temporal becomes geo-historical and bio-historical
Bottom-up biogeohistory displaces top-down geotheory

Strata have characteristic
fossil assemblages
in **orderly succession**

~late-18th
Century

Jean-Baptiste Lamarck
1744-1829

Organic change through deep time is naturalistic
Deep time is generally accepted
Natural history is detached from human history
Fossils are detached from mineralogy

**Antoine Laurent
Lavoisier**
1743-1794

Giovanni Arduino
1714-1795

Figure 10.1. Panorama of the founders of biostratigraphy.

This panorama is about the foundations of biostratigraphy, which is absorbing fossils in their strata into biogeohistory. In the six decades from Cuvier's establishing the fact of organic extinction to Darwin's establishing the fact of organic evolution, the biostratigraphers of Europe built the fossil-based geological time scale with no consensual notion of why species come and species go. But build it they did, bringing off Rudwick's Cuviero-Lyellian Revolution. Which shows that science, the progression of reliable knowledge, can flourish without the security of an overarching theory. Terse statements identifying the accruing memes of biogeohistorical knowledge are placed approximately, time running up the page.

Source: Author's representation, based on images via Wikimedia Commons (commons. wikimedia.org).

To the right, we have the English narrative, the best-known part of this panorama in English textbooks. Smith walked and walked during the canal-constructing, coal-mining and other earthworks as the landscape changed during the Industrial Revolution, and he walked into immortality by discovering that fossils characterised strata in orderly succession. Sedgwick and Murchison are best remembered for their roles in the great Cambrian–Silurian and Devonian controversies in the development of the geological timescale. Murchison knew his fossils and won more than he lost in those controversies; Sedgwick, who did not really know his fossils, lost more than he won. Phillips, nephew of Smith, was hugely knowledgeable on fossils and could write the far-sighted *Life on the earth* (1860) while remaining staunchly anti-Darwin.

But our story is more to the left and centre and focused on the 'Tertiary' (meaning almost all the Cenozoic Era), which is a gateway from the shallow-time of the modern world and those living in it back into the deep time and those long extinct. The three historical pronouncements listed below Cuvier are about him, confirming him as prime mover. Cuvier hauled the Eurasian and North American bones and teeth, collected over a century, into a coherent pattern of animals in what we now call the megafauna, confirming the fact of their extinction. Where Smith accomplished the orderly succession of fossils, Cuvier and Brongniart turned fossils and strata into biogeohistory. And Cuvier hauled the diversity of animals into four *embranchements*, or streams, or 'phyla'. Cuvier's biogeohistory changed the culture for the good, a point overlooked too easily while we are noting that his catastrophism was overthrown by Lyell's uniformitarianism, his rejection of organic evolution was overthrown by Darwin's *Origin*, and he was only half right in debating 'conditions of existence' against Geoffroy's 'unity of type', Geoffroy's themes having the upper hand for the next century (see below).

Left and centre are some of the conchologists of Europe—the monographers dealing in thousands of specimens and hundreds of species from fossiliferous strata. Lamarck demonstrated modern provincialism in the collections brought back from the South Seas. His successors had to explain differences among European faunas, especially as focused by Brocchi in Italy (or focused for Brocchi by Cuvier). Were they of different geography, or different environments, or different ages? Prévost progressed the facies concept in Brongniart's Parisian territory before Gressly did so in Switzerland. The key accomplishment, the synthesis, was Deshayes's, in confirming that there were three successional faunas in the Tertiaries of Europe. Lyell, well travelled and already very interested in the question of fossils and time, seized upon Deshayes's research (and supported it financially). Thus emerged from the shells the Eocene, Miocene and Pliocene epochs and correlation was added to fossil succession. Lyell believed that life on this earth was strongly cyclical, and that the three faunas were merely accidental slices of a temporal continuum. Deshayes knew better—the faunas were real, sure enough, but the idea of chronofaunas, large ecological units beyond the reach of shallow-time ecologists, had to wait a century.

Contemplate the trio in the upper left. D'Orbigny possibly was the intellectually most capacious palaeontologist of his time or of any time. He had a blazing vision of life as a long succession of stages punctuated by catastrophes. Several of his stages survive in the time scale and his actual palaeontology is not at all violent. For me, Moriz Hörnes in Vienna is special. Hörnes discovered that the marine faunas of the Cenozoic fall naturally into two super-faunas and that the natural break is between Lamarck's Eocene, below, and the Miocene–Recent (the living), above. So the younger group are the natural Neogene, and we are still living in a Neogene world. Oppel worked mostly in the Mesozoic (clearer and simpler than the Cenozoic), dividing and classifying fossil-bearing strata into zones. Zones, stages, chronofaunas—we have needed these divisions since biogeohistory began, but we can skip the wordy philosophical literature here.

The geological time scale essential for biogeohistory was built in the six decades between Cuvier's demonstrating the fact of organic extinction and Darwin's demonstrating the fact of organic evolution. It is noteworthy that the palaeontologists bringing off the Cuviero-Lyellian Revolution were neither biblical (young-earth) creationists nor organic evolutionists (Lamarck and an elderly and reluctant Lyell excepted). But seven were alive when Darwin's *Origin* was published.

The heroic age of geology?

Here are four intellectual streams in the late eighteenth and early nineteenth centuries (Figure 10.2).[1]

By the turn of the eighteenth to nineteenth century there were three cultural attitudes known as natural philosophy, natural history and natural theology. Natural philosophy included physics and astronomy, chemistry, crystallography and mineralogy, and the more intellectually respectable components of laboratory biology such as anatomy, physiology and embryology. It meant predictions and replicated experiments, it meant crunching numbers, it meant data leading to inductive conclusions; in modern terms it meant 'hard science' or 'real science'. Natural history included ecology and biosystematics, geography and biogeography. To your

---

1   The 'Heroic age' is from Karl Alfred von Zittel's great *History of geology and palæontology* (1901). It was Zittel's Third Period (1790–1820) after geological knowledge in the ages of antiquity and the beginnings of palaeontology and geology.

hard-nosed chemist it is descriptive, it is 'soft science'; and this mindset explains why Darwinian natural selection was and frequently still is not comprehended. And it helps explain why the possibility of a natural history in deep time dawned very late, two centuries after the rise of science. There is quite a literature on the origins and the rise of science. Did science as we know it come from the Greeks? Did Christianity help or hinder? Why did Chinese science and Islamic mathematics stall? The writings on these questions, from the scholarly to the pious, the populist and the polemic, are strewn with the social, religious and political special pleading of non-scientists. But there is an ingredient prominently missing from all of them, and that ingredient is historicity.

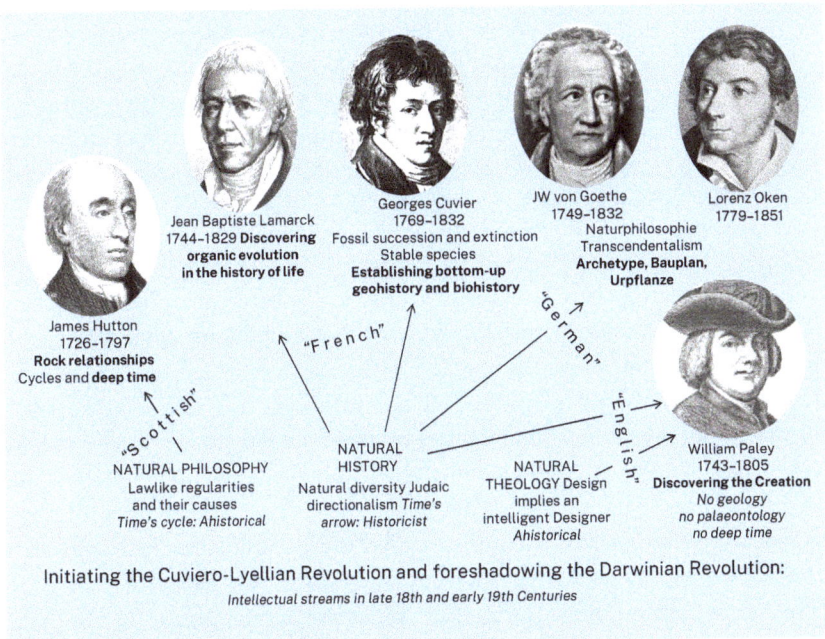

Initiating the Cuviero-Lyellian Revolution and foreshadowing the Darwinian Revolution:
Intellectual streams in late 18th and early 19th Centuries

Figure 10.2. Initiating the Cuviero-Lyellian Revolution in the late eighteenth and early nineteenth centuries.

The three intellectual categories of the perceived natural world were natural philosophy, natural history and natural theology. They gave rise to the various traditions or intellectual streams with some degree of regional or national foundation. But the vision of a recoverable history of the natural world in deep time, the historicist vision, was prominently French.

Source: Author's depiction, based on images via Wikimedia Commons (commons. wikimedia.org).

From Plato and Aristotle onwards there are recurring tussles, such as dreamy mysticism versus robust observation of nature, religion versus science, or idealism versus empiricism—contrasting world views struggling down the centuries for the soul of Western civilisation. Or so goes a hallowed narrative[2]—but where's the history? For history, another binary is available, Greek eternalism and their oracles predicting versus Jewish directionalism and their prophets warning. But it was all of two centuries after the rise of modern science before history of all stripes, human history, prehistory, biohistory and geohistory, got underway. Science, from the physics and astronomy of Kepler, Galileo and Newton to medically oriented anatomy and physiology, required no notion of deep time and no sense of earthly change. The vertebrate palaeontologist Henry Fairfield Osborn in the 1890s traced the 'evolution idea' back to the Greeks, meaning the pre-Socratic philosophers. He took pains to explain that by the evolution idea he really meant the cluster of ideas that came together down the centuries, to the time when evolution in any modern sense of the word was discovered by Lamarck, and the powerful and successful theory was forged at last by Darwin. So long as we don't forget those points, and acknowledging that there has been a century's science and scholarship since Osborn, his exuberantly written lectures are worth a look.[3]

The eighteenth century was the time of the Enlightenment Project, this project being the description and classification of the lush and diverse organic world in what we now call the terrestrial, neritic and pelagic realms. Linnaeus's objective was to know the mind of the Creator by way of the so-called Natural System, and he and all the others were thinking and working in biblical or shallow time. They were three-dimensional neontologists. It was late in the century when the Cuviero-Lyellian Revolution began, with the possibility of a history in the rocks recording the passage of deep time. They were palaeontologists, the four-dimensional biologists.

Our four streams of biostratigraphy, biogeography, transformational evolution and variational evolution arose out of natural history. Hutton clarified unconformity, igneous, metamorphic and sedimentary rocks,

---

2   An example in the popular genre is *The cave and the light: Plato versus Aristotle, and the struggle for the soul of Western Civilization* by Arthur Herman (2014).

3   Osborn (1894, 1902) was candid enough in using 'the evolution idea' to mean change through geological time. More confusing is 'evolution' in titles such as this: *Darwin's ghosts: In search of the first evolutionists*, by Rebecca Stott (2012). There is a literature on Darwin's 'forerunners', another of the alleged victims of Darwin's intellectual thieving, and a third on evolution and biblical creation. I find most of this profoundly unrewarding: Lamarck, Darwin and Wallace were the first organic evolutionists and of course there is a discernible narrative behind them, all the way back to the Jews and the Greeks.

and uplift in a conceptual rock cycle, a major achievement of the Scottish Enlightenment. Rock relationships and deep time are utterly fundamental. But Hutton's geological science in his ahistorical, earth-as-machine mindset exemplified time's cycle in natural philosophy, not time's arrow in natural history. Paley, on the far side of the display, was the climactic figure in the quintessentially English endeavour of natural theology (a forerunner of the modern intelligent design). Paley was like Hutton in being ahistorical (if in few other respects). In fact those times are remarkable for their lack of historical awareness. Hence this forceful quote from the highly influential 1950s book *Genesis and geology*:

> There is no historicism in [Paley's] Moral and Political Philosophy and no geology in his Natural History; and the two books are good illustrations that a sense of history was as uncharacteristic of utilitarian political philosophy as a sense of evolution was of eighteenth-century natural philosophy.[4]

Lamarck was discovering organic evolution when his taxonomic duties and responsibilities in the Museum in Paris were shifted from terrestrial plants to marine and fossil shells. His idea of life continuously changing ('transmuting') implied that species were unstable. But in his magnificent systematic studies, in his day-by-day work of description, classification and identification, he treated species as real, stable, discoverable entities, thereby laying the foundations for both biogeography and biostratigraphy. Cuvier discovered extinction and the vision of biogeohistory in fossil succession, all without evolution. Cuvier and Lamarck could hardly have been more at odds—yet clearly belong together in bringing off the splendid French foundation to the Cuviero-Lyellian Revolution. Which brings us to the Germans in this four-part pattern. The word 'German' here evokes such notions as transcendental anatomy, or the movement known as *Naturphilosophie*, a vision and a program to comprehend nature in her totality. Three words help: '*Urpflanze*', '*Bauplan*' and 'archetype'. Contemplating Linnaeus's botanical system, the philosopher and creative one-man cultural force JW Goethe developed the notion of a basic or primordial flowering plant, the *Urpflanze*. The comparably versatile Car Gustav Carus sketched a vertebrate animal skeleton, a Bauplan or primordial type, forerunner of Owen's archetype (about which more below).

Our story so far has neglected neptunism, the theory that minerals and rocks precipitated from an initially universal ocean which has been receding during geological time. Neptunism, usually associated with the name of

---

4   *Genesis and geology* (1959, p. 39).

Werner, was a historical theory overthrown by the ahistorical theory of our surge #1. But not in Edinburgh, Hutton's town. Wernerian historical thinking supplied the seedbed for Lamarck's historical thinking to survive and flourish in Edinburgh.[5]

Change was everywhere. Palaeontologists were piling up massive evidence for coherent patterns of change during deep time. We can see in the early decades of the nineteenth century a useful clarification emerging in discussions of the change. Was it the Bauplan concept of an internal structural control explaining succession through time, namely fish then four-legged tetrapods, reptiles then mammals, then humans? Or was adaptation to environmental change the external control? Do we have five fingers or toes per limb so as to do all the things that we use them for, or simply because we inherited five digits from our ancestors? In a celebrated debate in Paris in 1832 the opposing slogans were unity of type (in which form trumps function; more structuralist), and conditions of existence (in which function trumps form; more ecological, more adaptationist). Lamarck's colleague Geoffroy advocated for the former side, Cuvier for the latter. The ageing Goethe was strongly on Geoffroy's side, England's natural theologians were with Cuvier.

Evolution down the decades: Darwinian Revolution, anti-Darwin decades, Darwinian Restoration

By mid-century (Figure 10.3) that either/or debate had clarified somewhat (something of both structure and function) and the parameters had also shifted, most apparently in the mind of Richard Owen. Owen was a disciple of Cuvier, and we have seen that Cuvier's death opened for Owen the great opportunity of the newly discovered Australian megafauna. And Owen 'owed', in political and other ways, the English natural theologians, especially the influential and well-connected Buckland and Murchison. But the transcendental anatomy of the Germans, the *Urpflanze* and the

---

5   There were several 'Edinburgh Lamarckians' or transmutationists and the key figures were the geologist Robert Jameson (1774–1854) and the biologist Robert Grant (1793–1874). This was the thin red line from Lamarck's discovering organic evolution to Darwin's convincing the populace of its truth. Werner, Lamarck and Jameson were time's-arrow people, which explains why Lyell, a time's-cycle person promoting Hutton and his eternal cycles, spent some effort dismissing Lamarck. It was never just evolution vs creation; it was yet another clashing of world views. See Secord (1991); Jenkins (2016) and Jenkins (2019), *Evolution before Darwin: Theories of the transmutation of species in Edinburgh, 1804–1834.*

Bauplan, were taking hold in Owen via his contacts among the medical anatomists in Edinburgh. His intellectual migration from French–English adaptationism to German structuralism produced the famous archetype. In one perspective of Owen's intellectual journey, he was in the closet as a transformational evolutionist for virtually all of the two decades before Darwin published the *Origin*. In another perspective of this complex character, Owen saw the evolution of the horse and other modern animals as planned to benefit humans, newly arrived in the Creation. This vision identifies him as one of the last natural theologians.[6]

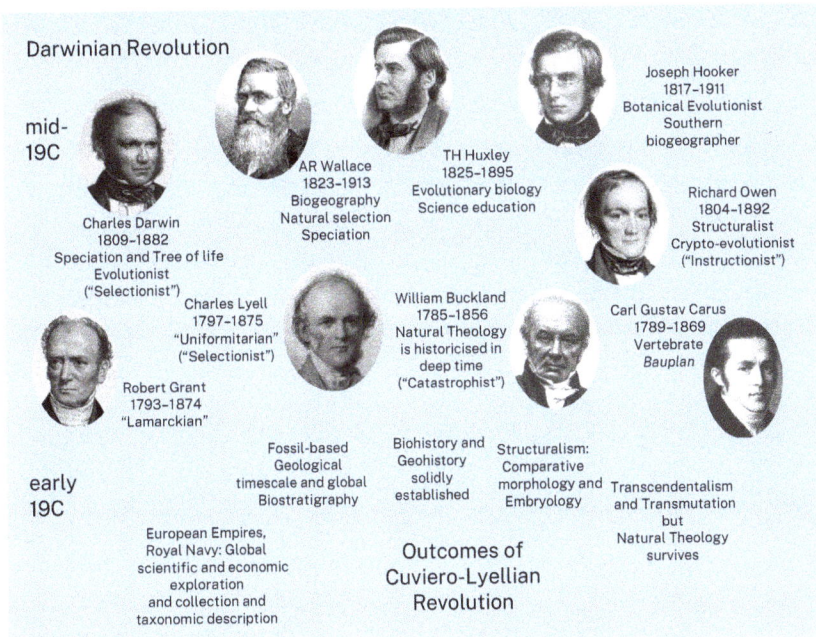

Darwinian Revolution

mid-19C

Joseph Hooker
1817–1911
Botanical Evolutionist
Southern
biogeographer

AR Wallace
1823–1913
Biogeography
Natural selection
Speciation

TH Huxley
1825–1895
Evolutionary biology
Science education

Richard Owen
1804–1892
Structuralist
Crypto-evolutionist
("Instructionist")

Charles Darwin
1809–1882
Speciation and Tree of life
Evolutionist
("Selectionist")

Charles Lyell
1797–1875
"Uniformitarian"
("Selectionist")

William Buckland
1785–1856
Natural Theology
is historicised in
deep time
("Catastrophist")

Carl Gustav Carus
1789–1869
Vertebrate
Bauplan

Robert Grant
1793–1874
"Lamarckian"

early
19C

Fossil-based
Geological
timescale and global
Biostratigraphy

Biohistory and
Geohistory
solidly
established

Structuralism:
Comparative
morphology and
Embryology

Transcendentalism
and Transmutation
but
Natural Theology
survives

European Empires,
Royal Navy: Global
scientific and economic
exploration
and collection and
taxonomic description

Outcomes of
Cuviero-Lyellian
Revolution

Figure 10.3. Legacy of the Cuviero-Lyellian Revolution to Darwin's generation.

It was a legacy, not merely the old about to be swept away by the new. Historicity expanded early in the nineteenth century, and also discernible were adaptation by divine plan, structuralism, functionalism and *Naturphilosophie*. After the Napoleonic wars and expanded global exploration, the biogeohistorical disciplines had shifted in the direction of biodiversity and systematics and biogeography. They sprouted the Darwin–Wallace selectionism, Owen's success in structuralism notwithstanding. For catastrophist/uniformitarian and instructionist/selectionist, see Table 10.1.

Source: Author's depiction, based on images via Wikimedia Commons (commons. wikimedia.org).

---

6   For Owen, see especially Nicolaas Rupke's *Richard Owen: Biology without Darwin* (2009).

Meanwhile Darwin and Wallace, via biogeography and ecology, founded the theory of evolution—adaptation and natural selection generated the tree of life, the metaphor for the pattern of speciation and extinction through deep time. Everything produced by the natural theologians as evidence of the glory of the Creator was available to Darwin as evidence for organic evolution. The anatomy of the structuralists' vertebrate animal, culminating in Owen's archetype, became prime evidence for the tree of life. The deep-time palaeontology synthesis was the most compelling in its need of a scientific explanation and the most frustratingly elusive in all its missing bits. Darwin could trawl the entire legacy of the Cuviero-Lyellian Revolution and make use of everything. The outcome was a fusion, an integration, a synthesis of the five theories of Mayr, or the two theories of tree of life and natural selection. This was the Darwinian Revolution (Part I).

Darwin's biosphere resided on stable continents and oceans. He knew and recruited the facts that his diverse organisms had lots of variation, as revealed in the domesticating of plants and animals. He knew that variation was inherited but inherited discretely, not like, say, mixing two inks—but the black box of transmission genetics was in the future. He knew about and recruited from the expanding fields of morphology and embryology—but the black box of developmental genetics was still further in the future. There has been much scrutiny of what Darwin did not know and of what he got wrong. This comment in 1973 by the philosopher of evolutionary biology, David Hull, is still relevant:

> The truly amazing feature of Darwin's intellect was the frequency with which he was able to 'guess' correctly, even though he lacked the requisite data and anything like an adequate theory governing the phenomena. Modern evolutionary biology is closer to the original Darwinian formulation today than it has ever been.[7]

The two cultures of evolution, then the tripod supplants the bell curve

But the theory of organic evolution did not stay fused, and this was not due to glib debating of binaries such as evolution versus special creation, or contingency versus intelligent design. The integrated theory split when

---

7   Hull (1973, p. 77).

the structuralism that became transformational evolution separated from the ecology and taxonomy that engendered variational evolution. With varying levels of promptness and degrees of enthusiasm, most taxonomists, ecologists and biogeographers on the one hand and most anatomists, physiologists and embryologists on the other accepted that evolution was here to stay. But not natural selection! There was the thin red line of the true believers, and there were very few palaeontologists among them in the last decades of the nineteenth and the first decades of the twentieth century, which have been called the anti-Darwin decades. That is, the internal drive of transformational evolution dominated over the external drive of environmental change, adaptation and selection. However, vertebrate palaeontology, comparative anatomy and embryology all flourished during the anti-Darwin decades (Figure 10.4).

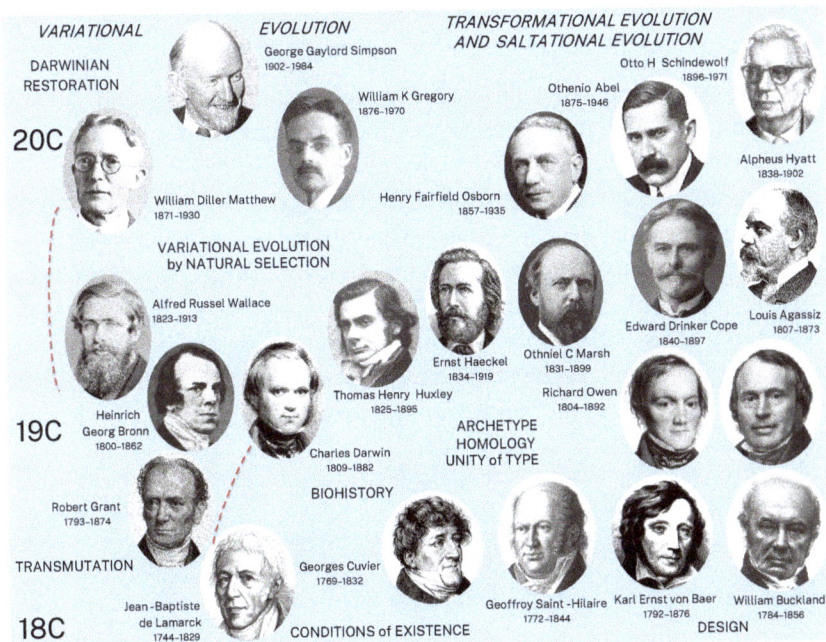

Figure 10.4. Panorama of palaeontologists in two evolutionary streams.

A panorama of palaeontologists (with Grant, Wallace and the embryologist von Baer as honoured inclusions) arranged to portray the lopsided streams of evolution. Matthew and Simpson appear as rather lonely on the variational side. On the transformational side, there were two anti-Darwinian nodes. The North American node centred on Cope, Hyatt and Osborn. In the German node, the school of typostrophism culminated in the triumph of the anti-Darwinian, anti-uniformitarian, orthogenetic and saltationist theories that dominated German palaeontology for part of the twentieth century (Schindewolf and Abel). Shown here are the streams reaching the Simpson–Schindewolf generation, but actually they continue through the rise of evo-devo and the expansion of the restoration, including what sociologists have called the palaeobiological revolution. The first thin red line acknowledges Lamarck's taxonomy-based vision of evolution. The lopsidedness was more than we see here, because omitted are the English Sedgwick, Murchison and Phillips and all the European conchologists except Lamarck. The second thin red line acknowledges the lonely theory of natural selection.

Source: Author's depiction, based on images via Wikimedia Commons (commons. wikimedia.org).

When population thinking in taxonomy, palaeontology and transmissional genetics (and more specifically population genetics) reinvigorated natural selection and produced the Darwinian Restoration in the 1930s–1940s, structuralism or transformational evolution felt marginalised, rich though it became in evidence for constructing the tree of life. It was simply the absence of a coherent discipline of developmental genetics, made all the more stark by the rise of population genetics. With the innovation of developmental genetics, evolutionary developmental biology, evo-devo, expanded.

And evo-devo surely did expand. The black box of ontogeny was closed for long enough after the black box of inheritance was opened. Here is an example of the pressure that our awareness of deep time and deep biogeohistory have been applying. Recall (Figure 3.8) that in 1555 Pierre Belon graphically displayed the detailed similarity of bird and human skeletons—splendid evidence of intelligent design. During all the millennia of plant and animal domestication, to be sure, ontogeny was as much a mystery as was phylogeny. In 2011 John Long brilliantly rescaled the bones of a Devonian fish to human dimensions to create *Gogonasus* Man, thus demonstrating relative differences overlying fundamental detailed similarities across 375 million years—differences generated by the tweaking capacities of ever-alert natural selection.

Our fifth chart (Figure 10.5) sums up this thesis of a 'natural history' and 'natural philosophy' proceeding through 250 years.

Table 10.1 is offered to cope with the jargon.[8]

---

8   The table can be seen as an extension of the Glossary. It assembles concepts and jargon from the biological side of biogeohistory, some of them quite esoteric. Worth more than a footnote are three richly challenging books by Peter Munz (1921–2006), *Our knowledge of the growth of knowledge* (1985), *Philosophical Darwinism* (1993) and *Beyond Wittgenstein's poker* (2004). Munz's prose contains comments (2004, p. 2) like this: 'It was Charles Darwin who finally put teeth into both Plato and Kant … He replaced, to put it bluntly, Plato's speculative metaphysics with hard-nosed biology.'

Expanded Paradigm: **Population Thinking, Tree Thinking and Homology Thinking:**
Species & Homologs are Metaphysical Individuals united by Individuality and Common Descent

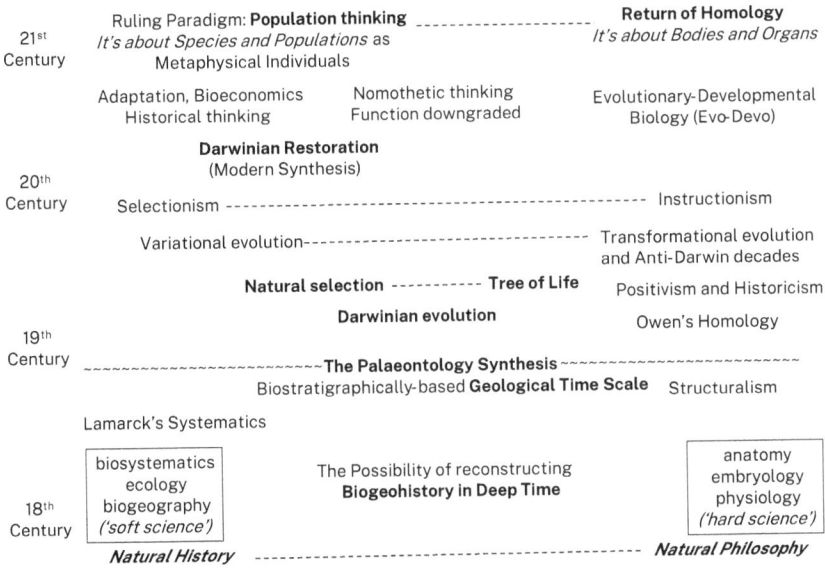

21st Century	Ruling Paradigm: **Population thinking** - *It's about Species and Populations* as Metaphysical Individuals	**Return of Homology** *It's about Bodies and Organs*

Adaptation, Bioeconomics Nomothetic thinking   Evolutionary-Developmental
  Historical thinking   Function downgraded   Biology (Evo-Devo)

20th Century

Darwinian Restoration
(Modern Synthesis)

Selectionism - Instructionism

Variational evolution - Transformational evolution
and Anti-Darwin decades

Natural selection - - - - - - - - - - **Tree of Life**   Positivism and Historicism

Darwinian evolution   Owen's Homology

19th Century - **The Palaeontology Synthesis** ~~~~~~~~~~~~~~~~~~~~~~~~~
Biostratigraphically-based **Geological Time Scale** Structuralism

Lamarck's Systematics

18th Century	biosystematics ecology biogeography *('soft science')*	The Possibility of reconstructing **Biogeohistory in Deep Time**	anatomy embryology physiology *('hard science')*

Natural History - *Natural Philosophy*

Figure 10.5. Historicity and evolutionism through a quarter-millennium.

The sweep of historicity and evolutionism through a quarter-millennium. The cleavage between natural history and natural philosophy is rooted in the Enlightenment and has resurfaced from time to time down the decades. Concerning organic evolution, the way forward at any given period has tended to be found more to the left of this diagram than to the right, from Lamarck's systematics through Darwin's selectionism to Ghiselin's and Vermeij's strongly historical thinking. But it is too easy to be simplistic in a two-dimensional chart. For example, Cuvier and Darwin were comparably strong in historicity and functionalism (or conditions of existence), and yet their stances on evolution were diametrically opposed.

Source: Author's depiction.

Table 10.1. Natural history and natural philosophy: Some binaries and other terms.

Down from the ancients		
Deep time	Biblical time	The balance of informed opinion on earth's great age shifted decisively during the eighteenth century.
Natural history	Natural philosophy	Eighteenth-century categories, spawning the separation of geohistory and biohistory from ahistorical science. The modern informal categories 'soft science' and 'hard science' are not dissimilar.
Aristotelian	Platonist	Versatile and elusive terms. They can mean, respectively, paying more attention to the tangible world and its life and meaning, or teleology, or paying more attention to ideal forms and mathematics.
Realist	Idealist	Likewise and naively, it can mean 'reality out there' versus ephemeral, impermanent shadows of reality.
Time's arrow	Time's cycle	Hebrew beginning to end vis-à-vis Greek eternalism. Hebrew prophetic vis-à-vis Greek oracular. The 'time's cycle' vision for the earth was developed by Hutton then Lyell (who soon had to abandon it). Among the numerous examples of cyclical behaviour perceived in biogeohistory, strict periodicity has been disproved for all cycles except astronomically forced climate (Vermeij, 2011). The spectacular rise of cyclostratigraphy notwithstanding, 'time's arrow' is the stronger drive, as in macroevolution and evolvability (Erwin, 2009).
The rise of historicity		
Historical; big history	Ahistorical	Human history, prehistory, biohistory, geohistory, solar system history, cosmic history: together they comprise big history, essential to high-quality education and in fuelling dreams of a Grand Unified History (Christian, 2011, 2018).
Historicity (good)	Historicism (bad)	The good: Awareness of history and its importance was a later eighteenth-century shift (Rudwick, 2014), in natural history seen most prominently in Cuvier and Darwin. Historicity spans the arts/science cultural gap.
		The bad: History is controlled by a developmental law or an iron succession (Popper, 1957; Munz, 2004).
		Transformational evolution has long been distorted by historicism, as in sloppy analogies with embryological growth. (Mea culpa: all approving references to historicism in my (2013b) should be to historicity.)

The rise of historicity		
Internalist history	Externalist history	Emphasis on the evidence-based science itself versus 'external' emphasis on the social and philosophical context. Examples: social and religious pressures or revolutions, Emma Darwin's faith, Platonist bishops, Marxist historicism, swathes of postmodernist sociology. The present essay is strongly internalist.
Configurational, contingent	Immanent	The twofold cosmos (Simpson, 1960): inherent characteristics of physical and experimental science with its laws, compared to historically derived structure and organisation and contingent events. Before Darwin, life on earth was seen as an orderly progression of immanent stages, like the stages in an organism's life span. Historical contingency is central to Darwinism. It is history, all the way down. It is narrative, not law.
Function	Form	The 1830 debate in Paris was function (Cuvier) versus form (Geoffroy) as the chief organising principle of life. Cuvier was deemed to have won that encounter; but Owen to have won the longer game for form (structuralism) in the 1840s.
Uniformitarian	Catastrophist	Whewell's dualist classification of influential geologists in the 1830s; the former with a membership of just one (Lyell). The dualism was not considered very important at the time, except as a debating tool, or for dismissing biblical creationists.
	Positivism	In its extreme form, positivism states that everything that is known is known scientifically; science displaces magic, religion and philosophy, all else becomes superficial or superfluous.

Darwin and after		
Variational evolution	Transformational evolution	Variational evolution: populations change, generation by generation, by differential selection of the available variation. Variational evolution occurs in and only in organic evolution including the origin of species. All other meanings of 'evolution', organic and inorganic, from cosmology to geology, from sociology to tabloid journalism, are transformational.
Evolution: ontogeny	Evolution: phylogeny	Pre-Darwin, these were often indistinct; in some languages a word for 'unfolding' was used for both. Post-Darwin, they tended to become entangled in such notions as 'ontogeny recapitulates phylogeny'.

Darwin and after		
Evolution: Process: Natural selection	Evolution: Pattern: Outcome: Tree of life	The Darwinian Restoration focused especially on process, the origin of species. Evo-devo inspects the pattern of the outcome. As a cogent advocate for evo-devo said, 'They're interested in species, we're interested in bodies' (Amundson, 2007).
Populationist	Typologist (~Essentialist)	This is Mayr's formulation in 1959 (Mayr later accepted Karl Popper's essentialist as much the same thing): for the typologist, the type (eidos) is real, variation is an illusion; for the populationist, the type is an abstraction, only the variation is real.
Gradual	Punctuated	These are two recurring views of biogeohistory. Lyell the lawyer and Darwin the gentleman naturalist successfully conflated punctuated change with biblical Creation. Saltations, macromutations and catastrophic mass extinctions all exemplified reactions to perceived extreme gradualism and ultra-selectionism in the Darwinian Restoration.
Individuals: e.g. the species taxon	Classes: e.g. the species category	Ghiselin (1997, pp. 304–305) illustrated the two meanings of 'species', the concrete historical entity and the abstract, thus: 'Biological species are populations within which there is, but between which there is not, sufficient cohesive capacity to preclude indefinite divergence.' 'The biological species is, roughly speaking, the most incorporative populational taxonomic category.'
Adaptationists	Structuralists	In a modern dialectic, neo-Darwinists are adaptationists, and evo-devo is structuralist (Amundson, 2007, 2014). But Breidbach and Ghiselin (2007, p. 167) do better with this: 'Adaptationist: Individuals that are organisms do not evolve. Individuals that are populations composed of individuals that are organisms do. Populations evolve by natural selection. Structuralist: Individuals that are organisms do not evolve. Individuals that are ontogenies do. Individuals that are ontogenies evolve by modifications of ontogeny ... [However,] what gets selected is organisms throughout their entire life cycles, and what evolves is populations thereof.'

Darwin and after		
Selectionist	Instructionist	There is some ambivalence of emphasis in references to Darwinism (Munz, 1985, 1993, 2004): on the one hand, the organism is perceived as adapted to pick up information, as if passively receiving instruction from the environment. On the other hand, options are presented to the organism for active selection. Not coincidentally, Lyell, the solitary uniformitarian of his times, was a selectionist whereas the others were instructionists, and he, Lyell, was a much more fertile intellectual influence on Darwin than were those high-achieving 'catastrophists' (Buckland, Sedgwick, Murchison, Owen, Phillips). In more recent times, Simpson was a selectionist and Schindewolf was an instructionist. (In an example from the narrative of discovering the structure of DNA, Munz identified Crick and Watson as selectionists and Franklin as the instructionist.)
Homology, homologue		'Homologue ... the same organ in different animals under every variety of form and function' (Owen, 1843, in Wagner, 2014).
		'Homology. — The relation between parts which results from their development from corresponding embryonic parts, either in different animals, as in the case of the arm of man, the fore-limb of a quadruped, and the wing of a bird; or in the same individual, as in the case of the fore and hind legs in quadrupeds, and the segments or rings and their appendages of which the body of a worm, a centipede, etc., is composed' (Darwin, *Origin*, 6th Edition, 1898, Glossary, p. 480).
		'Homology is resemblance due to inheritance from a common ancestor' (Simpson, 1961)
		'Attributes of two organs are homologous when they are derived from an equivalent characteristic of the common ancestor' (Mayr, 1982).
Homology thinking		'Homology thinking explains the properties of a homologue by citing the history of a homologue' (Ereshefsky, 2012, p. 381).
		'Homology thinking is a form of historical explanation that draws upon information about the phylogenetic origins and developmental underpinnings of body parts that evolve in lineages. It is a complement to functionalist explanations that seek to explain organismal diversity from the point of view of functional need' (Wagner, 2014, p. 425).
		Homology thinking joins population thinking and tree thinking in the three-legged stool of evolutionary biology (Wagner, 2016).

Source: Author's summary: see sources throughout table.

The Darwinian Restoration is shown simplistically as twofold. 'Nomothetic thinking' refers to the urge to be quantitative, to search for laws, to derive insights from counting and tabulating the fossil organisms found in the rocks and described. Neither ecology and adaptation (Cuvier's 'function') nor our expanding insights into ancient environments are prominent. In contrast, natural-historical thinking and Darwinian thinking and the pervasive reality of natural selection have spread far beyond natural history in recent decades, giving rise to such concepts as bioeconomics.

However, the rise of evo-devo clearly takes us beyond the Restoration.[9] The advance can be encapsulated like this: the concept of homology has been immensely useful since the days of Owen and Darwin, but that usefulness has been on the tree-of-life side of evolution (and the right-hand side of Figures 10.4 and 10.5). For the population geneticists and the field naturalists, the micro-evolutionists, it was all about populations and species, not about bodies. And now it is all about bodies too, and so there was call for a new paradigm identifying not two but three styles of research. Wagner responded with the three-legged stool of evolutionary biology. Population thinking is about micro-evolutionary mechanisms leading up to the origin of species. Tree thinking is about the macro-evolutionary study of biodiversity. Homology thinking is about the development of organisms, their structure and their function and their place in the evolving biosphere (Figure 10.6).[10]

---

9   We can pause here and present a lineage of transformational evolution in the form of strong, long-form and readable statements. In the nineteenth century there were Richard Owen's *The archetype and homologies of the vertebrate skeleton* (1848) and *On the nature of limbs* (1849). In the twentieth century we have ES Russell's *Form and function: A contribution to the history of animal morphology* (1916), a splendid effort by, so help me, a Lamarckian and something of an anti-Darwinist. There is also Stephen Jay Gould's *Ontogeny and phylogeny* (1977), written while his colossal literary talents were still well disciplined. We have in the twenty-first century Ron Amundson's eminently readable *The changing role of the embryo in evolutionary thought: Roots of evo-devo* (2007), and Sean B Carroll's *Endless forms most beautiful: The new science of evo devo and the making of the animal kingdom* (2005). An essay by Amundson is as challenging as its inviting title, *Charles Darwin's reputation: How it changed during the twentieth-century and how it may change again* (2014).

10   Two books by Michael Ghiselin capture the metaphysical essence of Darwinism: *The triumph of the Darwinian method* (1969) and *Metaphysics and the origin of species* (1997). For including homology thinking in the new paradigm, see Carroll (2005) and Amundson (2007), Ereshefsky (2012), Wagner (2016) on the tripod, and Günter Wagner's formidable *Homology, genes, and evolutionary innovation* (2014).

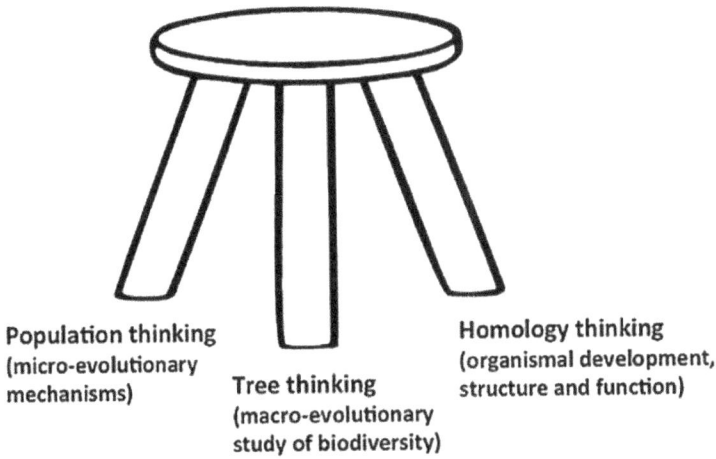

Population thinking
(micro-evolutionary
mechanisms)

Tree thinking
(macro-evolutionary
study of biodiversity)

Homology thinking
(organismal development,
structure and function)

Figure 10.6. Wagner's three-legged stool of evolutionary biology.

Darwin's tree of life and natural selection crystallised the two streams originating in eighteenth-century natural philosophy and natural history, which became transformational evolution and variational evolution, respectively. Embryology and homology have been in transformational evolution and have felt excluded from the Darwinian Restoration. With the rise of developmental genetics and evo-devo, a two-stream culture has become three-stream or better, a tripodal culture. And we are deep into the highly consilient biogeohistory of our surge #VIII.

Source: From Wagner (2016).

Consilience of inductions

In the 1970s I went to some pains to demonstrate that the biogeohistorical records of the terrestrial, neritic and pelagic realms ought to fit together, and that southern Australia for all of our troubles in constructing the narrative was part of the big picture, more global than local. The available evidence seemed to confirm that extratropical excursions by the benthic tropical foraminifera with photosymbionts were more or less consistent with the newly emerging oxygen-isotopic profiles from the oceanic plankton. Both implied warming and things were starting to fit together. But for a broader survey, inspect this chart of various trends through the Cenozoic Era (Figure 10.7).

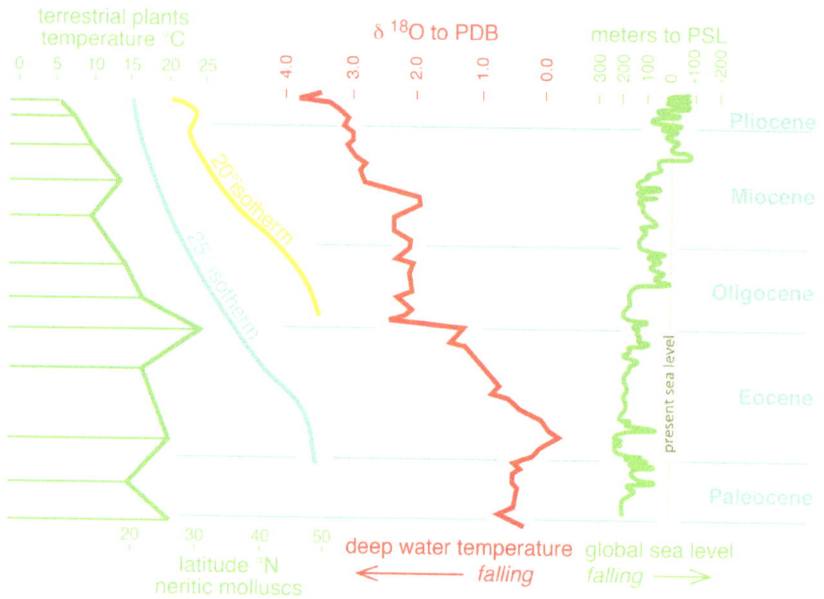

Figure 10.7. Consilience of induction in the Cenozoic Era.

Matched chronologically, trending through the Cenozoic Era, are several decades-old figures from the terrestrial, neritic and pelagic realms — terrestrial plants, marine molluscs, deep-ocean benthic foraminifera and a global sea level curve from sequence stratigraphy. By *consilience of induction* the theory of a fall in temperatures and sea levels from the Early Eocene onwards was much stronger than it would be for any one of the items by itself. The match is not perfect — which immediately presents worthwhile problems for further investigation. Thus, why did sea level fall in mid-Oligocene, well after the deep ocean cooled so strongly?

Source: From McGowran (2005a, Figure 8.3).

Lyell's generation knew that the Eocene was a time of global warmth. How? By the presence of palms, crocodiles, molluscs and corals in the Arctic lands, at tens of degrees higher latitudes than their modern tropical or subtropical counterparts—and the assumption that their very different physiologies hold true in a uniformitarian way. We have seen this strategy known to botanists as 'nearest living relatives'. In that way, Erling Dorf in 1955 could employ terrestrial floras as 'thermometers of the ages' to produce a curve of climate change for 40–50°N latitude in the western US; the thermometer has been falling since Eocene times. Since neritic molluscs are distributed according to water temperatures, Wyatt Durham in 1950 could reconstruct two marine isotherms, for 20° and 25°, and show them retreating equatorwards since Eocene times. Two sets of data, terrestrial and marine, assembled and dated independently, each open to legitimately sceptical interrogation as to the stability of the assumptions—and the leaves and shells tell the same story of long-term cooling in the respective environmental realms. We add Shackleton's 1985 deep-ocean-bottom curve of $\delta^{18}O$ values from benthic foraminifera. As we have noted, those numbers are susceptible to alteration of the shell, 'vital effects' during biocalcification, and salinity changes. And yet the three curves have a powerful mutual similarity! If quite disparate data from the terrestrial, neritic and pelagic realms of the biosphere show such a good mutual match through geological time, then the chances that we are seeing real broadscale climatic change are suddenly much better than they were for each of the three data sets in isolation. The mutual reinforcement of evidence from neritic shells, terrestrial leaves and oceanic isotopes far exceeds the sum of the parts. Adding a 1980s curve of global sea level, we see still more strengthening by persuasive correlations. Sea level has fallen as temperature has fallen, from the same high point in the Early Eocene.

It was the same William Whewell who coined 'catastrophists' and 'uniformitarians' who, in his *Philosophy of the inductive sciences* (1840), coined 'consilience of inductions' for this strategy of coordinating different lines of evidence to form a highly coherent pattern. The ultimate example of consilience is *On the origin of species*, a grand integrated brief for evolution by consilience. Only Stephen Jay Gould could have put these grand Victorian sentences into Darwin's mouth:

> I present you, in this book, with thousands of well-attested facts drawn from every sub-discipline of the biological sciences— from the transitory and vestigial teeth of embryonic whales, to transitional forms in the fossil record, to the invariant order of life in geological strata throughout the world, to documented cases of small-scale change in agriculture and domestication, to the use of the same bones for such different functions as a horse's run, a bat's flight, a whale's swim, and my writing of this manuscript, to the observation that faunas of isolated oceanic islands always resemble faunas from nearby mainlands, but only include creatures that can survive transport across the waters, et cetera, ad infinitum, though thousands of equally firm and disparate facts. Only one conclusion about the causes and changes of life—the genealogical linkage of all forms by evolution—can possibly coordinate all these maximally various items under a common explanation. And that common explanation must, at least provisionally, be granted the favour of probable truth.[11]

---

11   Gould (2003), see next note.

But, irony upon irony, Whewell as master of Trinity College at Cambridge rejected Darwin's thesis of consilience in its entirety, and even banned the *Origin* from the college library. Whewell took his prime examples of consilience from Newton's science of gravity and light. Not for the first time and not for the last time by a long chalk, a distinguished reductionist scientist failed to grasp the depth and strength of expansionist historical science.[12]

'Any concept is only as good as the research program it inspires.' What were we doing in the 1950s and 1960s?

The Modern Synthesis, or Darwinian Revolution (Part II), better still called the Darwinian Restoration, was in place by the 1950s and it made for exciting times for a wide-eyed acolyte discovering biogeohistory. But what did it do for our research program in Cenozoic southern Australia? For that matter, what was that 'research program'? Postwar Australia needed expanded geological mapping, geophysical surveying and exploratory drilling and it benefited greatly from the immigration program. So too did I: I was profoundly influenced by three immigrants immensely knowledgeable about fossils and strata, Armin Öpik, Curt Teichert and Martin Glaessner. Rereading Glaessner and Teichert—both gradualists and both anti-continental drifters—makes one feel the intellectual power now as it did six decades ago. They were searching, collecting and describing fossils in the process of correlation and age determination, labours underpinning everything from tectonics and mountain-building to evolution and biogeography, and labours that would have been appreciated immediately by their predecessors 150 years ago. This was what one did then and does now (however differently), when employing organic diversity in research and the systematics are not in good order (as the documentation of the biosphere never quite is). The seven surges have accreted down the decades and the context of systematic palaeontology has changed enormously in both its geological and its biological modes, and we do well in our rush to progress not to overlook that deep continuity.

---

12   We pushed this understanding of consilience in examining the Miocene biogeohistorical record (McGowran and Li, 1994). Meanwhile Edward O Wilson (1998) wrote a popular book under the title *Consilience* with the objective of bridging the sciences with the social sciences and other humanities. Wilson shifted the meaning of 'consilience' hard towards what I would call 'reductionism'. In *The hedgehog, the fox, and the Magister's pox* (2003), Stephen Jay Gould also found Wilson's consilience to be largely dressed-up reductionism.

My Palaeocene and Eocene microfaunas in southern Australia contained numerous specimens characterised by a shell of aragonite, not the usual calcite, and by peculiar structures within the chambers called toothplates, revealed by acid dissection or chipping. The group have been around for 100 million years and more, usually as minor components of fossil assemblages. Martin Glaessner realised in the 1930s that the toothplates in common could unite several genera as a new family. I assembled a plausible family tree (question marks notwithstanding) (Figure 10.8). The tree was plausible but the branching was in jumps, not gradual transitions. Once arisen, the genera changed very little from the Cretaceous to the Palaeogene to the Neogene and there was virtually no transition from the respective apparent ancestor to its descendant. That is, new taxa arose in innovation and remained in stasis.[13]

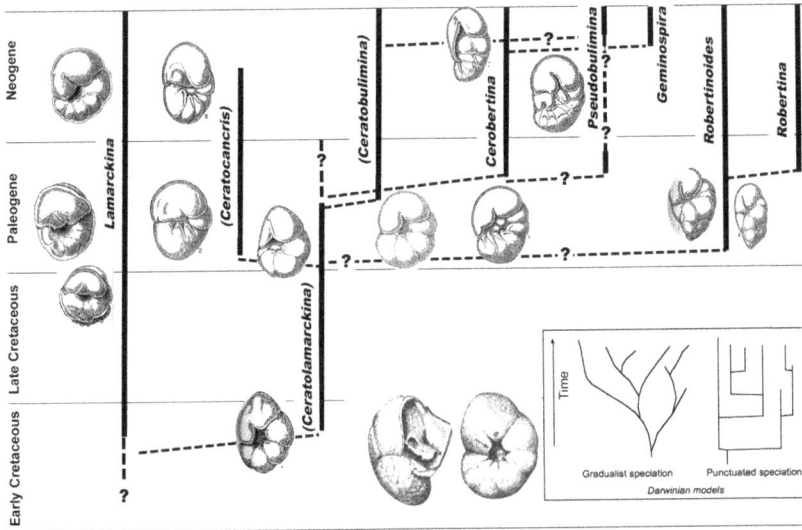

Figure 10.8. Family tree of the foraminiferal family Robertinidae, modest but distinctive twig on the great foraminiferal branch of the tree of life.

The shells are aragonite in these microbes, all less than 1 millimetre in largest dimension. Each subgenus (in brackets) and genus changes very little during its tenure (solid vertical lines), whereas there is sudden change at its origin. Hence the queries in the links — they are cover for our lack of detail about the transitions which are highly likely at the levels shown, strong alternatives not evident. This reconstruction was assembled a decade before sketches appeared contrasting gradualist speciation with punctuated speciation, and can be read as consistent more with punctuated speciation and stasis

---

13   The dissected *Ceratobulimina* displaying the toothplate in the last chamber is from Plummer (1936); for Glaessner (1937a) this character could unify a new family of foraminifera, the Ceratobuliminidae (which later became the Robertinidae), which I displayed again (2012b, Fig. 3).

than with the other 'Darwinian model' shown here. The isolated specimen at bottom is a *Ceratobulimina* with the last chamber dissected by Helen Jeanne Plummer in 1934 to show the internal toothplate which, Glaessner realised, was the characteristic uniting a new family of foraminifera.

Source: Adapted from McGowran (1966).

But I was more interested in translating the tree, constructed bottom-up, into an evolutionary classification, for at the same time the industry-driven resurgence in studying the planktonic foraminifera was exposing (to this callow youth anyway) a taxonomic tangle. The more influential workers including Hans Bolli and Walter Blow focused on industrial biostratigraphy and became interested in the evolutionary relationships thereby uncovered but not, unlike Glaessner, making that dipole into the tripole of biostratigraphy/evolution/taxonomy. The most influential in the classification of the foraminifera in the decades after Glaessner's *Principles of micropalaeontology* (1945) were Al Loeblich and Helen Tappan, assembling the treatise on the foraminifera. Their philosophy of taxonomy was less about evolutionary classification, building an edifice bottom-up, and more about constructing a top-down key, that is, an aid to rapid identification. Philosophically they were Aristotelian and pre-evolutionary, and I was crass enough to characterise them in print as that, and to recommend Glaessner's evolutionary philosophy of foraminiferal taxonomy, which had a lot in common with Simpson's mammalian taxonomy, namely Darwinian philosophy. My homily submerged without a whimper. It was not that people did not believe in variational evolution. Of course they believed. Instead, foraminiferal micropalaeontology was too busy with geological biostratigraphy and palaeoenvironmental problems and scientific and industrial opportunities to actually address the biological Darwinian Restoration.[14]

And then Eldredge with trilobites and Gould with sub-modern snails introduced the theory of punctuated equilibria, in which change was crammed into the speciation event, after which the species was typically in stasis. Punctuated speciation was conceived and advocated by palaeontologists comfortable in deep time. It was rejected by neontologists more comfortable in shallow time, people who 'owned' the processes of evolutionary biology—for how could specimens of long-extinct species say anything about the actual dynamic processes called 'evolution'?

---

14   The publications referred to were Glaessner (1945), Loeblich and collaborators, especially Bolli (1957), Blow (1969), Loeblich and Tappan (1964), Simpson (1961) and McGowran (1971).

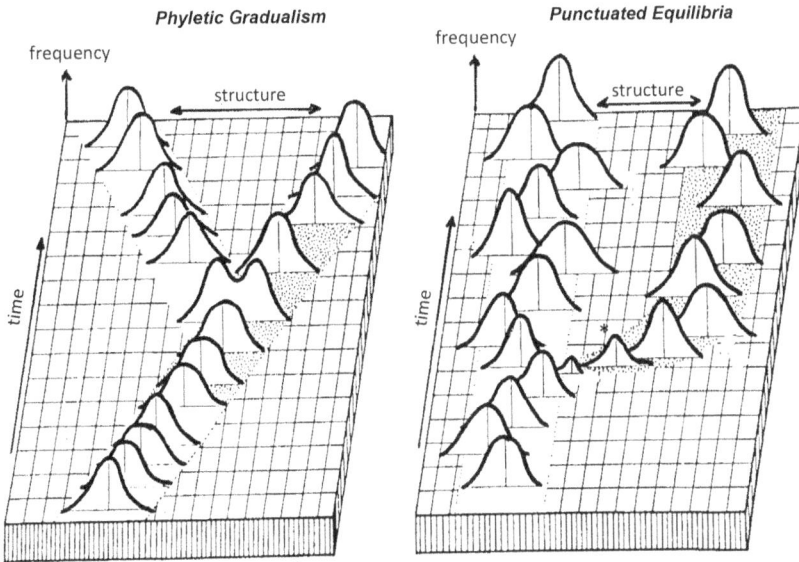

Figure 10.9. Phyletic gradualism and punctuated equilibrium are alternative models of evolutionary change.

Populations are signified by the bell curves and different species by alternate stippling. Gradual change begins with populations and accumulates more or less steadily thereafter. In the other model, a species varies in the short term but remains in stasis in the long term. Peripheral populations wither, or return to the mainstream — or once in a while (star) make the adaptive breakthrough to become a new species. Stasis is more important and cumulative change less. Time is unscaled here, but is in thousands not millions of years.

Source: After Vrba (1980).

Recall that the population thinking in Simpson's palaeontology symbolised the variable species as a bell curve (Figure 3.12). Central to Darwin's evolutionary strategy was gradual change, brought about by natural selection acting upon generously available variation through time. It is easy to imagine the curve for scallop shells shifting gradually through time. Thus a sudden apparent jump through the succession of strata would invoke the Lyellian fallback position of hiatus, of gaps in the sampled section; and the apparent evolutionary saltation could be explained away by the patchy fossil record. Although time is not scaled in Simpson's diagrams, their thrust is of changing adaptive zones, forcing gradual change in phyletic evolution. Simpson also estimated that most evolution was in that mode; speciation by splitting was less common; less common still but critically necessary was the straying of small populations to the margins of the adaptive zone, where most were extinguished or resorbed in the mainstream, and only the very few populations made a breakthrough—which is how bats were derived from mice and whales evolved from deer, for example. In its extreme form, this gradualism would have a species shifting steadily in one direction while a daughter species shifted in another. In the alternative model, species certainly shifted quite markedly through time, but the overall profile of the species was stasis. These diagrams have an unscaled dimension of time, but stasis seems to mean several million years or more and the speciation event a few thousand or less.

Where Simpson's diagrams imply ongoing shifts, punctuated speciation emphasises stasis (Figure 10.9). There are shifts and reversals between successive populations, presumably responding to short-term environmental shifts, but in the longer term neither the parent nor the daughter species is actually going anywhere in terms of observable characteristics ('structure'). The content of punctuated equilibrium is that a species arises rapidly, it exists for a time, it might or might not bud off a daughter species; it resists change, responding instead to serious stress (environmental or competitive) by moving to more congenial quarters or going extinct; it shifts, but not irrevocably. Punctuated equilibrium stimulated detailed scrutiny of morphological change in lineages, and there was to-ing and fro-ing between deep-time palaeontologists and shallow-time neontologists. The upshot is that we are more aware of a species possessing a beginning and a termination. That is, it is more than just a slice of a continuum. It is a real, distinct entity. Speciation in animals, once seen as happening more by phyletic evolution, is now seen as occurring more by splitting a lineage or clade.[15]

This shift in evolutionary world view occurred in the 1970s. At about the same time there was a philosophical shift. A species can be thought of as belonging to a class of species defined on the basis of its properties. Or, an individual species can be sought and discovered to be a missing link, as it were, to be restored to its place as a real twig on the real genealogical tree of life. Define or seek? Class or individual? Biology and philosophy both clarified the reality, the concreteness, of species in organic evolution. And down there at the modest level of the origin of species, there was a shift from gradualist to episodic. And a species was more than an assembly

---

15   The foundation text for punctuated equilibrium is Eldredge and Gould (1972). Figure 10.9, from Vrba (1980), clarified Eldredge and Gould more than they clarified themselves. Simpson's bell curve thinking on speciation was spelled out in 1937. In recent years the evolutionary potential of the large photosymbiotic foraminifera, fossil and modern, is being realised. On bell curves, see two splendid discussions of ecological species in Hohenegger (2014) and Hohenegger et al. (2022). Renema (2015) finds that populations of *Cycloclypeus* in south-east Australia, marginal to the great Indo-West Pacific province, were central to the speciation of *C. carpenteri* from *C. eidae*.

of individuals. It was itself an individual, it was bounded in space and time. It was a historical entity. Although some went unenthusiastically, we underwent a cultural shift.[16]

We can go further. The evolutionary drama takes place in the environmental theatre of the earth's crust. By about 1950 the theatre's shifting foundations were being visualised in two ways. A 'German' ('Stillean') view held that tectonic disturbances in the earth's crust tended to be intensive, short-lived, very widespread and even global, and separated by quiet intervals—tectonic quiescence. The countervailing 'American' ('Gillulyan') view was that there was no such pattern discernible in space and time in such phenomena as rock deformation and metamorphism, the intrusion of granites and the extrusion of volcanics, or mountain building. Entangled in this ontological jousting was the question of continental drift, of fixed or mobile continents and oceans, but by and large the Gillulyan view prevailed over the Stillean; processes were gradual over deep-time scales and episodic jumps were largely an artefact of a spotty preservation. Oceanfloor spreading and plate tectonics changed everything. Oceanfloor spreading was an elegant machine-like process in its steady state, but that steady state could not be sustained, most obviously because lower-density granitic crust could not be forced down a subduction zone. There had to be collisions necessitating global rearrangements of the convection cells, the prime example in our region being the India–Asia collision. The forces underlying the lithosphere, hydrosphere and biosphere were operating episodically. A distinguished geologist of mountain belts came to see a distinct pattern in time and space—episodic synchronism—in the tectonic evolution of the Alpine mountain chain:

> When the author set out to gather information on the timing of orogenic events, he started as a convinced Gillulyan; to his own surprise, he has ended up as a moderate Stillean.[17]

---

16 Biogeohistorians have long had problems with philosophy, problems crystallised by Ernst Mayr's observing that the philosophers of science actually served up philosophy of physics. Simpson thought that Aristotle was not helpful; Mayr thought that Plato was a disaster for biology. Mayr did something about it (*Towards a new philosophy of biology; observations of an evolutionist*, 1988). So too did the evolutionary biologist Michael Ghiselin (*Metaphysics and the origin of species*, 1997) with his advocacy for species as metaphysical individuals every bit as real as individual organisms. Arguments for species as individuals are not arguments for species in stasis between punctuations, and both differ from the physical arguments for an episodic world. But all three came together in the cultural shift as surge #VII accreted with surges #V and #VI.

17 Trümpy (1973), also quoted in McGowran (1978) under the heading 'Episodic history', reviewing evidence far distant from Trümpy's Alpine geology for supporting his 'neo-Stillean' *Weltanschauung*. The next section is about the K/Pg boundary, the mass extinction and the discovery of the Chicxulub bolide, and the shift of the *Weltanschauung* away from uniformitarianism and Lyellian gradualism. This note is to point out that that shift was underway for more than a decade before the bolide arrived.

Rudolf Trümpy was not alone in undergoing a cultural shift in the early 1970s.

So much for the biosphere and the lithosphere. What about the record of ancient surface environments, the stratigraphic record? The gradualism of Simpson and Gilluly was matched in their contemporary, Hollis Hedberg, driving force of the International Stratigraphic Guide. Hedberg wanted an agnostic framework anatomised into rocks, biozones and time, internationally agreed, and avoiding any formal commitment to natural breaks. In this he was sharply opposed especially to the Russian philosophy of stratigraphy, which was holistic and dedicated to finding and building the natural breaks into the geological time scale. Hedberg's gradualist philosophy in the 1950–1960s resembles Lyell's in the 1830s. We have seen that sequence stratigraphy is very different. Its rise in the 1970s was yet another a major cultural shift of those times.[18]

It's time to address Wagner's claim: *Any concept is only as good as the research program it inspires.*[19] Wagner's assertion holds for sharply focused concepts. For example, the bolide theory of widespread extinction is very sharp and looked to be testable in the best scientific manner, thereby appealing to mineralogists and experimental neontologists. But many palaeontologists remained underwhelmed until the relevant evidence began rolling in (see below). The theory of plate tectonics and its biogeohistorical implications stranded a parade of intellectual giants of biogeohistory—Simpson, Glaessner, Teichert, Mayr—as well as geophysicists. The principles of cladistics upended much of evolutionary taxonomy, and together with the molecular clock are bridging the heuristic gulfs between poorly fossilised organisms (those without a mineralised skeleton) and those with an extensive fossil record. These breakthroughs might seem to have left the Darwinian Restoration as a noble ruin and the work of yesterday's leaders as rubble.

But I think not. Instead, I agree with the view that Simpson's great synthesis of genetics, evolution and palaeontology in *Tempo and mode* (1944) formed the core of modern palaeontology.[20] We have some answers to three questions recurring down the centuries about biogeohistory.

---

18 McGowran and Li (2007).
19 Wagner (2014, p. 245). Italics in original.
20 Jackson and Erwin (2006). There was some celebrating of a 'palaeobiological revolution' not arising from Simpson but post-dating and effectively marginalising him (Sepkoski and Ruse, 2009; Turner, 2011; Sepkoski, 2015). This action downgraded ecology, bioeconomics and adaptation, and promoted chaos and chance at levels from catastrophic impacts downwards.

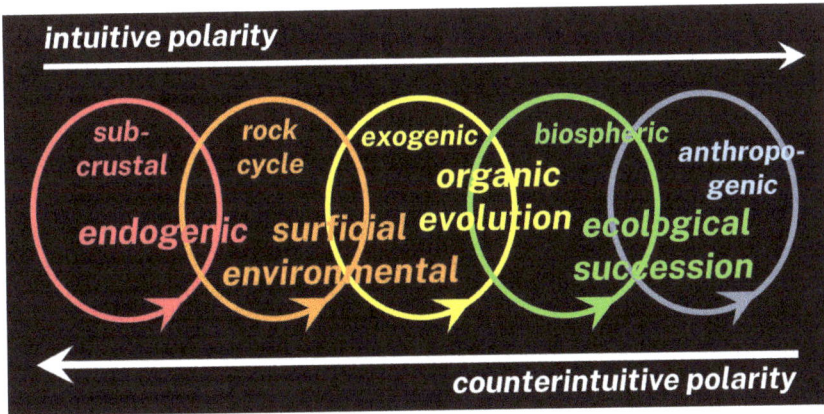

Figure 10.10. One way of contemplating time's arrow and time's cycle.

It seems reasonable to emphasise the arrow to the right, driven by plates moving at centimetres per year. Omitted are the only 'true' cycles, the astronomical sun–earth–moon, strongest to the right. The counterintuitive polarity acknowledges that in feedbacks the biosphere is by no means a passive recipient, but an increasingly active player as we come to understand anthropogenic disasters.

Source: Author's depiction.

Time's arrow versus time's cycle? Judaic historicity has delivered more to the cause of biogeohistory than has Greek eternalism. That is, time's arrow leads. But in going further by looking at conceptual 'cycles' from the deep-crustal to the shallow-time ecological (Figure 10.10), we see that time's arrow and time's cycle are still-healthy recurring metaphors; biogeohistory is not either/or, neither is it all or nothing.

The longest-known cycle is 'Hutton's rock cycle', tying together igneous, sedimentary and metamorphic processes driven by internal and uplift (the cause of uplift was obscure but the effect was plain to see in youthful-looking marine fossils occurring in folded rocks, high up in mountains). 'Endogenic' plate tectonics clarified the notion of the making and the breaking of supercontinents and the birth and death of oceans. This is the Wilson Cycle extending over several hundred million years. In the 'exogenic' direction from Hutton's cycle, which is into the biosphere, we have at scales of millions to tens of million years the theories of evolutionary direction, of evolutionary cycles, of evolutionary episodes and saltations. And these grade into the Milankovich cycles and human scales and shallow-time ecological theories. One arrow indicates the common-sense notion that things start deep and slow and that the biosphere is on the receiving end, not least through environmental impact and natural selection. The reverse arrow acknowledges feedback, such as photosynthesis constructing the oxygenated atmosphere and hydrosphere. Such ideas burst into prominence in the notion of Gaia and now the Anthropocene. It seems grimly appropriate to recognise human impact in a shade of grey.

Is evolution's engine an internal or external dynamic? There were geneticists who could entertain, as a thought experiment, a biosphere evolving without needing any environmental nudges along the way. Less extreme than this ultimate internal dynamic are all theories of transformation or saltation. Darwin's theory of natural selection was ecological, an intimate involvement of organism with environment. This has to be an external dynamic, but it builds an increasing complex biosphere with a strong signature of time's arrow.[21]

What is the tempo: gradual or episodic? The world of Lyell and Darwin was uniformitarian and gradualist, and so too was the world of Simpson, Hedberg and Gilluly a century later. From plate-tectonic reorganisation to punctuated evolution, we are all episodic now.

Mysterious Priabonian and the return of the chronofaunas

Late Eocene times, the Priabonian Stage, are distinctive, as we saw in the St Vincent Basin (Chapter 8) and in the Australo-Antarctic Gulf (AAG) and around its shores. What of the wider scene in the oceanic (pelagic), neritic and terrestrial environmental realms? And can we say something interesting about fossils and environment and evolution during this slice of geological time, beyond the greenhouse–icehouse theme?

Among the cascading ideas and challenges in his *The dinosaur heresies* (1986), Robert Bakker developed the notion of megadynasties in 300 million years of biogeohistory. There were four of these megadynasties in large land herbivores, thus: I, the Permian pelycosaurs etc.; II, the Permian–Triassic protomammals; III, the archosaurs of the Late Triassic–Cretaceous; and IV, the mammals of the Cenozoic. The megadynasties comprised dynasties, such as a real entity in the Cenozoic called 'Eocene mammals', say, or 'brontosaurs

---

21 Time's arrow and time's cycle are evocative metaphors, but Calcott and Sterelny (2011) could compare two world views within evolutionary palaeontology, namely time's arrow and no arrow. For Gould (2002), the most vocal participant in the 'revolution', there is a multitude of local histories for all the twigs on the tree of life, and there is an increase in complexity mostly because life began simple, but there is no unified history of life on earth as a whole. But for Vermeij (2009, 2011) and Vermeij and Leigh (2011), the pace of life increases through time and life's history is dominated by an ecological arrow of time.

+ stegosaurs' in the Cretaceous. Likewise, Bakker immediately grasped the Eldredge–Gould notion of punctuated equilibrium in organic evolution, of stasis in a lineage, meaning that the species changes little through its time range until something happens, and then there is rapid speciation. He supported the anti-Lyellian thesis that the record of the fossils in the rocks was basically chunky, and that chunkiness is intrinsic, not merely an artefact of the broken-ness and scatter of the archives of deep time. Bakker was acutely aware of the physiology and ecology behind his fossil bones, but he was more aware still that these deep-time phenomena were mostly beyond the reach of modern (shallow-time!) ecology.

We were long used to such labels as 'The age of reptiles' or 'The age of mammals', but Bakker's dynasties recalled for me Everett Olson's chronofauna. Painstakingly reconstructing the communities of vertebrate animals living in Permian–Triassic times, Olson recognised that community types persisted through deep time and across local variations. Individual species would be coming and going in speciation and extinction, but still the overall structure of the community remained perceptible. We have seen how speciations and extinctions are invaluable in stratigraphy, where fossils could organise strata in space and time to create the very possibility of biogeohistory. We have seen also the identification of ancient environments from the simple marine and nonmarine to the grander oceanic, neritic and terrestrial realms. And thirdly we have seen the reconstructions of relationships and genealogies of organisms called phylogenies. Chronofaunas and chronofloras are apart from that schema. And so researching the patterns of ecological organisation through time came to be called evolutionary palaeoecology, and chronofaunas do fit into that integrating discipline.[22]

But these ideas came closer to home when Lukas Hottinger began highlighting them from his magisterial position in foraminiferal micropalaeontology. Refer back to the great spread of the large, photosymbiotic species across the warm and sunlit ramps and platforms of Eocene Tethys (Figure 6.9). We saw that the foraminiferal lineages partitioned the available well-lit habitat into four big communities. Each community continued to change, to morph, species-budding-off-species, to evolve, meanwhile the overall pattern holding together. It held together for 18 million years, from the Palaeocene–Eocene thermal maximum (PETM) through the Middle Eocene climatic optimum (MECO), surviving Chill I and into MECO

---

22   References throughout this section are to Bakker (1986), Olsen (1952), Wing et al. (1992). Evolutionary palaeoecology in foraminifera: McGowran and Li (2000).

and a bit beyond. Those five biofacies retained their integrity for that time, notwithstanding the more or less continuous overturn in speciations and extinctions. And now refer to Figure 10.11.

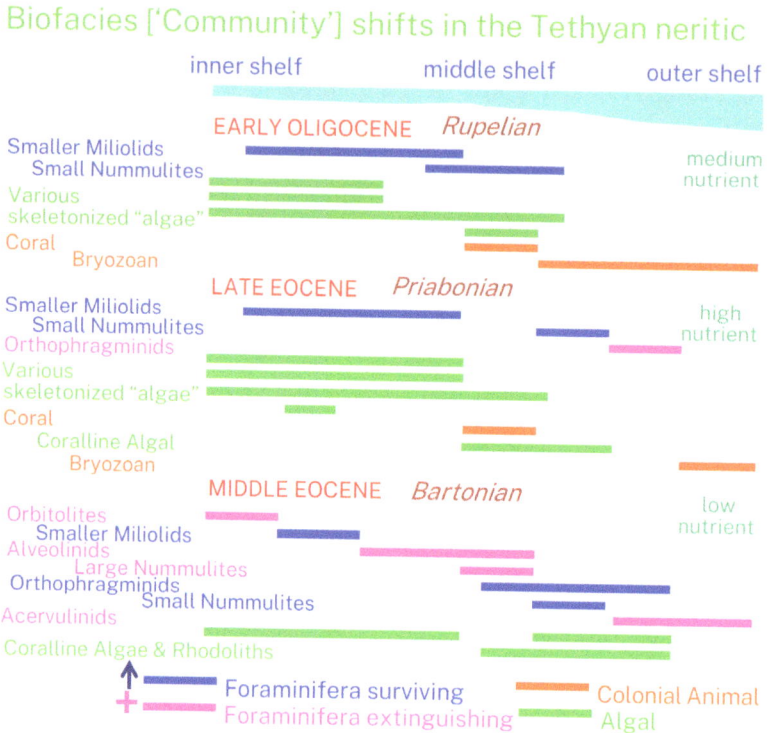

Biofacies ['Community'] shifts in the Tethyan neritic

Figure 10.11. Biofacies shifts and ecological partitioning in the Tethyan neritic.

The fossilised remains of communities, the *biofacies*, are distributed across the sunlit ramps and platforms in the neritic realm of Tethys. In the Middle Eocene there are seven associations of photosymbiotic foraminifera along with corals, algae and bryozoans. By Late Eocene times all the dominant forams except the Orthophragminids have gone, to be replaced mainly by algae. This implies a massive shift from low nutrient to high nutrient, from oligotrophy to eutrophy! And then by Early Oligocene time the Orthophragminids have gone too, and we have moved from the old to the new or modern faunas and facies, signalled not least by the advance of the bryozoans, which are to dominate the Neogene neritic where the corals are less comfortable.

Source: From McGowran (2009), adapted from Nebelsick et al. (2005).

Colonial animals with skeletons (corals and bryozoans) are part of the story along with the so-called 'algae' with skeletons. But we note the various groups of foraminifera keyed as to whether they survive or don't survive in this three-part succession of Middle Eocene (Bartonian), Late Eocene (Priabonian) and Early Oligocene (Rupelian).

What happened at 37.7 million years ago?
Coinciding turnovers in the three great environmental realms

Terrestrial Realm	Neritic Realm	Pelagic Realm	
North American Land Mammals	Tethyan photosymbiotic foraminifera	Planktonic Foraminifera	Deep-ocean benthic foram succession

(geological time scale on left: M Eocene — Late Eocene | Oligocene; values 30, 35, 40)

Markers on scale: Oi-1*, MLET*, MECO**

	Terrestrial	Neritic	Pelagic (planktonic)	Pelagic (benthic)
		duration 20 myrs	Neogene faunas	duration 15 myrs
	duration 18 myrs	Oligo-Miocene Chronofauna		Transitional Chronofauna
	White River Chronofauna	Priabonian Chronofauna	"transitional" assemblages	
			+	
	Eocene Chronofauna	Early Paleogene Chronofauna	Acarininid- Morozovellid assemblages	Paleogene Chronofauna
	duration 15 myrs	duration 18 myrs	duration 22 myrs	duration 17 myrs

* Onset "modern" icehouse
* Middle/Late Eocene turnover in the pelagic realm (**foraminifera** & **radiolarians**)
** Middle Eocene climatic optimum   (at Geomag Chron C17n.3n)

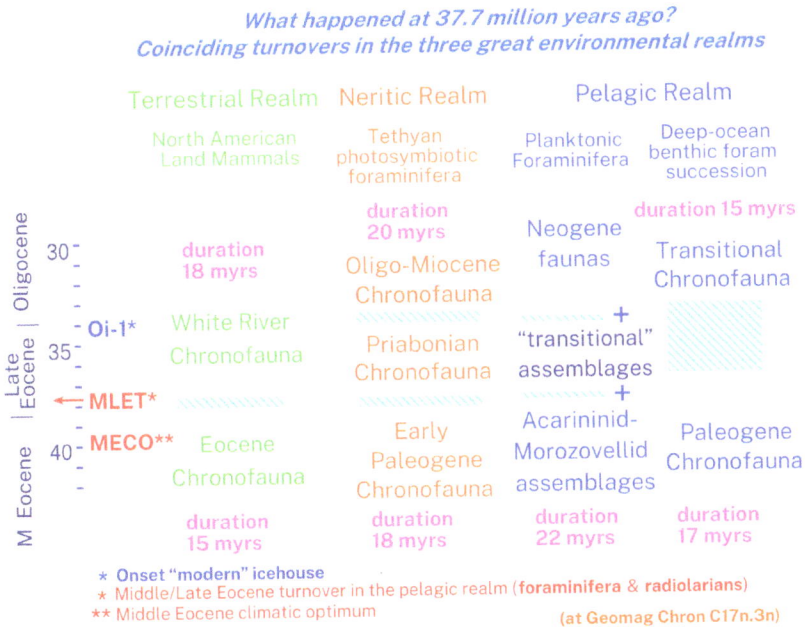

Figure 10.12. Successions of chronofaunas in the three environmental realms from the Eocene into the Oligocene—the most critical transition from the greenhouse world of the Cretaceous and Palaeogene to the icehouse world of the Neogene and today.

Excess nutrient caused the first crisis in planktonic foraminifera in the open ocean, coevally with a similar crisis in two successions in the neritic and coevally with the chronofaunal turnover in terrestrial mammals. The reverse happened at the second crisis which was caused by the onset of modern-type Antarctic glaciation (event Oi-1). The central point is that long-lived (14–15 million years) and deep-time 'community' entities turned over in concert in all realms. And the Priabonian, mysterious in southern Australia, turns out to be mysterious elsewhere.

Source: Adapted from McGowran (2009).

Three stages in geological time, two big changes between them. All five large-foram communities are hit hard in their detail after the Bartonian and note, in the broad picture, that only the large Orthophragminids plus two 'smaller' categories get through. In contrast, we see a blooming across the shelves and ramps of the algae. This shift signals a wholesale shift throughout the Tethyan neritic from low-nutrient to high-nutrient, from oligotrophy to eutrophy. For Hottinger, this event at the end of the Middle Eocene was the most significant in the history of the big photosymbiotic foraminifera. Where he marked it as a late-stage event in an evolutionary cycle, we marked the event as a major chronofaunal collapse, between the Early Palaeogene and the Priabonian Chronofaunas (Figure 10.12).

The changeover happened due to post-MECO cooling, invigorated circulation over the carbonate platforms and a flush of nutrient, giving the corals and various algae a strong advantage.[23]

In the pelagic realm, the planktonic foraminiferal chronofauna (calcareous shells; duration, 22 myrs) changed abruptly very near the end of the Middle Eocene, simultaneously with a turnover in the radiolarians, ecologically very different and with opaline shells. Yet another bunch of initials: MLET (Middle–Late Eocene turnover).[24] Meanwhile the deep-ocean benthos has the same structure, namely a Palaeogene chronofauna (17 myrs duration) succeeded by a Transitional chronofauna (15 myrs duration).

I once identified the silica window within the Late Eocene, meaning a window in time of a couple of million years within which sediments particularly rich in silica and poor in carbonate seemed to be concentrated—sediment ranging from the shallow seas marginal to the AAG to various places in the southern deep oceans. The fossils recurring in the silica window are the opaline skeletons of sponges, radiolarians and diatoms. More recently a glauconite horizon has been traced from Blanche Point and Browns Creek to the west in the AAG and out into the ocean, in the South Atlantic and the Weddell Sea. Glauconite when concentrated is long known as a greensand; it is associated with high levels of nutrient and dark organic muds.[25]

In the terrestrial realm, the best-known fossil succession is the splendid North America land mammal record, classified as chronofaunas. The Eocene chronofauna (duration 15 myrs) is succeeded by the White River Chronofauna (duration 18 myrs) at or near the Middle–Late Eocene boundary. Once again, two long periods of evolution within many lineages, separated by a very short interval of rapid change.

There are two deep-time ecological phenomena here, both beyond the reach of modern ecology. One is the coherence or stability of these entities perceived as chronofaunas, their duration estimated in an eight-figure number of years. The other is the synchrony in turnover or crisis. Three lots of single-celled marine organisms, shallow and deep, planktonic and benthic, suffered major change at the same time as land mammals. And we see this major global shift in the Priabonian manifested locally, in the strata on the north flank of the AAG, from carbonate-rich to carbonate-

---

23   Hottinger (1982, 2001) and some long conversations with Lukas in 1996 and 2001.
24   MLET: Kamikuri and Wade (2012); Newsam et al. (2017).
25   Two more recent references are McGowran (2009) and McGowran et al. (2016).

poor strata. Global cooling renewed after MECO, with evidence growing of some ice on Antarctica, and global-scale fluctuations in nutrient had a lot to do with it, whether as flushes in runoff from the luxuriantly vegetated land or as nutrient-charged upwellings from the deep.

Death of the dinosaurs at the K-Pg: Strangelove Ocean, Living ocean or heterogeneous ocean?

Biostratigraphy—correlation and age determination—comes first. Surely this notion is old hat, taking us back to the days of Cuvier? Well, consider the contrasting plates of fossils from the Chalk and the Lower Eocene (Figure 10.13).

These strata were regarded as contiguous in the days of Lyell and Darwin, who believed it feasible that an immense stretch of time was hidden in an huge hiatus at the changeover. But as stratigraphic knowledge progressed and the Palaeocene Epoch was erected and dated ever more precisely, the contrasts actually strengthened. Buffon knew in 1750 that the ammonites disappeared from the 'chalk seas' of the Mesozoic in Europe and elsewhere. And as it became necessary to be clear and precise and civilised about the divisions in the geological time scale, so too did it become obvious to some observers that the extinction of the ammonites should mark the close of the Mesozoic Era. We went one better with Martin Glaessner's 1937 chart of the planktonic foraminifera, those worldwide, small and abundant fossils being much more versatile markers (Figure 4.9). Planktonic foraminiferal biostratigraphy took hold in the 1950s and was joined by the coccoliths and radiolarians in the late 1960s in keeping deep-ocean drilling on course and solving the refractory stratigraphic problems of the AAG and southern Australia. Meanwhile the Cenozoic strata were being calibrated numerically as radiometric dates tumbled out of the geochemical laboratories. Forbidding environmental barriers were being breached, twice, the barriers between and within the pelagic, neritic and terrestrial environmental realms; one breach was by palynology, wherein sporomorphs blown out to sea could be matched directly with marine microfossils in the same samples; the other was the new backbone to the Cenozoic time scale in the geomagnetic record of earth's flip-flopping magnetic field. And this reversible geophysical signal reminds one of the various geochemical ratios that were investigated for environmental signals, which also became chronological signals.

Fossil Group No. 40.
Lower Eocene Fossils.

a. Nipadites umbonatus.
b. Paracyathus caryophyllus.
c. Pentacrinus sub-basaltiformis.
d. Ophiura Wetherellii.
e. Vermicularia Bognoriensis.
f. Hoploparia Bellii.
g. Zanthopsis tuberculata.

Fossil Group No. 41.
Lower Eocene Fossils.

a. Terebratulina striatula.
b. Pinna affinis.
c. Cyrena cuneiformis.
d. Cryptodon angulatum.
e. Voluta Wetherellii.
f. Aporrhais Sowerbii.
g. Nautilus imperialis.
h. Corbopoma Colei.
i. Lamna elegans.
j. Otodus obliquus.

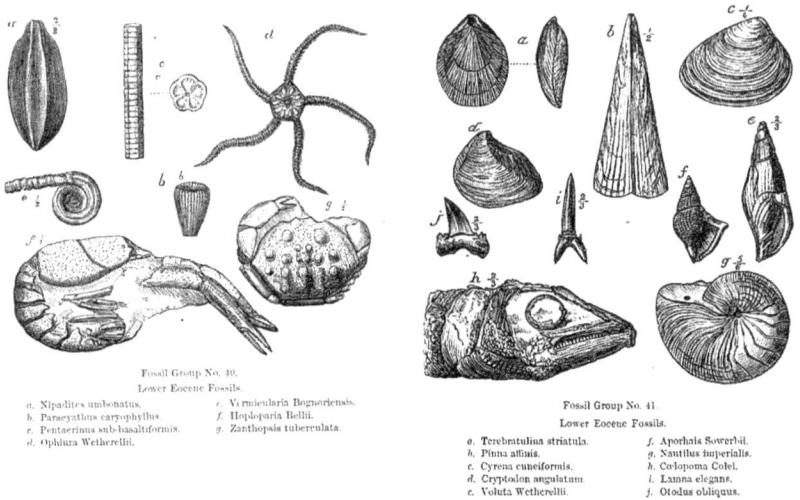

~~~~~~~~~~~~~~~~~~~Cretaceous/Paleogene boundary~~~~~~~~~~~~~~~~~~

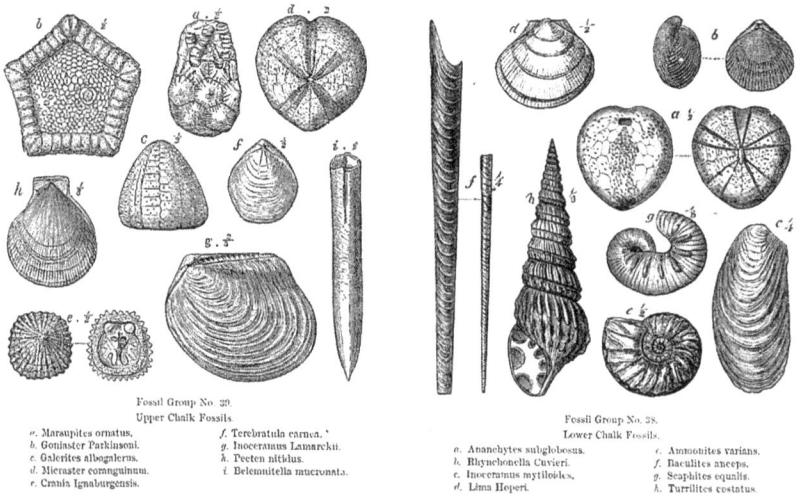

Fossil Group No. 39.
Upper Chalk Fossils.

a. Marsupites ornatus.
b. Goniaster Parkinsoni.
c. Galerites albogalerus.
d. Micraster cor-anguinum.
e. Crania Ignaburgensis.
f. Terebratula carnea.
g. Inoceramus Lamarckii.
h. Pecten nitidus.
i. Belemnitella mucronata.

Fossil Group No. 38.
Lower Chalk Fossils.

a. Ananchytes subglobosus.
b. Rhynchonella Cuvieri.
c. Inoceramus mytiloides.
d. Lima Hoperi.
e. Ammonites varians.
f. Baculites anceps.
g. Scaphites equalis.
h. Turrilites costatus.

Figure 10.13. Jukes's plates of Cretaceous and Eocene fossils in Darwin's time.

These plates from an outstanding textbook of Darwin's time illustrate the fossil contrast between the Cretaceous chalks and the Eocene sands and clays, especially the London Clay. Lyell and Darwin suspected a very large hiatus, missing time, at the Cretaceous–Palaeogene boundary. Note, in the Lower Eocene, *Nipadites umbonatus*, which is the fruit of *Nypa*, the mangrove palm, the pre-eminent botanical signifier of tropical conditions during the Early Eocene climatic optimum (EECO) at high northern latitudes and also in the AAG.

Source: Adapted from Joseph Beete Jukes, *The student's manual of geology* (1862).

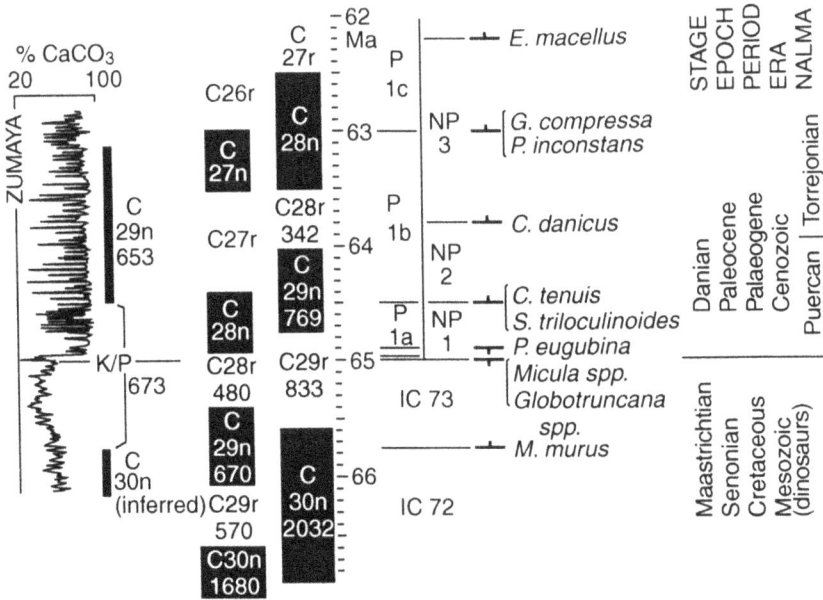

Figure 10.14. Nailing the K-Pg, the Cretaceous/Palaeogene boundary.

This is a very strong example of consilience. At the centre is the numerical scale, from 66 to 62Ma. To its right, planktonic foraminiferal (P zones) and coccolith (NP and IC zones) events are calibrated to numerical ages, which come from radiochronology. Further right are divisions of the geological time scale, from stage to era, plus the North American Land Mammal Ages, which are terrestrial zones. To the left are the magnetostratigraphic chrons recording reversals in Earth's magnetic field (n, normal; r, reversed) as perceived respectively in the 1980s and 1990s. Numbers in the latter are estimated durations in 10^5 years. Further left is a cyclostratigraphic record of limestone-shale rhythms in percent limestone, yielding estimates in thousands of years. It took careful integrating with geomagnetic stratigraphy and correlations between the pelagic, neritic and terrestrial realms to actually confirm that the dinosaurs disappeared at the same time, work that was still in progress when the bolide hypothesis arrived and the rigorous cyclostratigraphic confirmation of synchronicity was still to come.

Source: Adapted from McGowran (2005a, Figure 7.8).

Back to the central point, the K-Pg: many species of planktonic coccoliths were found to disappear at exactly the same level as all but two of the planktonic foraminifera and the last of the ammonites. The mass extinction event among microfossils in the photic zone in the pelagic realm seemed to match the extinction of the ammonites and marine reptiles in the neritic and pelagic realms. That coming-together soon fixed the paperwork on defining the Cretaceous–Palaeogene boundary.[26] And it seemed that the dinosaurs went out then, too. So, when the physicist Alvarez and his team in 1980 produced the theory of asteroidal impact based on the iridium anomaly at exactly the level in question, the literal and metaphorical bolt from the blue was as beautiful an event as even a physicist or chemist might dream of. However, in 1980 it was still being established that catastrophe struck simultaneously in the pelagic, neritic and terrestrial realms. The correlation chart that seems to indicate simultaneity between the realms was assembled years after the theory assumed it (Figure 10.14).

Hardly as exciting as the bolide of doom itself, the chart was not likely to feature in the mass media. But it is impossible to overestimate the importance of getting the ages right, or the impact of new techniques for establishing correlation and age determination.

'What happened to the dinosaurs?' is among the hardiest of perennial scientific and popular-scientific topics, not least due to the pulling power of the dinosaurs themselves. Theories of extinction probably run into the hundreds. They have ranged from the ecological (e.g. asthma, triggered by the newly arisen flowering plants; their eggs were eaten by the newly arisen mammals) through the terrestrial and environmental (extreme volcanism; climatic change, too hot or too cold, too dry or too wet) to the celestial (cosmic rays, comets, meteors). In the realm of the absurd, just one example: iron stains on a rock face imaged a brontosaur, implying human cave art produced before the dinosaurs missed Noah's boat. Two vertebrate palaeontologists summarised many of the theories: Dale Russell in 1979 and Michael Benton in 1990,[27] by which time all the theories and models had settled into one of two modes—predictably enough, William

---

26   The pale chalks of western Europe culminate in the Danian Stage in Denmark, in which no ammonites, belemnites or marine reptiles were found. But because the obvious boundary in the Danish strata was at the top of the Danian where the chalks suddenly stopped and the fossils were very different, the stratigraphers of western Europe placed the Danian Stage at the top of the Cretaceous System and Mesozoic Erathem, not at the base of the Cenozoic Erathem and its divisions. This made things difficult—clumsy and inconvenient and misleading—for us on the opposite side of the planet. We micropalaeontologists badly needed the K/Pg boundary clearly at the base of the Danian, not ambiguously at its top. As I pontificated in 1968: 'The stratigraphic distribution of the planktonic foraminifera and nannoplankton gives very strong support to the contention by de Grossouvre at the turn of the century, that the extinction of the ammonites marks the close of the Mesozoic' (1968a, p. 354). In an ideal world, solemn committees know that global beats parochial or regional. In this case they made the right decision. I have been on the right side and the wrong side in these decisions. Right feels better.
27   Russell (1971); Benton (1990). Benton has a chapter on mass extinction and a list of the hypotheses in his *The dinosaurs rediscovered: How a scientific revolution is rewriting history* (2019). That word 'revolution' again. What is meant is an array of new techniques being used on a stream of fossil discoveries in the fervent biogeohistorical culture of our times—that is, our surge #VIII.

Whewell's uniformitarianism or catastrophism, respectively more ecological and gradualist and more sudden and violent. There are numerous reasons for rejecting many of the suggestions down the decades, but I want to focus on one particular theme, and that is that almost all the theories could be downgraded or dismissed out of hand because they failed to address one simple question: what was happening in the neritic and pelagic realms during the dinosaurs' crisis on the lands?

Let me pose the follow-on question demanding high-precision chronology and interdisciplinary tools: what happened next?

Dynamic earth, or better, mobile earth's crust was a minority belief until the 1960s, when visionaries were glimpsing the making and breaking of supercontinents and the birth and death of oceans. This was the new paradigm of continental drift and plate tectonics, and it added a new dimension to the old question of global climatic change. While the Cenozoic Era was being sorted into its epochs in the earlier nineteenth century, it emerged that the times immediately antecedent to ours were much colder, and before that, much warmer. Initiating in Pleistocene research and seeping down into the Palaeogene were three transforming scientific and biogeohistorical innovations, each owing to the technology of sampling modern and ancient oceans. The first was biogeographic—oceanic organisms are distributed according to the global temperature gradient, so changes in distribution through time should inform us about environmental changes through time. The second was biogeochemical, meaning initially the stable isotopes of oxygen and carbon. The third was the cyclostratigraphy generated by the influence of earth–moon–sun cycles on earthly environments through time. Indeed, I would go so far as to say that the outstanding development down these decades has been a three-way rebalancing of our imbalances between the discovered fossils themselves, their refined and precise dating, and their revealed environmental context. From 1750 to 1950 our knowledge of the record of fossils in strata, the distribution of shells and bones in space and time, accumulated much faster than did its context of environments at the local, regional and global levels. One major reason was the central role of fossils in building the time scale and in economic exploration, in Europe, in the European empires and globally. The variational evolutionists Darwin, Matthew, Simpson were always conscious of the ecological context of organic evolution—that is, the environment—but what that actually meant was another matter. The ecological saga of the horses, browsing before shifting to grazing through the Neogene desiccation, was a great textbook exemplar of evolution in action but rather a lonely one. As we have seen, progress in understanding ancient environments accelerated in the 1950s–1960s.

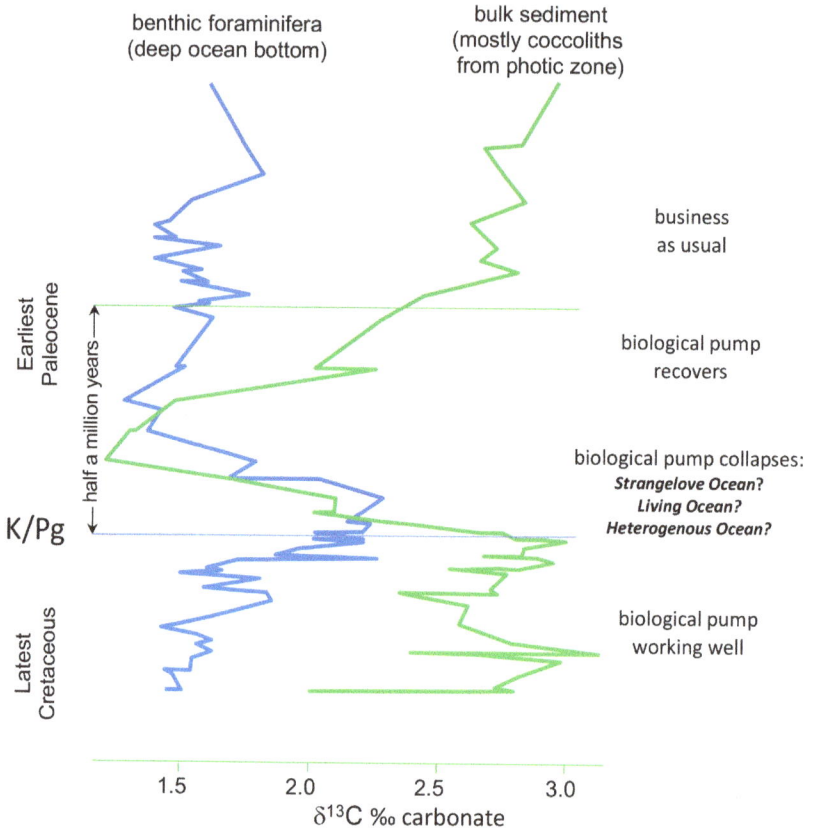

Figure 10.15. K-Pg collapse and restoration of the biological pump.

Resources known as food or nutrient grow in the photic zone of the global ocean and some is exported to the depths. The stable functioning of this biological pump is confirmed by the carbon numbers in shells of foraminifera — the isotopic carbon gap between surface calcifiers (heavier) and deep-bottom calcifiers (lighter). The overlap just above the K-Pg extinction level implied collapse, and a dead ocean ('Strangelove'). That theory turned out to be too extreme ('Living ocean'), and the global ocean turns out to be geographically variable in its response to the impact of the Chicxulub bolide ('Heterogeneous').

Source: Adapted from Alegret and Thomas (2013, Figure 5).

Arising from oceanic drilling was the notion of the Strangelove Ocean. Recall the Monterey isotopic profiles from the Ninetyeast Ridge illustrating the biological pump operating during the Miocene. Light carbon, C_{org}, was fixed photosynthetically in the photic zone, leaving a somewhat heavier reservoir for the shells' C_{carb}. C_{org} was exported to the depths, and that signal was duly preserved in the benthic foraminiferal C_{carb} down there. We see the same effect in the Late Cretaceous (Figure 10.15).

Below the K-Pg boundary the surface–bottom contrast implies that the pump is working. Immediately above the boundary the profiles merge and even cross over. This pattern of no carbon-isotopic gradient implied that there was no photosynthetic fixing of carbon, therefore no export, no profile of carbon isotopes from mixed layer through thermocline to bottom. The pump had collapsed, and the ocean as we know it was unrecognisable—no thermohaline circulation, no halothermal circulation, no photosynthesising coccoliths, no planktonic or benthic foraminifera with photosymbionts. Was this not the comprehensive devastation of the carbonate factories of the global ocean, an outcome of catastrophic impact, the global darkness and the fearsome nuclear winter? Hence, 'Strangelove Ocean'.

The benthic foraminifera deep down below depended on the biological pump. We had known since the late 1950s that the deep-water benthic foraminifera did not follow their planktonic cousins into the K-Pg extinction.[28] For example, I found in the deep Indian Ocean that the benthic change was at the end of the Palaeocene, not at its beginning. Characteristically Cretaceous species were spilling into the Palaeocene in the deep-oceanic realm. Deep-ocean benthic foraminiferal extinction is as little as 3 per cent in some places, much higher in others. Their food had to come from somewhere up in the light. This misfit became significant in clarifying the impact hypothesis and its aftermath. The ocean suffered a massive, serious but partial collapse not meriting the appellation 'Strangelove', and there arose a scenario of the 'Living ocean'. Robust, strongly calcified species especially *Nuttallides truempyi* and *Stensioeina beccariiformis* carried on through, except for a short and very sharp interruption (but not their extinction) in some places, just when two opportunistic species were the only planktonic survivors in the waters above.

---

28 Beckmann (1960).

Accumulating evidence suggests that the impact of the Chicxulub bolide caused a strong acidification of the global ocean as a major reason for ecological collapse in the neritic and pelagic realms. This would have discriminated between the planktonic foraminifera and the coccoliths on the one hand and pretty much all the other microbes in the surface waters on the other—the opaline diatoms and radiolarians, the organic walled dinoflagellates and numerous organisms without skeletons. Primary productivity was reduced perhaps by 50 per cent. Mass extinction occurred. But the catastrophe was not a uniform blanket; acidification varied across the world and affected the upper waters and not directly the depths.

The deep-sea biotas struggled on for 300,000 years as the biological pump was strengthening. The planktonic community structure, from the surface waters to the thermocline, reinvented itself with new species; the benthic community recovered with its old species. The impact surely affected the deepest waters via profound ecological disruption, but recovery from devastation was strong, and more rapid than had been thought.[29]

It is interesting to compare this reconstruction of the foraminifera (Figure 10.16) with the chart that pioneered all this, Glaessner's in 1937 (Figure 4.9). The extinction was there in plain sight; *Guembelitria* is there as one of the two survivors; *Hedbergella*, the other, was yet to be recognised as a distinct genus. Glaessner's work was repeated and extended into the 1960s,[30] when the even more spectacular demise of the coccoliths was added. All of this gave precision to the disappearance of the ammonites, rudistid bivalves and marine reptiles. But Glaessner's and Otto Renz's discovered extinction as a scientific problem remained in stasis during the Darwinian Restoration as it did in Darwin's day. Much has been made of how the Chicxulub bolide changed our culture by replacing Lyell's uniformitarianism with episodic thinking ('catastrophism') at about the time, 1980, that multiple mass extinctions were being discovered in the library. But episodic thinking had been making headway for a decade and more, as we have seen, and I prefer to point to the consilience of the eighth surge. Palaeobiological thinking was given both teeth and discipline by biostratigraphy and palaeoceanography. Biogeohistory lives!

---

29   This discussion is influenced strongly by Alegret and Thomas (2013), Birch et al. (2016), and Henehan et al. (2019).
30   See especially Renz (1936) and Luterbacher and Premoli Silva (1964).

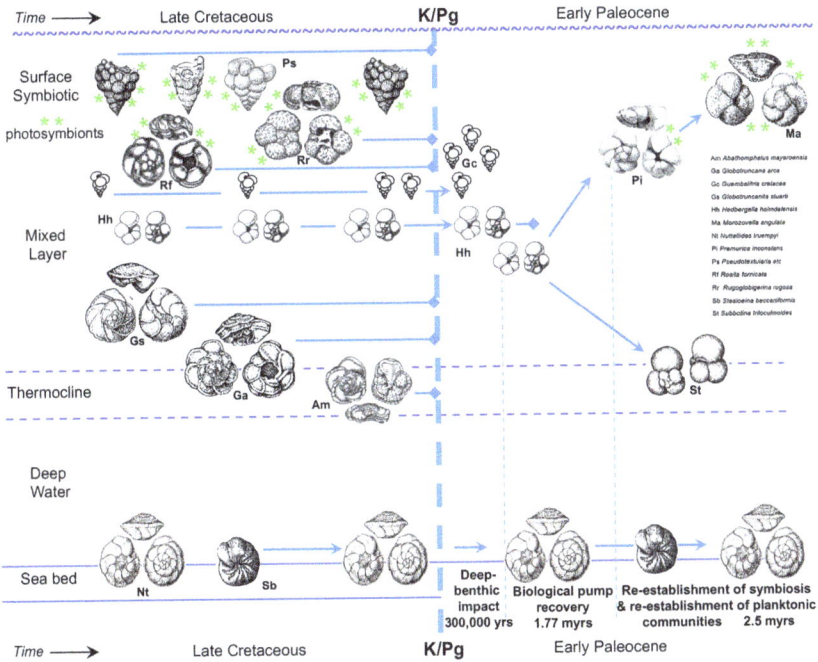

Figure 10.16. Foraminifera reconstructing K-Pg collapse and recovery.

The plankton sort into photosymbiotic at the surface (green symbols), the mixed layer and the thermocline; all but two inconspicuous planktonic lineages simultaneously went extinct (along with the coccoliths also with calcite skeletons). But much of the deep-water benthos did not go extinct. The two species shown, *Nuttallides truempyi* and *Stensioeina beccariiformis*, had robust calcitic skeletons showing no effects of the acidification occurring at shallow depths. And they still had food. The planktonic lineages in the Palaeocene recovery are not shown, but ecological recovery and evolutionary expansion are signified by photosymbiosis returning and the thermocline being reoccupied. That the deep-ocean foraminifera survived the (undoubtedly serious) impact meant that the mass extinction of the calcifiers in the photic zone was not the whole story: bio-production continued in the photic zone even if at half the normal rate or less.

Source: Author's depiction, inspired by Birch et al. (2016, Fig. 1); illustrations from numerous sources.

Epilogue: Eternal tensions revisited

Treating Palaeogene and Neogene as informal biochrons

The discovery of deep time and biogeohistory proceeds at various time scales in the neritic, terrestrial and oceanic or pelagic environmental realms. Therefore 'the' time scale is a work in progress, responding to the growth in reliable knowledge, notwithstanding that its stability down the decades is a highly desirable asset. Stability is maintained by the members of the commissions who ratify by vote the decisions in erecting and massaging the time scale.

We have two main traditions during these decades in managing and parcelling the time, strata and events of the Cenozoic Era at higher levels— *either* into the Tertiary and Quaternary periods *or* into the Palaeogene and Neogene periods.

The Cenozoic Era itself is stable, having been settled when mass extinction was discovered among the calcareous microplankton in the oceanic realm, thereby corroborating at global level the assertion by late-eighteenth-century palaeontologists in Europe that the extinction of the ammonites marked the end of the Mesozoic Era.

The Tertiary and Quaternary periods were more or less stable, perhaps by benign neglect during the plate tectonic revolution and the meteoric rise of palaeoceanography, and the sense that packing biozones, magnetozones and the numerical scale in Ma units into the epochs was sufficient, especially when the two periods were unbalanced so absurdly as 94 per cent and 6 per cent of the era.

That absurdity was one reason why the Palaeogene and Neogene periods came back into favour. More important was the belatedly recognised, 'natural' two-part Cenozoic perceived long ago by Moriz Hörnes in Vienna—'natural' meaning the message of the shelly faunas in Europe, where Hörnes saw that the living faunas were still clearly Neogene. We inhabit a Neogene world.

The 2004 edition of *The geologic time scale* omitted the Quaternary, thereby offending a population of shallow-time scientists, naturalists and environmentalists. The reaction in these times of woke identity politics was to restore the Quaternary Period by decapitating the Neogene. It was proposed that a three-part schema of Palaeogene, Neogene and Quaternary be imposed upon a two-part biogeohistory. In the overriding interests of stability, these changes succeeded. Penalties include the severe distortion of Neogene biogeohistory.

However, the ever-advancing importance of the oceanic realm and deep-ocean drilling are reinforcing a natural two-part Cenozoic with a still-living Neogene biome; and the Quaternary[1] is a perfectly good name for the natural Late Neogene. These are evidence-based statements. The essence of the Quaternarians' position remains that Quaternary's status as sub-Cenozoic instead of sub-period is important to them, as they become more anthropocentric with such outcomes looming as an 'Anthropocene Epoch' and the imbalance in period durations increasing from 6:94 to 4:96.

Biochrons are divisions of time based on organic-evolutionary events. In all three environmental realms we perceive two superbiochrons in the Cenozoic Era, informal entities validating Hörnes' vision of long ago. The hinge is between the Eocene plus Palaeocene and the Miocene plus Pliocene to Holocene.

There have long been indications that the Oligocene belongs as it were partly or entirely with the Miocene, not with the Eocene as it now does, in the Palaeogene. Note, however, that the scenario of Cenozoic climatic states, from hothouse to icehouse, strongly identifies the tipping point of the Cenozoic as very close to the present Eocene–Oligocene boundary. But note too that the Palaeogene and Neogene periods are not included in this grand scenario. They were not missed!

---

1   Gibbard (2019).

Heresy to stratigraphic puritans and stabilists it well may be, but I have interpolated informal and unratified periods between formal and ratified eras above and epochs and ages below. Some diagrams lack a Palaeogene–Neogene boundary. Informality can make good use of fuzzy boundaries.[2]

Consensus lacking regionally on the Eocene–Oligocene boundary

In a somewhat prolonged chain of events, we came to recognise the Eocene–Oligocene boundary at the base of the Chinaman Gully Formation in the Adelaide district, a correlation stiffened by the discovery of geomagnetic Chron C13n in the Chinaman Gully and immediately above it in the Aldinga Member of the Port Willunga Formation, whereas C13r was found immediately below it in the topmost Member of the Blanche Point Formation. The boundary is a ~50 m downcut and backfill in places identifying that the onset of glaciation Oi-1 (Oligocene glaciation-1) on East Antarctica had a pronounced glacioeustatic effect near-field, just across the waters of the Australo-Antarctic Gulf. My Victorian colleagues identified the 'Oi-1 disconformity' above and cutting into the T0 seam, the last of the great Traralgon Coals in Gippsland and the last surge of the conifer forests before the great chill. So far, so good—seemingly a useful regional perception of the most profound change in global climatic states for the entire Cenozoic Era.

However, my colleagues now have shifted their identification of the Oi-1 event from above the T0 coal seam to below it and up into it. The 'Oi-1 disconformity' is shifted 2–3 million years downscale and has become the 'zone of Oi-1 glaciation' encompassing underlying sand and a substantial chunk of the coal seam itself.[3]

The basis for the shift is nearest living relative (NLR), the technique whereby the environmental parameters of living plants might inform us about an ancient flora. According to fluctuations in the NLR numbers, mean annual temperatures fell sharply then rose again. The inference is that

---

2   The geologic time scale is updated from time to time. See Gradstein et al. (2004, 2012, 2020). From simple to complex, from old to young, deep time to shallow time, Wikipedia has answers on topics of time and its classification. But I would add this, on the matter of rigid definitions: a definition can be a fine servant but an oppressive master.

3   Holdgate and Sluiter (2021); Sluiter et al. (2022).

the T0 coal seam preserves *within itself* the change from Warmhouse II to Coolhouse I.[4] No downcut at that time, no apparent glacioeustasy, and the T0 seam becomes of Early Oligocene age.

Eustasy and isostasy — again

In 2007 I wrote, somewhat ebulliently:

> Chronostratigraphy enfolds or pervades all the other stratigraphies. There was a time when 'dynamic' sedimentology pushed aside 'static' and 'dry-as-dust' and 'layer-cake' stratigraphy—or so the young turks of those days would have had us believe. It has been an abiding joy to watch sequence stratigraphy and cyclostratigraphy reassert stratigraphy's rightful place in the scheme of things.[5]

Figures 8.6 and 9.8 in this book display plausible transcontinental (south-east–south-west) matches between microfossil changes in neritic strata, implying changes in sea level in the Late Eocene to Late Miocene in southern Australia, and deep-oceanic strata archiving oxygen-isotopic signals, implying episodes of glaciation. That is the strong basis for suggesting that glacioeustasy was operating, Oligocene to present, on Coolhouse Earth. However, signs of tectonic activity from gentle local tilting to regional uplift were never far away.

In south-east Australia, a new study has detected a shift from a basically tensional regional tectonic regime (i.e. faulting mostly normal; uplift implied absent) to a basically compressional regime (i.e. faulting mostly reverse; uplift explainable).[6]

The changeover from one state to the other happened relatively quickly, roundabout the Eocene–Oligocene boundary. We were acutely aware that isostasy jangles with eustasy in producing transgression, regression and unconformity across continental margins; even so, this coinciding of change in regional tectonic state with change in global climatic state is remarkable.

---

4   The Eocene–Oligocene boundary has been bracketed but not yet captured in the Gippsland and Otway basins, unlike the St Vincent Basin in South Australia, which has both the geomagnetics and earliest Oligocene foraminifera. Holdgate and Sluiter call for (sporomorphic) biostratigraphic research targeting this problem; the boundary is also elusive in the long-studied sections at Browns Creek and Castle Cove (Gallagher et al., 2020).

5   McGowran (2007, p. 81).

6   Mahon and Wallace (2020, 2022).

Moreover, the authors found no relationship between the channels in the Gippsland Palaeogene, such as the Marlin channel, and the unconformities, such as the Marlin unconformity. There was no compression, therefore no uplift; and with depths in these long-lived channels approaching 500 metres, Palaeogene glacioeustasy was ruled out too.

In south-west Australia, a new study challenges the inferred influence of global eustasy on the stratigraphic development of the Dugong supersequence, the carbonates of the Eucla Basin and the Great Australian Bight.[7] Eustasy is relegated to third position, below tectonics and the newly emphasised, vigorous deep circulation believed to produce sedimentary drifts known as contourites. The study was of the images of seismic surveys. Since biostratigraphy based on hands-on data remains very sparse indeed in the Eucla Basin and there is no geomagnetic stratigraphy, seismic analysis of images can be high in plausibility while less than compelling in challenging the unconformities identifying glaciations. Seismic stratigraphy strengthens the scientific case for drilling in the Great Australian Bight at the very time that populist antipathy is against that drilling.

---

7   Stoker, Holford and Totterdell (2022).

Abbreviations

AAG	Australo-Antarctic Gulf
AFS	Auversian Facies Shift
C13n, C13r	Geomagnetic chron, normal and reversed
CCD	calcite compensation depth in the ocean
DSDP	Deep Sea Drilling Project; succeeded by Ocean and International Ocean Drilling Projects
EECO	Early Eocene climatic optimum
EOT	Eocene–Oligocene transition
K-Pg	Cretaceous–Palaeogene boundary (previously K-T)
LOWE	Late Oligocene warming event
MECO	Middle Eocene climatic optimum
Mi-1 (etc.)	Miocene glaciation 1 (etc.) (oceanic signal thereof)
MICO	Miocene climatic optimum
MLET	Middle–Late Eocene oceanic planktonic transition
MMCT	Middle Miocene climatic transition
MPU, IOU, ILU, IMU	the hiatuses separating the four packages of strata in Cenozoic southern Australia
NLR	nearest living relative (of an extinct plant taxon)
ODP, IODP	see DSDP
Oi-1 (etc.)	Oligocene glaciation 1 (etc.) (oceanic signal thereof)
PETM	Palaeocene–Eocene thermal maximum
PrOM	Priabonian oxygen isotope maximum

Glossary

abyssal plain The floor of the ocean.

active continental margin The leading edge as the continent approaches a subduction zone or a continental collision, hence more igneous activity or compressive deformation. The passive margin or trailing edge is tensional, and less deformed. The active/passive binary is rarely used now.

adaptationist In evolutionary thinking, emphasises process especially selection, contrasting with structuralist, which used to emphasise history.

aeolianites Coastal ridges and dunes, limestones winnowed from the limey muds exposed during low sea levels during the Pleistocene. Calcretes mostly are soil limestones derived from the fine fraction blown further.

affinity A noncommittal term pointing to similarities among organisms or organs, useful at one time for avoiding the traps of creation on the one hand or transmutation (evolution) on the other. Also, 'aff.' is useful for pointing to a near relative when uncertain about the firm identification.

ahistorical Thinking about how something is made or how it works, its history in time or place being irrelevant.

albedo The degree to which solar energy is reflected, not absorbed.

algae Used for green organisms ranging from cyanobacteria ('blue-green algae') to single-celled with a nucleus (eukaryotes), some with robust skeletons ('calcareous algae').

allochronous See *synchronous*.

allostratigraphic See *stratigraphy*.

ammonite A mollusc, member of the cephalopods (along with squid, octopus, nautiloid) with chambered shells of aragonite. Abundant and diverse in later Palaeozoic and Mesozoic seas. Together with the trilobites, the ammonites have long been the brand symbols of palaeontology.

anaerobic Environments lacking oxygen, such as black muds, or strata showing no signs of disturbance. Anoxic means much the same thing. A less extreme diagnosis is dysaerobic or dysoxic.

anagenesis Evolution, where speciation, change from ancestor to descendant, occurs without splitting. Previously regarded as the dominant mode of speciation, anagenesis is now believed to be less common than cladogenesis, where the lineage splits.

angiosperms Flowering plants.

Anthropocene The term identifies the time during which human activities have changed the world. Widely used informally and thus useful, but its suggested formal addition to the geological time scale is contested and less useful.

anti-Darwin Not questioning deep time or organic evolution, but not accepting the mechanism natural selection.

aragonite The less common mineral of calcium carbonate (the other is calcite). Coral and most molluscan skeletons are aragonite.

Araucarians A large group of conifers, mostly in the Southern Hemisphere; prominent in the Palaeogene forests around the Australo-Antarctic Gulf (AAG). Colloquially called 'pines'. Australian examples are the bunya-bunya and *Wollemia*.

archetype Pre-evolutionary notion of the basic plan, the mother figure, the central idea, the prototype. Richard Owen's archetypal vertebrate animal is the most famous (Owen 1848, 1860). Archetypes are embedded in homology thinking (see Table 10.1).

Auversian Facies Shift In oceanic stratigraphy, the shift from dominant Palaeogene facies to Neogene facies, roughly the 40–30 Ma span. I expanded it to (what later was named) the warmhouse–coolhouse shift, but the term is obsolescent.

basalt Igneous rock, dense, dark, fine-grained, forming the floor of the deep oceans.

basin, sedimentary Receptacle for the products of erosion and transport, and for precipitation including shells.

bauxite See *laterite*.

benthos, benthic Living on the ocean bottom or in the muds.

biodiversity The number of kinds of organisms in the region, community or sample.

biofacies See *facies*.

biogeography The spatial distribution of organisms and communities, giving clues to changes through time of, for example, changing climate or changing lands and seas.

biogeohistory Historical biology and historical geology became biohistory and geohistory which, in their intertwining, become biogeohistory. Such triple-barrelled terms signify the increasing breakdown of the separate silos of the sciences.

biome, biota Biome is a community held together by environmental controls, such as climate. Biota is a looser term for an assemblage of organisms.

biosphere Collectively, life on earth. It is used along with atmosphere, hydrosphere (ice sheets and glaciers can be calved off as the cryosphere) and reactive lithosphere (the latter meaning the parts of the earth's crust reactive to weathering or hot water).

biostratigraphy See *stratigraphy*.

bolide hypothesis That the extinctions at the end of the Cretaceous Period were caused by the impact on earth of an asteroid or comet.

brackish lid See *estuarine circulation*.

calcite The most common mineral of calcium carbonate, $CaCO_3$.

calcite compensation depth (CCD) The calcite shells of foraminifera and coccoliths fall into the deep ocean, which is acidic due to accumulated CO_2. The shells are attacked chemically. There develops a surface in the ocean below which dissolution exceeds input and no carbonate can accumulate on the floor. This equilibrium surface is the CCD, preserved in drill cores. Its reconstruction in space and time is a valuable tool in palaeoceanography.

calcrete See *aeolianites*.

Cambrian The first Period in the Palaeozoic Era.

carbon, organic Carbohydrate 'CH_2O' in the carbon cycle, shown as C_{org} compared to carbonate carbon, C_{carb}.

carbon isotopes Carbon-12, carbon-13 and carbon-14 (^{12}C, ^{13}C, both stable, and ^{14}C, unstable, radioactive) are the same chemical element. However, the difference in neutron number causes very slight but measurable fractionation of carbon in photosynthesis (plant formation and decay) and in calcification (shell formation and dissolution) as carbon is moved between the ocean, the atmosphere, the biosphere and coalfields and oil and gas reservoirs.

carbon ratios The slight and chemically insignificant shifts in the $^{12}C/^{13}C$ ratio in growing green plants gives lopsidedly clumsy ratios, expressed more elegantly as departures per thousand from an agreed standard set at zero. This is delta-carbon-13 ($\delta^{13}C$). As a shell grows, its calcium carbonate takes on the $\delta^{13}C$ ratio in the water at that place and that time.

catastrophist The mindset that change in nature is sudden, or violent.

chalk Blackboard chalk is, or was, the mineral gypsum. The chalk as in the 'White cliffs of Dover' is mostly fossil skeletons in the mineral calcite.

chemostratigraphy See *carbon isotopes* and *oxygen isotopes*. Chemical changes preserved in the calcium carbonate of fossils, most notably of microscopic foraminifera, is plotted through successive strata, most notably recovered in drilling the ocean floors. Changes in oxygen and carbon ratios can be traced around the planet and shown to be of both environmental and chronological significance. Numerous other chemical series are being compiled as analytical tools are developed.

chronicles What happened and when, from births, deaths and marriages, to village records of the annual harvests, to the range chart of microfossils through a sample section of strata; all are chronicles, the evidence of history.

chronofauna An ecological entity in deep time. As communities evolve, the inhabitants speciating and extinguishing, a recognisable coherence can be maintained for millions of years; until a turnover occurs, the most extreme example being a mass extinction.

chronology Erecting a time scale, getting the dates and correlations correct.

chronostratigraphy See *stratigraphy*.

clade An evolving lineage, a twig or branch on the tree of life, a group with common ancestor (monophyletic) and all descendants.

cladogenesis Speciation, the origin of species, by splitting.

classification See *systematics*.

climatic optimum Episodes of warming, or warm spikes as seen on a chart of oxygen isotopic readings. 'Optimum' on the assumption that warming = good, cooling = bad.

climatic state Earth's average global surface temperature has varied during the Cenozoic Era by more than 20°. The 65-million-year history can be divided at tipping points into meaningful climatic states. See *hothouse, warmhouse, coolhouse* and *icehouse*.

coeval See *synchronous*.

cold ring ocean, warm ring ocean The Neogene and modern ocean and climate are dominated by the Southern Ocean and the Circum-Antarctic Current. This is a cold ring. In Cretaceous times and shrinking in the later Palaeogene, there was Tethys, an ocean in the lower latitudes. This was a warm ring.

conchologist Student of molluscs and their science.

consilience, consilient When different disciplines, theories or lines of evidence point to the same conclusion. The whole is markedly stronger than the sum of the parts.

continent Large area of land geographically outlined by the sea, but defined more meaningfully as a thick slab of earth's crust comprising mostly granitic and metamorphic rocks. A supercontinent has accreted various smaller pieces of continental crust.

convergent evolution Different unrelated clades, pursuing the same lifestyles, come to look remarkably similar. In organisms with a limited range of shapes, such as planktonic foraminifera, this adaptive explanation is probably blurred by chance.

coolhouse Antarctica is icecap-prone at least some of the time; earth's climatic state for most of the Neogene including Quaternary.

correlate In geology, to correlate means believed to have the same age.

cosmopolitan Organisms distributed widely.

craton The oldest and most stable central part of the continent, made up of igneous and metamorphic rocks with thin cover.

cross-section Geologically reconstructing the third dimension, depth, to a two-dimensional map.

crust The outer shell of the earth. Oceanic crust, 5–10 kilometres thick, is mostly basalt or basaltic. Granitic crust, 30–50 kilometres thick, is mostly granite and metamorphic rocks, such as gneiss.

cyanobacteria (aka blue-green algae) Bacteria photosynthesising, as they have done for more than three billion years.

Darwinian Meaning organic evolution and the origin of species by natural selection.

Darwinian Restoration Darwin's supertheory of evolution can be thought of as five theories in, most usefully, two groups: the tree of life and natural selection. Natural selection was widely rejected during the seven 'anti-Darwin decades' until it became widely understood in the 1930–1940s, causing what Ghiselin called the Darwinian Restoration.

deep time Time in millions of years ago or billions. Lately the human-historians and sociologists have begun using 'deep time' for what to me is clearly covered by the neologism 'shallow time'. Thus the educational topic 'Deep Time history of Australia' is about millennia, not mega-anna.

deformation Forces in the earth's crust bend and break into folds and faults, and raise temperatures to cause metamorphism.

desiccation Drying out, of a turd or a continent.

diachronous, diachrony See *synchronous.*

dinocysts The cyst stage of some dinoflagellates, single-celled eukaryotes, some marine and planktonic. Typically 15–100 microns in diameter, dinocysts are made of very tough acid-resistant 'dinosporin', and have an excellent microfossil record in fine-grained sediments out of the reach of oxygen in groundwater.

dip-and-strike In geological mapping, dip is the angle in degrees of a tilted bed below the horizontal; strike is the compass direction of that horizontal line.

disconformity See *unconformity.*

dispersal and vicariance See *vicariance and dispersal.*

duricrust Deep weathering of the continent reduces rocks to clays, releasing silica, iron and calcium (and others). These can precipitate as crusts known variously as silcrete, ferricrete and calcrete.

dysaerobic See *anaerobic.*

ecological gradient Fossils in lateral and vertical arrangements through strata reflect gradients in temperature, salinity, oxygen, etc., pointing to (e.g.) marine transgressions.

ecological partitioning The land and sea and their resources are parcelled out to communities. This is a spectrum, from communities finely divided into specialists to communities coarsely divided into generalists or opportunists. The specialists are said to be in K-mode ecological strategy and the generalists are in r-mode strategy.

ecostratigraphy See *stratigraphy.*

endemic Locally present and persistent.

endogenic, exogenic Forces respectively from within the earth, or in the environment.

endosymbiont See *symbionts*.

environmental realm The big three realms are the terrestrial (the nonmarine), the neritic (shallow marine spilling across the continental margins) and the pelagic (the deep oceans).

epifaunal and infaunal Benthic animals and communities living respectively on the seafloor (or on rocks or plants) and in the mud. Their ratio might indicate nutrition; rich infauna indicates that food is being buried before it can be consumed on or above the surface of the muds.

epoch Informally, a slice of time marked by some event or characteristic. Formally, a division of geological time—for example, the Eocene Epoch.

erosion The wearing away of rock surfaces by water, wind and ice (physical erosion), and by rotting and dissolving (chemical erosion).

erratic A rock abruptly separated from its source, transported by ice.

estuarine circulation Density differences controlled by salinity differences: brackish above normal marine salinity produces surface water outflow. A brackish lid can form at scales up to the oceanic, inhibiting exchange of gases between ocean and atmosphere.

eukaryote An organism's chromosomes are contained within a nucleus.

eustasy Global changes in sea level. Glacioeustasy is due to waxing and waning of ice caps. Tectonoeustasy is due to changes in the volume of the global ocean basin.

eutrophic Rich in nutrients, potentially developing a crisis in oxygen levels.

evo-devo Evolutionary developmental biology.

evolution Most broadly: incessant changing in the biosphere throughout deep time. Most specifically: changes in organisms through speciation (origin of species) and extinction.

evolution, transformational See Table 10.1.

evolution, variational See Table 10.1.

exogenic See *endogenic*.

extinction Removal of a species or taxon, termination of a lineage.

facies Environmental aspect. Attributes that clump, generalise and distinguish, for example, molluscan facies, biofacies, pelagic facies, chemofacies, seismic facies. A most useful concept because of its fuzzy boundaries.

faults Breaks in rocks, due to compression (reverse and thrust faults), or stretching (normal faults) or spreading in the ocean floor in opposite directions (transform faults). Faults clustered under sustained or intense pressure form a shear zone.

ferricrete See *duricrust*.

fluviatile To do with rivers, running water.

foraminifera (forams) Single-celled phylum in the kingdom Protozoa, mostly with shells ('tests') forming a half-billion-year fossil record. The bearers of signals from ancient oceans.

genus (genera) Taxon, next level above species.

geomagnetic chron Repeated reversals of earth's magnetic field are recorded in the rocks in all realms as intervals of normal and reversed polarity, yielding a powerful time scale between the pelagic, neritic and terrestrial realms.

glacial To do with glaciers and glaciations.

glacioeustatic See *eustasy*.

Gondwana (Gondwanaland) The supercontinent giving rise to Australia, Antarctica and India. The short form is more common now.

gradualism See *punctuated equilibrium*.

granite, granitic intrusion Igneous rock, coarse-grained (being intruded in continental crust and cooling slowly). Composed mostly of quartz and the feldspar minerals, with micas.

greenhouse (glasshouse) Environment warmed by atmospheric entrapment of carbon dioxide, methane, water vapour and other gases.

gymnosperms Seed plants without flowers, such as conifers.

gyre In oceanography, a large system of circulating ocean currents.

halothermal See *thermohaline*.

hiatus The time gap marked in strata by the rock relationship called an unconformity.

historical science See Table 10.1.

historicity See Table 10.1.

homology See Table 10.1.

hothouse Earth's climatic state for 9 million years during the Early Eocene.

hotspot (Geology) A focused source of heat below the crust. (Biology) A localised particularly high diversity of plants or animals.

ice effect See *oxygen isotopes*.

icehouse Earth's climatic state for the past 3.3 million years with large icecaps at both poles.

igneous Rocks crystallised from molten material (magma).

infaunal See *epifaunal*.

instructionist See Table 10.1.

isochronous At the same time, spanning the same time.

isostasy Buoyancy, as earth's crust floats in the mantle.

isotherms Lines connecting points with the same temperature.

isotopes Chemical elements in two or more forms, depending on the number of neutrons in the nucleus.

***K*-mode ecological strategy** See *ecological partitioning*.

lagoonal circulation Density differences controlled by salinity differences: normal marine above hypersaline produces surface water inflow, opposite to estuarine circulation.

Lamarckian Inheritance of use and disuse ('soft' inheritance, as opposed to genetics-based 'hard' inheritance). It is a travesty, using like this the name of the discoverer of organic evolution in a deep-time world where the balance of nature is constantly in flux.

land bridge, transoceanic bridge Explaining the dispersal of terrestrial and freshwater animals between continents across deep oceans.

large foraminifera Foraminifera, large (diameters can be over a centimetre, not fractions of a millimetre like the vast majority) with numerous chambers and cavities packed with algal photosynthesisers.

laterite Iron oxides and the clay mineral kaolin, concentrated by intense and deep chemical weathering of the landscape. Bauxite forms where aluminium content is high.

Leeuwin Current Off the western margin and into the AAG then the Bight, this current runs against the anticlockwise oceanic gyre. On the strong evidence of tropical-type organisms immigrating, it has had 'proto-Leeuwin' antecedents from the latest Cretaceous throughout the Cenozoic Era.

lineage A line of descent.

lineaments Lines and alignments across the landscape or seascape with an underlying cause.

lithology The mineral composition of a rock.

lithostratigraphy Distinguishing and arranging bodies of rock.

Lutetian Gap In southern Australia, the critical changes in biogeohistory across the Lower/Middle Eocene boundary are not resolved with clarity and precision in timing.

magma The molten material or mush from which igneous minerals and rocks are precipitated.

Mammerickx microplate 68,000 km^2 piece of the Antarctic plate, detached during the India–Asia collision at about 47 Ma.

marine transgression Where marine facies are found overlying nonmarine facies or an unconformity, we infer that oceanic water spilled on to the continent, expanding the neritic realm. A marine regression would seem to be the reverse, but actually the two are not mirror images.

megafauna Most used specifically for the large animals on most continents and some islands in the Pleistocene but extinct.

megathermal These divisions are used in palaeobotanical reconstructions of ancient climatic states: Megathermal, above 24°C, Meso-Megathermal 20–24°C, Mesothermal 14–20°C, Microthermal 0–14°C.

mesotrophic Nutrient levels, intermediate between eutrophic (high) and oligotrophic (low) levels.

metamorphic Minerals and rocks transformed by heat and pressure.

microbe and macrobe Microbes are usually called microorganisms, microscopic and usually unicellular. Macrobes are multicellular. The terms are potentially useful but rarely used.

microevolution and macroevolution Respectively evolution up to the species level and evolution above the species level.

microflora, marine Refers in palaeoceanography to the fossilisable photosynthesisers, namely coccoliths and discoasters (calcite skeletons), diatoms and silicoflagellates (opal skeletons) and dinocysts (acid-resistant cysts of dinoflagellates).

microfossils Usually 'hard parts', shells of microbes, fragments of colonies, teeth of small mammals, and resistant parts, acid-resistant spores, pollen grains and dinocysts.

midocean ridge The circumglobal mountain system formed by oceanfloor spreading at plate boundaries.

Milankovitch fluctuations Cycles in the motions of the earth–moon–sun recorded in microfossils and strata.

Modern Synthesis The 'Modern Synthesis' is the reconciliation in the 1930–1940s of genetics, palaeontology and taxic biology, aka the Darwinian Restoration.

Monterey hypothesis As recorded in an oceanic carbon and oxygen isotope profile, CO_2 in the Miocene environment is drawn down by carbon burial, triggering the expansion of the West Antarctic icecap.

morphology Studying the form and structure and relationships of organisms and their parts.

moulds For a fossil shell: the external and internal 'negative prints' are the external and internal moulds. Replacing the shell itself forms a cast.

mountain building In plate-tectonic collision, involving deformation, igneous episodes and metamorphism.

Murravian Gulf The Late Oligocene – Middle Miocene seaway in the Murray Basin, reaching not far south of Broken Hill.

natural classification See *systematics*.

natural history Evidence-based discipline, included minerals, rocks, fossils, animals and plants.

natural philosophy Evidence-based discipline, included the experimental disciplines.

natural selection 'This preservation of favourable variations and the rejection of injurious variations, I call Natural Selection' (Darwin, *On the origin of species*, 1859, p. 81).

natural theology The study of nature celebrating the Creation.

nautiloids The first group of the cephalopods, the molluscs with chambered shells.

neontologist Neontology is to palaeontology as shallow time is to deep time. Rarely used by the neontologists themselves.

neotectonics Current or recent movements in earth's crust (but 'recent' is usefully unclear).

neptunism The eighteenth-century theory that rocks, including granites and basalts, precipitated from a universal ocean.

neritic Shallow seas spilling across the continental margins; the realm between the terrestrial and the pelagic.

new systematics A 1940s name for systematics responding to the Darwinian Restoration.

oceanic About the ocean, from the photic zone to the deep-ocean floor.

oceanic front The boundary between two distinct water masses.

oceanography Studying the ocean, its features and its history.

oligotrophic In the ocean, low levels of the nutrients required by phytoplankton, especially nitrate, phosphate and silica.

ontogeny The origin, development and lifespan of an organism.

ooze Biogenic sediment in the deep ocean. Calcareous ooze comprises mostly coccoliths and foraminifera. Siliceous ooze comprises mostly diatoms and radiolarians.

opportunistic Ecologically, rapid responders to environmental stresses, from minor and local to catastrophes.

ordination Geologically, getting the succession correct, of strata and events. Ecologically, statistically extracting a signal from a large sample. Theologically, accepting a priest into holy orders.

organic carbon C_{org}, carbon fixed in photosynthesis. Ancient C_{org} is observed via its effect on C_{carb}, carbonate carbon in shells.

organic evolution All life on earth is related in genealogies, ancestor to descendant, as the tree of life grows for perhaps 4 billion years.

orogenic, Orogeny, Delamerian Late Cambrian – Early Ordovician events, now exposed in the eastern Mt Lofty Ranges and Kangaroo Island.

orthogenetic Organic evolution thought to be driven by some guiding principle, or internal momentum. At one time popular, long obsolete.

oxygen isotopes Oxygen-16, oxygen-17 and oxygen-18 (^{16}O, ^{17}O, ^{18}O) are the same chemical element. However, the difference in neutron number causes very slight but measurable fractionation of oxygen from the reservoir in calcification (shell formation and dissolution) in response to temperature and salinity (the ice effect).

oxygen ratios The slight and chemically insignificant shifts in the $^{16}O/^{18}O$ ratio in growing shells in the sea gives lopsidedly clumsy ratios, expressed more elegantly as departures per thousand from an agreed standard set at zero. This is delta-oxygen-18 ($\delta^{18}O$).

palaeobiological revolution A term used to raise the profile of modern palaeontology.

palaeoceanography Discovering the birth, life and death of ancient oceans.

palaeosol Ancient soil.

palynology Investigating two great groups of microfossils surviving strong-acid digestion of sedimentary rocks; dinocysts, the cysts of some dinoflagellates; and sporomorphs, spores and pollen grains.

parallel evolution Where different lineages have evolved in similar ways.

Paratethys A shallow sea extending from central Europe to western Asia in the Oligocene and Miocene.

passive continental margin See *active continental margin.*

pelagic zone (pelagial) The water column of the open ocean. The pelagic realm is used more commonly here.

photic zone The oceanic waters receiving sunlight. Phytoplankton free-living and in symbiotic partnership photosynthesise in the euphotic zone.

photosymbionts The photosynthesising partners of protists and animals in the sea. The best known is the dinoflagellate *Symbiodinium.*

phyletic evolution Organic evolution.

phylogeny Development of a group of organisms. A branch of the tree of life.

physiographic About landscapes and seascapes.

plankton (**planktonic**; vulgarly, **planktic**) Floating organisms. Among those with significant fossil records, phytoplankton includes the diatoms (opal shells) and coccoliths (calcite shells); zooplankton includes the radiolarians (opal shells) and planktonic foraminifera (calcite shells).

plate tectonics Discovering the plates with their basaltic parts (oceans) and granitic parts (continents); the making and breaking of supercontinents and the birth and death of oceans.

podocarps A large group of conifers, mostly in the Southern Hemisphere; prominent in the Palaeogene forests around the AAG. Colloquially called 'pines'.

polar amplification Global warming and cooling produces, respectively, flatter and steeper longitudinal gradients in temperature. Fluctuations are stronger at the poles than in the tropics.

Primary, Secondary, Tertiary, Quaternary The first evidence-based geological timescale, initiated by Arduino in the mid-eighteenth century.

prokaryote An organism's chromosomes not enveloped within a nucleus in the cell.

proto–Leeuwin Current See *Leeuwin Current*.

provenance In geology, reconstructing the origin of sedimentary materials.

punctuated equilibrium The theory of evolution wherein the event of speciation packs in virtually all the change, ancestor and descendant meanwhile remaining in stasis (unchanged). The opposing theory is gradualism. Informed that PE was evolution by jerks, SJ Gould retorted: rather than evolution by creeps.

***r*-mode ecological strategy** See *ecological partitioning*.

radiometric Dating rocks using radioactive isotopes in ancestor/descendant ratios.

range, range chart Compiling the ranges of fossil taxa, at first through the strata, then in maturing discovery through a geological time scale.

ratites Birds, mostly large and flightless, distributed on most of the fragments of the former Gondwanaland.

reactive lithosphere Those parts of the earth's crust within reach of the reactive agents in the hydrosphere and biosphere. Reactive agents include CO_2, water and heat, and plant penetration.

regolith The zone above fresh bedrock: fragmenting, unconsolidated, decomposing rock and dust, and the soils. The processes in the regolith are physical weathering (e.g. expansion and contraction, wetting and drying, root penetration) and chemical weathering (CO_2, plant decay) since the main rock-forming minerals are high-temperature silicates inherently unstable at earth's surface.

revolution A profound and far-reaching change. For example: the discovery of biogeohistory, the discovery of Darwinian evolution and the discovery of plate tectonics.

rift, rift valley A pull-apart rock structure due to tension, producing normal faults.

rock relationships Fundamental examples are: igneous intrusion and extrusion; faults, tensional and compressional; and unconformity (in the rocks) and hiatus (in time).

rocks, kinds of The three classes of rock are igneous, sedimentary and metamorphic.

Rodinia Neoproterozoic supercontinent.

Sahul The land as it appeared during the Pleistocene ice ages, comprising Tasmania, Australia and New Guinea, facing the land of Sunda across the Wallacean archipelago.

saltation, saltationist In biology, an apparently sudden jump in a lineage, postulated as due to a macromutation among the genes.

savant A learned person, including those now called scientists.

science Robust organised knowledge; conjecture and refutation; evidence-based theories of the universe.

selectionist See Table 10.1.

sequences Unconformities separate packages of strata which have time significance. This is allostratigraphy, which unifies the specialist stratigraphies when we construct biogeohistory. Unconformities used to be the dumping ground for our ignorances, as in the 'imperfections' of the fossil record in Darwin's day. No longer.

shallow time See *deep time*.

silcrete See *duricrust*.

silicates Rock-forming minerals based on silicon and oxygen: quartz, feldspars, micas, the dark minerals and clays.

siliceous, siliciclastics A general term for non-carbonate sediments: clay, silt, sand, etc.

Southern Ocean The ocean around Antarctica is oceanographically distinct and indeed the engine room of the global ocean and environment. But its geographic distinctness from the Atlantic, Indian and Pacific Oceans is legalistic and arbitrary.

spatial, temporal, spatiotemporal Words to do with space and time.

speciate To form a new species. See *anagenesis*.

species The basic unit of classification and the most inclusive unit that exists (lives) in nature, although populations rather than species are the functioning ecological unit.

sporomorphs See *palynology*.

spreading, oceanfloor pattern The expansion of crustal plates by upwelling basaltic magma.

stasis The theory of evolution stating that species are not gradually changed by environmental selective pressures.

stratigraphy The study of layered rocks, mostly sediments, as documents of space and time and events in earth history and life history. The foundation documents were fossils, thus biostratigraphy, and sediments, lithostratigraphy. Other names followed as their subject matter was discovered and put to use: thus magnetostratigraphy, chemostratigraphy, cyclostratigraphy. As well as their classical use in indicating time and geological age, fossils order and arrange environmental shifts (ecostratigraphy). Developing and repairing the geological time scale is chronostratigraphy. See also *chronology* and *sequences*.

stratum, strata Layers of sediment.

striations Lines and grooves and scratches on pavements and boulders due to glacial transport. The boulders become faceted in the process.

structuralist See Table 10.1.

structure, structural evolution Faults, folds, any other deformation or configuration; nowadays 'tectonics' has more impact.

succession, stratigraphic Compiling strata, fossils and interpreted events in the correct order in time. See also *ordination*.

Sunda See *Sahul.*

superposition The principle that strata lower are older, higher are younger. (Deceitful pseudo-exceptions are downcuts and backfills, or upending of strata in mountains. Beware also the activities of wombats, rabbits and termites.)

symbionts, symbiosis The partnerships such as between corals and algae, or between benthic and planktonic foraminifer on the one hand and the photosynthesising algae (diatoms or dinoflagellates) on the other. This photosymbiosis has been reinvented repeatedly, as has been detected in the fossil record on carbon isotopes.

synchronous Happening at the same time, like coeval (being at the same time), unlike allochronous (at different times), or diachronous (through time). Lines joining points of same time are isochrons.

systematics, taxonomy, classification Systematics is the study of diversity of organisms and relationships among them. Classification is ordering and arranging organisms in groups based on their relationships. Taxonomy is the theory of classification.

taxon A taxonomic group, big or small.

taxonomy See *systematics*

tectonic Movements in the earth's crust and their effects.

tectonoeustatic See *eustasy*.

temperature Gradient change in temperature quantified through space (such as equator to pole) or time.

temporal To do with time.

temporal entities Subdivisions of a stretch of time.

terrestrial The land environment, nonmarine; contrast with neritic and oceanic.

Tethys, Tethyan Ocean The ancient ocean at low latitudes, destroyed by continental collisions.

thermocline The oceanic boundary between warmer surface waters and the bulk of the deep dark cold ocean.

thermohaline, halothermal Thermohaline, density-driven by cold Antarctic bottom water, as today; contrasts with halotherml, density-driven by warm brines in the days of Tethys.

three-legged stool See Table 10.1.

thrust fault See *faults*.

time's arrow, time's cycle See Table 10.1

tipping point Critical point of no return, applicable to earth's Cenozoic climatic states.

topographic Physical features on the surface.

transform fault Boundary between two crustal plates moving in opposition.

transgression See *marine transgression*.

tree of life The metaphor for organic evolution: all life is related.

trench, oceanic Trenches at subduction zones, the deepest parts of the ocean.

trophic Feeding relationships among members of a community. See eutrophic, mesotrophic and oligotrophic.

two-way time In seismology, the time taken for a seismic wave to travel from source to a reflector and return to a reader.

unconformity A break in a succession of rocks indicating a hiatus in time. Tilting or folding during the hiatus produces an angular unconformity. Without the tilting, sometimes called a disconformity. Strata draped over igneous or metamorphic rocks produce a nonconformity (not to be confused with an intrusion!).

uniformitarian Natural laws and processes have operated in the past as they operate in the present. The shibboleth 'The present is the key to the past' gets less attention than the equally valid 'The past is the key to the present'.

uplift Buoyancy (e.g. of an iceberg or a continent) responding to unloading (by erosion or ice melting) or to thickening by tectonic compression.

upwelling Deeper waters brought up at an oceanographic boundary. Its significance is to bring rich nutrient-laden waters within reach of photosynthesising organisms (phytoplankton).

vicariance and dispersal Dispersal is migration into a new habitat; in vicariance the population is split by a new barrier, for example a change in climate and vegetation, a rise in sea level, or even a continent fragmenting.

Wallacea, Wallacean archipelago The biogeographically rich zone of islands between Australian Sahul and Asian Sunda.

Walther's law of facies Disruptions aside, a vertical progression of facies is caused by a lateral succession of depositional environments.

warm ring ocean See *cold ring ocean*.

warmhouse Earth's climatic state for much of the Palaeogene.

water mass A body of oceanic water with temperature, salinity and CO_2 load distinct from its neighbours.

weathering, chemical and physical See *regolith*.

Whiggish Believing in inevitable progress, and judging the past in terms of the present. In biogeohistory, focusing on the growth of reliable knowledge, not on the ebb and flow of ideological fashion.

Wilson cycle The making and breaking of supercontinents and the birth and death of ancient oceans.

younging The direction in which the rocks, especially strata, or their characteristics are decreasing in age.

Zealandia The continent, mostly submerged, east of Australia.

zone In stratigraphy, strata characterised by their fossils or magnetic signature, or other consistent and useful properties. Strictly or a touch pedantically (but the distinction has its champions), the rocks are the biozone or magnetozone and the equivalent time is the biochron or magnetochron.

References

Abele, C & McGowran, B. (1959). The geology of the Cambrian south of Adelaide (Sellick Hill to Yankalilla). *Transactions of the Royal Society of South Australia*, 82, 301–320.

Abreu, VS, Hardenbol, J, Haddad, GA, Baum, GR, Droxler, AW & Vail, PR. (1998). Oxygen isotope synthesis: A Cretaceous ice-house? In P-C de Graciansky, J Hardenbol, T Jacquin & PR Vail (eds), *Mesozoic and Cenozoic sequence stratigraphy of European basins* (pp. 75–80). SEPM (Society of Sedimentary Geology) Special Publication No. 60. doi.org/10.2110/pec.98.02.0003.

Alegret, L & Thomas, E. (2013). Benthic foraminifera across the Cretaceous/Paleogene boundary in the Southern Ocean (ODP Site 690): Diversity, food and carbonate saturation. *Marine Micropaleontology*, 105, 40–51. doi.org/10.1016/j.marmicro.2013.10.003.

Alley, NF & Lindsay, JM. (1995). Tertiary. In JF Drexel & WV Preiss (eds), *The Geology of South Australia. Volume 2: The Phanerozoic* (pp. 150–217). South Australia Department of Mines and Energy, Bulletin 54.

Almond, DO, McGowran, B & Li, Q. (1993). *Late Quaternary foraminiferal record from the Great Australian Bight and its environmental significance.* Memoirs of the Association of Australasian Palaeontologists 15.

Amundson, R. (2007). *The changing role of the embryo in evolutionary thought: Roots of evo-devo.* Cambridge University Press. doi.org/10.1017/CBO9781139164856.

Amundson, R. (2014). Charles Darwin's reputation: How it changed during the twentieth-century and how it may change again. *Endeavour*, 38, 257–267. doi.org/10.1016/j.endeavour.2014.10.009.

Archibald, JD. (2009). Edward Hitchcock's pre-Darwinian (1840) 'tree of life'. *Journal of the History of Biology*, 42, 561–592. doi.org/10.1007/s10739-008-9163-y.

Baatsen, M, von der Heydt, AS, Huber, M, Kliphuis, MA, Bijl, PK, Sluijs, A & Dijkstra, HA. (2020). The Middle-to-Late Eocene greenhouse climate, modelled using the CESM 1.0.5. *Climate of the Past.* doi.org/10.5194/cp-2020-29.

Bakker, R. (1986). *The dinosaur heresies.* Penguin Books.

Bartlett, ME. (2006). Crespin, Irene (1896–1980). *Australian Dictionary of Biography*, online edition. National Centre of Biography, Australian National University. Retrieved 22 September 2022 from: adb.anu.edu.au/biography/crespin-irene-9863.

Beckmann, JP. (1960). *Distribution of benthonic foraminifera at the Cretaceous/Tertiary boundary of Trinidad (West Indies).* Report of the International Geological Congress, 21st Session, Norden, Copenhagen. Part 5, pp. 57–69.

Belon, P. (1955). *L'Histoire de la nature des oyseaux, avec leurs descriptions et naïfs portraicts retirez du naturel, escrite en sept livres* [*The history of the nature of birds, with their descriptions and naïve portraits taken from the natural, written in seven volumes*]. Paris.

Belperio, AP (compiler). (1995). *Quaternary.* In JF Drexel and WV Preiss (eds), *The Geology of South Australia. Vol. 2, The Phanerozoic*, 219–280. Geological Survey of South Australia, Bulletin 54.

Benton, MJ. (1990). Scientific methodologies in collision: The history of the study of the extinction of the dinosaurs. *Evolutionary Biology*, 24, 371–400.

Benton, MJ. (2019). *The dinosaurs rediscovered: How a scientific revolution is rewriting history.* Thames & Hudson.

Berger, WH. (1979). Impact of deep-sea drilling on paleoceanography. In M. Talwani, WW Hay, and WBF Ryan (eds), *Deep drilling results in the Atlantic Ocean: Continental margins and paleoenvironment* (Maurice Ewing Series Vol. 3, pp. 297–314). American Geophysical Union.

Berger, WH. (2009). *Ocean: Reflections on a century of exploration.* University of California Press. doi.org/10.1525/9780520942547.

Berger, WH. (2011). Geologist at sea: Aspects of ocean history. *Annual Review of Marine Science*, 3, 1–34. doi.org/10.1146/annurev-marine-120709-142831.

Berger, WH. (2013). On the beginnings of palaeoceanography: Foraminifera, pioneers and the *Albatross* expedition. In AJ Bowden, FJ Gregory & AS Henderson (eds), *Landmarks in foraminiferal micropalaeontology: History and development* (pp. 159–179). The Micropalaeontological Society, Special Publications, Geological Society of London. doi.org/10.1144/TMS6.13.

Berger, WH & Vincent, E. (1986). Deep-sea carbonates: Reading the carbon-isotope signal. *Geologische Rundschau*, 75, 249–269. doi.org/10.1007/BF01770192.

Berger, WH & Wefer, G. (1996). Expeditions into the past: Paleoceanographic studies in the South Atlantic. In G Wefer, WH Berger, G Siedler & DJ Webb (eds), *The South Atlantic: Present and past circulation* (pp. 363–410). Springer-Verlag. doi.org/10.1007/978-3-642-80353-6_21.

Berggren, WA. (1960). Paleogene biostratigraphy and planktonic foraminifera of the SW Soviet Union: An analysis of recent Soviet publications. *Stockholm Contributions in Geology*, 6, 63–125.

Berggren, WA. (1968). Phylogenetic and taxonomic problems of some Tertiary planktonic foraminiferal lineages. *Tulane studies in Geology and Paleontology*, 6, 1–22.

Berggren, WA, Olsson, RK & Reyment, RA. (1967). Origin and development of the foraminiferal genus *Pseudohastigerina* Banner and Blow, 1959. *Micropaleontology*, 13, 265–288. doi.org/10.2307/1484830.

Berggren, WA, Pearson, PN, Huber, BT & Wade, BS. (2006). Taxonomy, biostratigraphy, and phylogeny of Eocene *Acarinina*. In PN Pearson, RK Olsson, BT Huber, C Hemleben & WA Berggren (eds), *Atlas of Eocene planktonic foraminifera* (Cushman Foundation Special Publication 41, pp. 257–326). Allen Press.

Bernecker, T, Partridge, AD & Webb, JA. (1997). Mid–Late Tertiary deep-water carbonate deposition, offshore Gippsland Basin, southeastern Australia. In NP James & J Clarke (eds), *Coolwater carbonates in space and time* (pp. 21–236). Society of Economic Paleontologists and Mineralogists Special Volume No 56. doi.org/10.2110/pec.97.56.0221.

Bijl, PK. (2011). *Environmental and climatological evolution of the Early Paleogene Southern Ocean* [PhD thesis, Laboratory of Palaeobotany and Palynology, Utrecht University]. LPP Contributions Series No 4.

Bijl, PK, Houben, AJP, Hartman, J, Pross, J, Salabarnada, A, Escutia, C & Sangiorgi, F. (2017). Oligocene–Miocene paleoceanography off the Wilkes Land Margin (East Antarctica) based on organic-walled dinoflagellate cysts. *Climate of the Past Discussions*. Manuscript under review. doi.org/10.5194/cp-2017-148.

Bijl, PK, Sluijs, A, & Brinkhuis, H. (2013). A magneto- and chemostratigraphically calibrated dinoflagellate cyst zonation of the Early Palaeogene South Pacific Ocean. *Earth Science Reviews* 124, 1–31. doi.org/10.1016/j.earscirev.2013.04.010.

Birch, HS, Coxall HK, Pearson, PH, Kroon, D & Schmidt, DN. (2016). Partial collapse of the marine carbon pump after the Cretaceous–Paleogene boundary. *Geology*, 44(4), 287–290. doi.org/10.1130/G37581.1.

Black KH, Archer, M, Hand, SJ & Godthelp, H. (2012). The rise of Australian marsupials: A synopsis of biostratigraphic, phylogenetic, palaeoecologic and palaeobiogeographic understanding. In JA Talent (ed.), *Earth and life* (pp. 983–1078). Springer Science + Business Media BV. doi.org/10.1007/978-90-481-3428-1_35.

Blow, WH. (1956). Origin and evolution of the foraminiferal genus Orbulina d'Orbigny. *Micropaleontology*, 2, 57–70. doi.org/10.2307/1484492.

Blow, WH. (1969). Late Middle Eocene to recent planktonic foraminiferal biostratigraphy. In P Brönnimann & HH Renz (eds), *Proceedings of the First International Conference on Planktonic Microfossils* (vol 1, pp. 199–421). EJ Brill.

Bohaty, SM & Zachos, JC. (2003). Significant Southern Ocean warming event in the late Middle Eocene. *Geology*, 31, 1017–1020. doi.org/10.1130/G19800.1.

Bohaty, SM, Zachos. JC, Florindo, F & Delaney, ML. (2009). Coupled greenhouse warming and deep-sea acidification in the Middle Eocene. *Paleoceanography*, 24(2), PA2207. doi.org/10.1029/2008PA001676.

Bolli, HM. (1957). The genera *Globigerina* and *Globorotalia* in the Paleocene–lower Eocene Lizard Springs Formation of Trinidad, BWI. *United States National Museum Bulletin*, 215, 51–81.

Bourman, RP. (1995). A review of laterite studies in South Australia. *Transactions of the Royal Society of South Australia,* 119, 1–28.

Bourman, RP. (2007). Deep regolith weathering on the summit surface of the Southern Mount Lofty Ranges, South Australia: A contribution to the 'Laterite' debate. *Geographical Research*, 45, 291–299. doi.org/10.1111/j.1745-5871.2007.00461.x.

Bourman, RP & Ollier, CD. (2002). A critique of the Schellmann definition and classification of 'laterite'. *Catena*, 47, 117–131. doi.org/10.1016/S0341-8162(01)00178-3.

Bowler, JM, Kotsonis, A & Lawrence, CR. (2006). Environmental evolution of the Mallee region, Western Murray Basin. *Proceedings of the Royal Society of Victoria* 118(2), 161–210.

Brady, HB. (1884). Report on the Foraminifera dredged by H.M.S. *Challenger* during the Years 1873–1876. *Report on the Scientific Results of the Voyage of H.M.S. Challenger during the years 1873–76. Zoology* 9 (part 22), i–xxi, 1–814; pl. 1–115.

Breidbach, O & Ghiselin, MT. (2007). Evolution and development: Past, present and future. *Theory in Biosciences* 125, 157–171. doi.org/10.1016/j.thbio.2007. 02.001.

Briguglio, A, Hohenegger, J & Less, G. (2013). Paleobiological applications of three-dimensional biometry on larger benthic foraminifera: A new route of discoveries. *Journal of Foraminiferal Research*, 43(1), 72–87. doi.org/10.2113/ gsjfr.43.1.72.

Browne, J. (2007). *Darwin's origin of species: A biography*. Atlantic Books.

Buckland, W. (1837). *Geology and mineralogy, considered with reference to natural theology*. London: William Pickering. doi.org/10.5962/bhl.title.25353.

Calcott, B & Sterelny, K. (2011). Introduction: A dynamic view of evolution. In B Calcott & K Sterelny (eds), *The major transitions of evolution revisited* (pp. 1–14). The MIT Press. doi.org/10.7551/mitpress/9780262015240.003. 0001.

Callen, RA. (2020). Neogene Billa Kalina Basin and Stuart Creek silicified floras, northern South Australia: A reassessment of their stratigraphy, age and environments. *Australian Journal of Earth Sciences*, 67(5), 605–626. doi.org/ 10.1080/08120099.2020.1736630.

Campana, B. (1955). The structure of the eastern South Australian ranges: The Mt. Lofty-Olary arc. *Journal of the Geological Society of Australia*, 2, 47–61. doi.org/10.1080/08120095409414131.

Campana, B. & Wilson, RB. (1953). *The geology of the Jervis and Yankalilla military sheets: Explanation of the geological maps*. Department of Mines, South Australia, Geological Survey, Report of Investigations 3.

Cann, JH & Clarke, JDA. (1993). The significance of *Marginopora vertebralis* (Foraminifera) in surficial sediments at Esperance, Western Australia, and in last interglacial sediments in northern Spencer Gulf, South Australia. *Marine Geology*, 111, 171–187. doi.org/10.1016/0025-3227(93)90195-2.

Carozzi, AV. (1965). Lavoisier's fundamental contribution to stratigraphy. *Ohio Journal of Science*, 65, 71–85.

Carpenter, RJ, Truswell, EM & Harris, WK. (2010). Lauraceae fossils from a volcanic Palaeocene oceanic island, Ninetyeast Ridge, Indian Ocean: Ancient long-distance dispersal? *Journal of Biogeography*, 37, 1202–1213. doi.org/ 10.1111/j.1365-2699.2010.02279.x.

Carroll, SB. (2005). *Endless forms most beautiful: The new science of evo devo and the making of the animal kingdom*. WW Norton.

Carter, AN. (1964). *Tertiary foraminifera from Gippsland, Victoria, and their stratigraphical significance*. Geological Survey of Victoria Memoir 23. Department of Mines.

Chapman, F. (1915). New or little known fossils in the National Museum, Part 17— Some Tertiary Cephalopoda. *Proceedings of the Royal Society of Victoria*, new series, 27, 350–361.

Christian, D. (2011). *Maps of time: An introduction to big history*. 2nd Edition. Berkeley: University of California Press.

Christian, D. (2018). *Origin story: A big history of everything*. Penguin: Random House UK.

Clode, D. (2006). *Continent of curiosities: A journey through Australian natural history*. Cambridge University Press.

Clode, D. (2007). *Voyages to the South Seas: In search of Terres Australes*. The Migunyah Press.

Coblentz, D, Sandiford, M, Richardson, R, Zhou, S & Hillis, R. (1995). The origins of the intraplate stress field in continental Australia. *Earth and Planetary Science Letters*, 133, 299–309. doi.org/10.1016/0012-821X(95)00084-P.

Colbert, EH. (1973). *Wandering lands and animals*. Dutton.

Colbert, EH. (1993). *William Diller Matthew, paleontologist: The splendid drama observed*. Columbia University Press.

Contreras, L, Pross, J, Bijl, PK, Koutsodendris, A, Raine, JI, van de Schootbrugge, B & Brinkhuis, H. (2013). Early to Middle Eocene vegetation dynamics at the Wilkes Land Margin (East Antarctica). *Review of Palaeobotany and Palynology*, 197, 119–142. doi.org/10.1016/j.revpalbo.2013.05.009.

Contreras, L, Pross, J, Bijl, PK, O'Hara, RB, Raine, JI, Sluijs, A & Brinkhuis, H. (2014). Southern high-latitude terrestrial climate change during the Palaeocene–Eocene derived from a marine pollen record (ODP Site 1172, East Tasman Plateau). *Climate of the Past*, 10, 1401–1420. doi.org/10.5194/cp-10-1401-2014.

Corfield, R. (2004). *The silent landscape: The scientific voyage of HMS* Challenger. Joseph Henry Press.

Cracraft, J. (1974). Continental drift and vertebrate distribution. *Annual Reviews of Ecology and Systematics*, 5, 216–251. doi.org/10.1146/annurev.es.05.110174.001243.

Cramwinckel, MJ, Woelders, L, Huurdeman, EP, Peterse, F, Gallagher, ST, Pross, J, Burgess, CE, Reichart, G-J, Sluijs, A & Bijl, PK. (2019). Surface-circulation change in the Southern Ocean across the Middle Eocene Climatic Optimum: Inferences from dinoflagellate cysts and biomarker paleothermometry. *Climate of the Past Discussions.* Manuscript under review. doi.org/10.5194/cp-2019-35.

Crouch, EM, Shepherd, CL, Morgans, HEG, Naafs, BDA, Dallanave, E, Phillips, A, Hollis, CJ & Pancost, RD. (2020). Climatic and environmental changes across the Early Eocene Climatic Optimum at mid-Waipara River, Canterbury Basin, New Zealand. *Earth-Science Reviews*, 200, 102961. doi.org/10.1016/j.earscirev.2019.102961.

Darragh, TA. (1985). Molluscan biogeography and biostratigraphy of the Tertiary of southern Australia. *Alcheringa* 9, 83–116. doi.org/10.1080/03115518508618960.

Darragh, TA. (1986). The Cenozoic Trigoniidae of Australia. *Alcheringa*, 10, 1–34. doi.org/10.1080/03115518608619039.

Darragh, TA. (1994). Paleocene bivalves from the Pebble Point Formation, Victoria, Australia. *Proceedings of the Royal Society of Victoria*, 106, 71–103.

Darragh, TA. (1997). Gastropoda, Scaphopoda, Cephalopoda and new Bivalvia from the Pebble Point Formation, Victoria, Australia. *Proceedings of the Royal Society of Victoria*, 109, 57–108.

Darwin, C. (1959). *The voyage of the* Beagle. JM Dent & Sons (Original work published 1845).

Darwin, C. (1964). *On the origin of species* [Facsimile of first edition] (E Mayr, ed.). Harvard University Press. (Original work published 1859.) doi.org/10.2307/j.ctvjf9xp5.

Davies, TA, Luyendyk, BP, Rodolfo, KS, Kempe, DRC, McKelvey, BC, Leidy, RD, Horvath, GJ, Hyndman, RD, Thierstein, HR, Herb, RC, Boltovskoy, E & Doyle, P. (1974). *Initial reports of the Deep Sea Drilling Project* (Vol. 26). US Government Printing Office.

Dawson, L. (1985). Marsupial fossils from Wellington Caves, New South Wales: The historic and scientific significance of the collections in the Australian Museum, Sydney. *Records of the Australian Museum, 37*(2), 55–69. doi.org/10.3853/j.0067-1975.37.1985.335.

Dobzhansky, T. (1937). *Genetics and the origin of species*. Columbia University Press.

D'Onofrio, R, Luciani, V, Dickens, GR, Wade, BS & Turner, SK. (2020). Demise of the planktic foraminifer Morozovella during the Early Eocene climatic optimum: New records from ODP Site 1258 (Demerara Rise, western equatorial Atlantic) and Site 1263 (Walvis Ridge, south Atlantic). *Geosciences*, 10, 88. doi.org/10.3390/geosciences10030088.

Douglas, RG & Savin, SM. (1978). Oxygen isotope evidence for the depth stratification of Tertiary and Cretaceous planktic foraminifera. *Marine Micropaleontology*, 3, 175–196. doi.org/10.1016/0377-8398(78)90004-X.

du Toit, AL. (1937). *Our wandering continents*. Oliver and Boyd.

Dugan, KG. (1980). Darwin and Diprotodon: The Wellington Caves fossils and the law of succession. *Proceedings of the Linnean Society of New South Wales*, 104(4), 265–272.

Dunbar, CO & Rodgers, J. (1957). *Principles of stratigraphy*. New York: John Wiley.

Dutton, A, Lohmann, KC & Leckie, RM. (2005). Insights from the Paleogene tropical Pacific: Foraminiferal stable isotope and elemental results from Site 1209, Shatsky Rise. *Paleoceanography*, 20(3), PA3004. doi.org/10.1029/2004 PA001098.

Eggleton, RA (ed.). (1998). The state of the regolith. *Proceedings of the Second Australian Conference on Landscape Evolution & Mineral Exploration*. Geological Society of Australia Inc., Special Publication No 20.

Eiseley, L (ed.). (1967). *On a piece of chalk, by Thomas Henry Huxley*. Charles Scribner's Sons. (Original work published 1868.)

Eldredge, N & Gould, SJ. (1972). Punctuated equilibria: An alternative to phyletic gradualism. In TJM Schopf (ed.), *Models in paleobiology* (pp. 82–115). Freeman, Cooper.

Ereshefsky, M. (2012). Homology thinking. *Biology and Philosophy*, 27, 381–400. doi.org/10.1007/s10539-012-9313-7.

Erwin, DH. (2009). Macroevolution and microevolution are not governed by the same processes. In FJ Ayala & R Arp (eds), *Contemporary debates in philosophy of biology*, (pp. 180–199). Hoboken, NJ: Wiley-Blackwell. doi.org/10.1002/ 9781444314922.ch10.

Escutia, C, Brinkhuis, H & the Expedition 318 Scientists. (2014). From greenhouse to icehouse at the Wilkes Land Antarctic Margin: IODP Expedition 318 synthesis of results. *Developments in Marine Geology*, 7, 295–328. doi.org/ 10.1016/B978-0-444-62617-2.00012-8.

Evans, D, Sagoo, N, Renema, W, Cotton, LJ, Müller, W, Todd, JA, Saraswati, PK, Stassen, P, Ziegler, M, Pearson, PN, Valdes, PJ & Affek, HP. (2018). Eocene greenhouse climate revealed by coupled clumped isotope-Mg/Ca thermometry. *PNAS* 115, 1174–1179. doi.org/10.1073/pnas.1714744115.

Exon, NF, Kennett, JP, and Malone, MJ. (eds). (2004). *Proceedings of the Ocean Drilling Program. Vol. 189: Scientific results, the Tasmanian gateway: Cenozoic climatic and oceanographic development.* Ocean Drilling Program. doi.org/10.2973/odp.proc. sr.189.2004.

Ezard, THG, Aze, T, Pearson, PN & Purvis, A. (2011). Interplay between changing climate and species' ecology drives macroevolutionary dynamics. *Science*, 332, 349–351. doi.org/10.1126/science.1203060.

Falkowski, PG, Katz, ME, Knoll, AH, Quigg, A, Raven, JA, Schofield, O & Taylor, FJR. (2004). The evolution of modern eukaryotic phytoplankton. *Science*, 305, 354–360. doi.org/10.1126/science.1095964.

Feary, DA & James, NP. (1998). Seismic stratigraphy and geological evolution of the Cenozoic, Cool-Water Eucla Platform, Great Australian Bight. *American Association of Petroleum Geologists Bulletin*, 52, 792–816.

Findlay, E. (1998). *Arcadian quest: William Westall's Australian sketches.* National Library of Australia.

Flinders, M. (1814). *A Voyage to Terra Australis: Undertaken for the purpose of completing the discovery of that vast country, and prosecuted in the years 1801, 1802, and 1803, in His Majesty's Ship the* Investigator. G & W Nicol.

Frakes, LA. (1999). Evolution of Australian environments. In A Orchard (ed.), *Flora of Australia* (second edition, pp. 163–203), Australian Government Publishing Service.

Frakes, LA, Probst J-L & Ludwig, W. (1994). Latitudinal distribution of paleotemperature on land and sea from Early Cretaceous to Middle Miocene. *Comptes Rendus Académie Science Paris*, 318, sér II, 1209–1218.

Francis, JE, Marenssi, S, Levy, R, Hambrey, M, Thorn, VC, Mohr, B, Brinkhuis, H, Warnaar, J, Zachos, J, Bohaty, S & DeConto, R. (2009). From greenhouse to icehouse—the Eocene/Oligocene in Antarctica. In F Florindo & M Siegert (eds), *Antarctic climate evolution* (Developments in Earth & Environmental Sciences 8, pp. 309–368). Elsevier. doi.org/10.1016/S1571-9197(08)00008-6.

Frieling, J, Huurdeman, EP, Rem, CCM, Donders, TH, Pross, J, Bohaty, SM, Holdgate, GR, Gallagher, SJ, McGowran, B & Bijl, PK. (2018). Identification of the Paleocene–Eocene boundary in coastal strata in the Otway Basin, Victoria, Australia. *Journal of Micropalaeontology*, 37, 317–339. doi.org/10.5194/jm-37-317-2018.

Frieling, J & Sluijs, A. (2018). Towards quantitative environmental reconstructions from ancient non-analogue microfossil assemblages: Ecological preferences of Paleocene–Eocene dinoflagellates. *Earth-Science Reviews* 185, 956–973. doi.org/10.1016/j.earscirev.2018.08.014.

Gallagher, SJ, Wade, B, Li, Q, Holdgate, GR, Bown, P, Korasidis, VA, Scher, H, Houben, AJP, McGowran, B & Allen, T. (2020). Eocene to Oligocene high paleolatitude neritic record of Oi-1 glaciation in the Otway Basin southeast Australia. *Global and Planetary Change*, 191, 103218. doi.org/10.1016/j.gloplacha.2020.103218.

Gansser, A. (1966). The Indian and the Himalayas: A geological interpretation. *Eclogae Geologicae Helvetiae*, 59, 831–848.

Ghiselin, MT. (1969). *The triumph of the Darwinian method*. University of California Press.

Ghiselin, MT. (1997). *Metaphysics and the origin of species*. State University of New York Press.

Gibbard, PL. (2019). Giovanni Arduino—the man who invented the Quaternary. *Quaternary International*, 500, January, 11–19. doi.org/10.1016/j.quaint.2019.04.021.

Gibson, GM, Totterdell, JM, Moore, MP, Goncharov, A, Mitchell, CH & Stacey, AR. (2012). *Basement structure and its influence on the pattern and geometry of continental rifting and breakup along Australia's southern rift margin*. Geoscience Australia Record 2012/47. Geoscience Australia.

Gibson, GM, Totterdell, JM, White, LT, Mitchell, CH, Stacey, AR, Morse, MP & Whitaker, A. (2013). Pre-existing basement structure and its influence on continental rifting and fracture zone development along Australia's southern rifted margin. *Journal of the Geological Society*, 170, 365–377. doi.org/10.1144/jgs2012-040.

Gillispie, CC. (1959). *Genesis and geology: The impact of scientific discoveries upon religious beliefs in the decades before Darwin*. Cambridge: Harvard University Press (1951), reprinted as Harper Torchbook, Harper and Row (1959). All citations are to the 1959 edition.

Gischler, E. (2011). Rock stars: Johannes Walther. *GSA Today*, August, 14–15.

Glaessner, MF. (1937a). On a new family of foraminifera. *Studies in Micropaleontology*, 1(3), 19–38.

Glaessner, MF. (1937b). Planktonforaminiferen aus der Kreide und dem Eozän und ihre stratigraphische Bedeutung [Cretaceous and Eocene plankton foraminifera and their stratigraphic significance]. *Studies in Micropaleontology*, 1, 27–52.

Glaessner, MF. (1945). *Principles of micropalaeontology*. Melbourne University Press.

Glaessner, MF. (1951). Three foraminiferal zones in the Tertiary of Australia. *Geological Magazine*, 88, 273–283. doi.org/10.1017/S0016756800069600.

Glaessner, MF. (1955). Taxonomic, stratigraphic and ecologic studies of foraminifera and their interrelations. *Micropaleontology*, 1, 3–8. doi.org/10.2307/1484407.

Glaessner, MF, McGowran, B & Wade, M. (1960). Discovery of a kangaroo bone in the Middle Miocene of Victoria. *Australian Journal of Science*, 22, 484–485.

Gould, SJ. (1977). *Ontogeny and phylogeny*. Harvard University Press.

Gould, SJ. (1980). G.G. Simpson, paleontology, and the modern synthesis. In E Mayr & WB Provine (eds), *The evolutionary synthesis* (pp. 153–172). Harvard University Press. doi.org/10.4159/harvard.9780674865389.c23.

Gould, SJ. (2001, 4 October). The man who set the clock back. *The New York Review of Books*.

Gould, SJ. (2002). *The structure of evolutionary theory*. Cambridge & London: Harvard University Press.

Gould, SJ. (2003). *The hedgehog, the fox, and the Magister's pox: Mending the gap between science and the humanities*. Vintage. doi.org/10.4159/harvard. 9780674063402.

Gradstein, FM, Ogg, JG, Schmitz, MD & Ogg, GM (eds). (2012). *The geologic time scale 2012*. Elsevier. doi.org/10.1016/C2011-1-08249-8.

Gradstein, FM, Ogg, JG, Schmitz, MD & Ogg, GM (eds). (2020). *Geologic time scale 2020*. Elsevier. doi.org/10.1016/C2020-1-02369-3.

Gradstein, FM, Ogg, JG & Smith, AG. (2004). *A geologic time scale 2004*. Cambridge University Press. doi.org/10.1017/CBO9780511536045.

Greenwood, DR, Moss, PT, Rowett, AI, Vadala, AJ & Keefe, RL. (2003). Plant communities and climate change in southeastern Australia during the Early Paleogene. In SL Wing, PD Gingerich, B Schmitz & E Thomas (eds), *Causes and consequences of globally warm climates in the Early Paleogene* (pp. 365–380). Geological Society of America Special Paper 369. doi.org/10.1130/0-8137-2369-8.365.

Haeckel, E. (1876). *The history of creation: On the development of the Earth and its inhabitants by the action of natural causes. A popular exposition of the doctrine of evolution in general, and of that of Darwin, Goethe, and Lamarck in particular* (trans. rev. ER Lankester, second edition). Henry S. King & Co. doi.org/10.5962/bhl.title.128563.

Haiblen, AM, Opdyke, BN, Roberts, AP, Heslop, D & Wilson, PA. (2019). Midlatitude southern hemisphere temperature change at the end of the Eocene greenhouse shortly before dawn of the Oligocene icehouse. *Paleoceanography and Paleoclimatology*, 34, 1995–2004. doi.org/10.1029/2019PA003679.

Haig, DW. (2003). Palaeobathymetric zonation of foraminifera from lower Permian shale deposits of a high-latitude southern interior sea. *Marine Micropaleontology* 49, 317–334. doi.org/10.1016/S0377-8398(03)00051-3.

Haig, DW, Griffin, BJ & Ujetz, BF. (1993). Redescription of type specimens of *Globorotalia chapmani* Parr from the Upper Paleocene, Western Australia. *Journal of Foraminiferal Research*, 23, 275–280. doi.org/10.2113/gsjfr.23.4.275.

Haig, DW, Mossadegh, K, Parker, JH & Keep, M. (2019). Middle Eocene neritic limestone in the type locality of the volcanic Barique Formation, Timor-Leste: Microfacies, age and tectonostratigraphic affinities. *Journal of Asian Earth Sciences*, 1, 1–20. doi.org/10.1016/j.jaesx.2018.100003.

Hall, TS & Pritchard, GB. (1902). A suggested nomenclature for the marine Tertiary deposits of southern Australia. *Proceedings of the Royal Society of Victoria*, 14, 75–81.

Haq, BU. (1983). Paleogene paleoceanography: Early Cenozoic oceans revisited. *Oceanologica Acta*, 4 (Supplement), 71–82.

Haq BU, Hardenbol, J, Vail, PR, Stover, LE, Colin, JP, Ioannides, NS, Wright, RC, Baum, GR, Gombos, Jr, AM, Pflum, CE, Loutit, TS, Jan du Chêne, R, Romine, KK, Sarg, JF, Posamentier, HW & Morgan, BE. (1988). Mesozoic and Cenozoic chronostratigraphy and cycles of sea-level change. In CK Wilgus, BS Hastings, CGSC Kendall, HW Posamentier, CA Ross & JC van Wagoner (eds), *Sea-level changes: An integrated approach* (pp. 71–108). Society of Economic Paleontologists and Mineralogists, Special Publication 42. doi.org/10.2110/pec.88.01.0071.

Harris, WK. (1965). Basal Tertiary microfloras from the Princetown area, Victoria, Australia. *Palaeontographica Abteilung B,* 115, 75–106.

Harris, WK. (1971). Tertiary stratigraphic palynology, Otway Basin. In H Wopfner & JG Douglas (eds), *The Otway Basin of Southeastern Australia* (pp. 67–87). Geological Surveys of South Australia and Victoria.

Harris, WK & McGowran, B. (1973). *South Australian Department of Mines Cootanoorina No. 1 Well. Part 2: Upper Palaeozoic and Lower Cretaceous micropalaeontology.* Report of Investigations No 40. Geological Survey of South Australia.

Henehan, MJ, Edgar, KM, Foster, GL, Penman, DE, Hull. PM, Greenop, R, Anagostou, E & Pearson, PN. (2020). Revisiting the Middle Eocene Climatic Optimum 'carbon cycle conundrum' with new estimates of atmospheric pCO2 from Boron isotopes. *Paleoceanography and Paleoclimatology*, 35, e2019PA003713. doi.org/10.1029/2019PA003713.

Henehan, MJ, Ridgwell, A, Thomas, E, Zhang, S, Alegret, L, Schmidt, SN, Rae, JWB, Witts, JD, Landman, NH, Greene, SE, Huber, BT, Super, JR, Planavsky & Hull, PM. (2019). Rapid ocean acidification and protracted Earth system recovery followed the end-Cretaceous Chicxulub impact. *PNAS* 116, 22500-4. doi.org/10.1073/pnas.1905989116.

Herman, A. (2014). *The cave and the light: Plato versus Aristotle, and the struggle for the soul of Western Civilization.* Random House.

Herold, N, Huber, M, Greenwood, DR, Müller, RD & Seton, M. (2011). Early to Middle Miocene monsoon climate in Australia. *Geology*, 39, 3–6. doi.org/10.1130/G31208.1.

Hill, RS (ed.). (1994). *History of the Australian vegetation: Cretaceous to recent.* Cambridge University Press.

Hill, RS, Tarran, MA, Hill, KE & Beer, YK. (2018). The vegetation history of South Australia. *Swainsona*, 30, 9–16.

Hillis, R & Reynolds, S. (2003). In situ stress field of Australia. In RR Hillis & D Muller (eds), *Evolution and dynamics of the Australian Plate* (pp. 101–113). Geological Society of Australia, Special Publication 22.

Hitchcock, E. (1851). *The religion of geology and its connected sciences.* Glasgow: William Collins.

Hohenegger, J. (2011). *Large foraminifera: Greenhouse constructions and gardeners in the oceanic microcosm.* Bulletin No. 5. The Kagoshima University Museum.

Hohenegger, J. (2014). Species as the basic units in evolution and biodiversity: Recognition of species in the recent and geological past as exemplified by larger foraminifera. *Gondwana Research* 25, 707–728. doi.org/10.1016/j.gr.2013.09.009.

Hohenegger, J & Briguglio, A. (2012). Axially oriented sections of Nummulitids: A tool to interpret larger benthic foraminiferal deposits. *Journal of Foraminiferal Research*, 42, 134–142. doi.org/10.2113/gsjfr.42.2.134.

Hohenegger, J, Torres-Silva, AI & Eder, W. (2022). Interpreting morphologically homogeneous (paleo-)populations as ecological species enables comparison of living and fossil organism groups, exemplified by Nummulitid Foraminifera. *Journal of Earth Sciences*, 20 (20), 1–16. doi.org/10.1007/s12583-021-1567-z.

Holdgate, GR. (2003). Coal, world-class energy reserves without limits. *Geology of Victoria*, 23, 489–518.

Holdgate, GR & Gallagher, SJ. (1997). Microfossil paleoenvironments and sequence stratigraphy of Tertiary cool-water carbonates, onshore Gippsland Basin, SE. Australia. In NP James & JDA Clarke (eds), *Cool and temperate water carbonates* (pp. 205–220). Society of Economic Palaeontologists and Mineralogists, Special Publication 56. doi.org/10.2110/pec.97.56.0205.

Holdgate, GR & Gallagher, SJ. (2003). Tertiary. In WD Birch (ed.), *Geology of Victoria* (pp. 299–335). Geological Society of Australia Special Publication 23.

Holdgate, GR, Kershaw, AP & Sluiter, IRK. (1995). Sequence stratigraphic analysis and the origins of Tertiary brown coal lithotypes, Latrobe Valley, Gippsland Basin, Australia. *International Journal of Coal Geology*, 28, 249–275. doi.org/10.1016/0166-5162(95)00020-8.

Holdgate, GR, McGowran, B, Fromhold, T, Wagstaff, BE, Gallagher, SJ, Wallace, MW, Sluiter, IRK & Whitelaw, M. (2009). Eocene–Miocene carbon-isotope and floral record from brown coal seams in the Gippsland Basin of southeast Australia. *Global & Planetary Change*, 65, 89–103, doi.org/10.1016/j.gloplacha.2008.11.001.

Holdgate, GR, Rodriquez, C, Johnstone, EM, Wallace, MW & Gallagher, SJ. (2003). The Gippsland Basin top–Latrobe unconformity, and its expression in other SE Australian basins. *APPEA Journal*, 43, 149–173. doi.org/10.1071/AJ02007.

Holdgate, GR, Sluiter, IRK. (2021). The T0 coal seam in the Latrobe Valley: A revised age and implications to Traralgon Formation stratigraphy. *Australian Journal of Earth Sciences*, 68(6), 763–781. doi.org/10.1080/08120099.2021.1876762.

Holdgate, GR, Sluiter, IRK, Clowes CD & Hannah, MJ. (2021). The spatial and temporal occurrence and significance of dinoflagellates and other marine fossils within onshore coal measures, Gippsland Basin, Australia. *Australian Journal of Earth Sciences*, 69(5), 630–649. doi.org/10.1080/08120099.2022.2002930.

Holdgate, GR, Sluiter, IRK & Taglieri, J. (2017). Eocene–Oligocene coals of the Gippsland and Australo-Antarctic basins—Paleoclimatic and paleogeographic context and implications for the earliest Cenozoic glaciations. *Palaeogeography, Palaeoclimatology, Palaeoecology*, 47, 236–255, doi.org/10.1016/j.palaeo.2017.01.035.

Holdgate, GR, Wallace, MW, Sluiter, IRK, Marcuccio, DM, Fromhold, TA & Wagstaff, BE. (2014). Was the Oligocene–Miocene a time of fire and rain? Insights from brown coal swamps of the southeastern Australian landscape. *Palaeogeography, Palaeoclimatology, Palaeoecology*, 411, 65–78. doi.org/10.1016/j.palaeo.2014.06.004.

Holford, SP, Hillis, RR, Hand, M & Sandiford, M. (2011). Thermal weakening localizes intraplate deformation along the southern Australian continental margin. *Earth & Planetary Science Letters*, 305, 207–214. doi.org/10.1016/j.epsl.2011.02.056.

Holford, SP, Tuitt, AK, Hillis, RR, Green, PF, Stoker, MS, Duddy, IR, Sandiford. M & Tassone, D. (2014). Cenozoic deformation in the Otway Basin, southern Australian margin: Implications for the origin and nature of post-breakup compression at rifted margins. *Basin Research*, 26, 10–37. doi.org/10.1111/bre.12035.

Hollis, C. (2014). Was the Early Eocene ocean unbearably warm or are the proxies unbelievably wrong? *Rendiconti Online Societa Geologica Italiana*, 31, 109–110.

Holmes, A. (2013). *The age of the earth*. Harper and Brothers.

Hottinger, L. (1982). Larger foraminifera, giant cells with a historical background. *Naturwissenschaft*, 69, 361–371. doi.org/10.1007/BF00396687.

Hottinger, L. (1997). Shallow benthic foraminiferal assemblages as signals for depth of their deposition and their limitations. *Bulletin de la Societé géologique de France*, 168(4), 491–505.

Hottinger, L. (2001). Learning from the past. In L Levi-Montalcini (ed.), *Frontiers of life. Volume 4(2): Discovery and spoliation of the biosphere* (pp. 449–477). Academic Press.

Hou, B, Frakes, LA, Sandiford, M, Worrall, L, Keeling, J & Alley, NF. (2008). Cenozoic Eucla Basin and associated palaeovalleys, southern Australia—Climatic and tectonic influences on landscape evolution, sedimentation and heavy mineral accumulation. *Sedimentary Geology*, 203, 112–130. doi.org/10.1016/j.sedgeo.2007.11.005.

Houben, AJP, Bijl, PK, Sluijs, A, Schouten, S & Brinkhuis, H. (2019). Late Eocene Southern Ocean cooling and invigoration of circulation preconditioned Antarctica for full-scale glaciation. *Geochemistry, Geophysics, Geosystems*, 20, 2214–2234. doi.org/10.1029/2019GC008182.

Houben, AJP, Quaijtaal, W, Wade, BS, Schouten S & Brinkhuis, H. (2019). Quantitative organic-walled dinoflagellate cyst stratigraphy across the Eocene–Oligocene Transition in the Gulf of Mexico: A record of climate- and sea level change during the onset of Antarctic glaciation. *Newsletters on Stratigraphy*, 52(2), 131–154. doi.org/10.1127/nos/2018/0455.

Howchin, W. (1888). The foraminifera of the older Tertiary of Australia (No. 1, Muddy Creek, Victoria). *Transactions of the Royal Society of South Australia*, 11, 1–20.

Howchin, W. (1929). *The geology of South Australia: With notes on the chief geological systems and occurrences in the other Australian States* (second edition). Gillingham and Co. Limited.

Huber, M, Brinkhuis, H, Stickley, CE, Döös, K, Sluijs, A, Warnaar, J, Schellenberg, SA & Williams, GL. (2004). Eocene circulation of the Southern Ocean: Was Antarctica kept warm by subtropical waters? *Paleoceanography*, 19(4), PA4026, doi.org/10.1029/2004PA001014.

Hull, DL. (1973). *Darwin and his critics: The reception of Darwin's theory of evolution by the scientific community*. Harvard University Press.

Hunt, P. (2011). *Molluscs of South Australia*. The Malacological Society of South Australia.

Hutton, J. (1788). Theory of the Earth. *Transactions of the Royal Society of Edinburgh*, 1, 209–305.

Hutton, J. (1795). *Theory of the Earth with proofs and illustrations. Edinburgh*: William Creech.

Huurdeman, E. (2017). *Late Paleocene – Early Eocene long and short-term environment and climate change in Southeast Australia*. [Unpublished Master of Science thesis]. Utrecht University.

Huurdeman, EP, Frieling, J, Reichgelt, T, Bijl, PK, Bohaty, SM, Holdgate, GR, Gallagher, SJ, Peterse, F, Greenwood, DR & Pross, J. (2020). Rapid expansion of paratropical rainforests into the southern high latitudes at the onset of the Paleocene–Eocene Thermal Maximum. *Geology*, 48(1), 40–44, doi.org/10.1130/G47343.1.

Huxley, J. (1942). *Evolution, the modern synthesis*. Harper Brothers.

Huxley, TH. (1878). *Physiography: An introduction to the study of nature*. Second edition. London: MacMillan and Co.

Imbrie, J & Imbrie, KP. (1979). *Ice ages: Solving the mystery*. Hillside, NJ: Enslow.

Jackson, JBC & Erwin, DH. (2006). What can we learn about ecology and evolution from the fossil record? *Trends in Ecology & Evolution*, 21, 322–328. doi.org/10.1016/j.tree.2006.03.017.

James, NP & Bone, Y. (2021). *Biogenic sedimentary rocks in a cold, Cenozoic ocean: Neritic southern Australia*. Springer. doi.org/10.1007/978-3-030-63982-2.

Jenkins, B. (2016). Neptunism and transformism: Robert Jameson and other evolutionary theorists in early nineteenth century Scotland. *Journal of the History of Biology*, 49, 527–557. doi.org/10.1007/s10739-015-9425-4.

Jenkins, B. (2019). *Evolution before Darwin: Theories of the transmutation of species in Edinburgh, 1804–1834*. Edinburgh University Press. doi.org/10.3366/edinburgh/9781474445788.001.0001.

Jennings, JT (ed.). (2009). Natural history of the Riverland & Murraylands. *Royal Society of South Australia*, Occasional Publications No. 9.

Jepsen, GL, Simpson, GG & Mayr, E (eds). (1949). *Genetics, paleontology & evolution*. Princeton University Press.

Jones, RW & Brady, HB. (1994). *The* Challenger *Foraminifera*. Oxford University Press.

Jukes, JB. (1862). *The student's manual of geology*. Adam & Charles Black.

Kamikuri, S & Wade, BS. (2012). Radiolarian magnetobiochronology and faunal turnover across the Middle/Late Eocene boundary at Ocean Drilling Program Site 1052 in the western North Atlantic Ocean. *Marine Micropaleontology*, 88–89, 41–53 doi.org/10.1016/j.marmicro.2012.03.001.

Katz, ME, Finkel, ZV, Grzebyk, D, Knoll, AH & Falkowski, PG. (2004). Evolutionary trajectories and biogeochemical impacts of marine eukaryotic phytoplankton. *Annual Review of Ecology, Evolution and Systematics*, 35, 521–556. doi.org/10.1146/annurev.ecolsys.35.112202.130137.

Kennett, JP & Exon, NF. (2004). Paleoceanographic evolution of the Tasman Seaway and its climatic implications. In NF Exon, JP Kennett & MJ Malone (eds), *The Cenozoic Southern Ocean: Tectonics, sedimentation, and climate change between Australia and Antarctica* (pp. 345–367). Geophysical Monograph No. 151. American Geophysical Union. doi.org/10.1029/151GM19.

Kennett, JP, Houtz, RE, Andrews, PB, Edwards, AR, Gostin, VA, Hajós, M, Hampton, MA, Jenkins, DG, Margolis, SV, Ovenshine, AT & Perch-Nielsen, K. (1974). Development of the circum-Antarctic Current. *Science*, 186(4159), 144–147, doi.org/10.1126/science.186.4159.144.

Kennett, JP & Stott, LD. (1991). Abrupt deep-sea warming, palaeoceanographic changes and benthic extinctions at the end of the Palaeocene. *Nature* 353, 225–229. doi.org/10.1038/353225a0.

Kominz, M, Browning, JV, Miller, KG, Mizinseva, S & Scotese, CR. (2008). Late Cretaceous to Miocene sea-level estimates from the New Jersey and Delaware coastal plain coreholes: An error analysis. *Basin Research*, 20, 2211–226. doi.org/10.1111/j.1365-2117.2008.00354.x.

Korasidis, VA, Wallace, MW, Wagstaff, BE & Hill, RS. (2019). Terrestrial cooling record through the Eocene–Oligocene transition of Australia. *Global and Planetary Change,* 173, 61–72. doi.org/10.1016/j.gloplacha.2018.12.007.

Lagoe, MB, Davies, TA, Austin, Jr, JA & Olson, HC. (1997). Foraminiferal constraints on very high-resolution seismic stratigraphy and Late Quaternary glacial history, New Jersey continental shelf. *Palaios*, 12, 249–266. doi.org/10.2307/3515426.

Lange, RT. (1978). Carpological evidence for fossil Eucalyptus and other Leptospermeae (subfamily Leptospermoideae of Myrtaceae) from a Tertiary deposit in the South Australian arid zone. *Australian Journal of Botany*, 26, 221–233. doi.org/10.1071/BT9780221.

Laurent, G. (2002). Alcide d'Orbigny entre Cuvier et Lamarck [Alcide d'Orbigny between Cuvier and Lamarck]. *Comptes Rendus Palevol*, 1(6), 347–358. doi.org/10.1016/S1631-0683(02)00066-0.

Li, Q, James, NP & McGowran, B. (2003). Middle and Late Eocene Great Australian Bight lithobiostratigraphy and stepwise evolution of the southern Australian continental margin. *Australian Journal of Earth Sciences* 50, 113–128. doi.org/10.1046/j.1440-0952.2003.00978.x.

Li, Q & McGowran, B. (1998). Oceanographic implications of recent planktonic foraminifera along the southern Australian margin. *Marine and Freshwater Research*, 49, 439–445. doi.org/10.1071/MF97196.

Li, Q & McGowran, B. (2000). *The Miocene foraminifera from Lakes Entrance Oil Shaft, southeastern Australia*. Memoirs of the Association of Australasian Palaeontologists 22.

Li, Q, Quilty, PG, Moss, GD & McGowran, B. (1996). Southern Australian endemic and semi-endemic foraminifera from southern Australia: A preliminary report. *Journal of Micropalaeontology* ,15, 169–185. doi.org/10.1144/jm.15.2.169.

Li, Q, Simo, AJ, McGowran, B & Holbourn, A. (2004). The eustatic and tectonic origin of Neogene unconformities from the Great Australian Bight. *Marine Geology*, 203, 57–81. doi.org/10.1016/S0025-3227(03)00329-3.

Lindsay, JM. (1981). *Tertiary stratigraphy and foraminifera of the Adelaide City area, St Vincent Basin, South Australia* [Unpublished Master of science thesis]. University of Adelaide.

Lindsay, JM & Alley, NF. (1995). St Vincent Basin. In JF Drexel & WV Preiss (eds), *The geology of South Australia. Volume 2. The Phanerozoic* (pp. 163–172). South Australia. Department of Mines and Energy, Bulletin 54.

Lindsay, JM & McGowran, B. (1986). Eocene/Oligocene boundary, Adelaide region, South Australia. In C Pomerol & I Premoli Silva (eds), *Geological events at the Eocene-Oligocene boundary* (pp. 165–173). Elsevier Science Publishing.

Lipps, JH. (1970). Plankton evolution. *Evolution*, 24, 1–22. doi.org/10.1111/j.1558-5646.1970.tb01737.x.

Lipps, JH. (2002). Alcide d'Orbigny and American micropaleontology. *Comptes Rendus Palevol*, 1, 461–469. doi.org/10.1016/S1631-0683(02)00069-6.

Loeblich, AR & Tappan, H. (1964). Sarcodina, chiefly 'Thecamoebians' and Foraminiferida. In RC Moore (ed.), *Treatise on invertebrate paleontology* (Part C, vols 1–2, pp. 900). Geological Society of America.

Long, J. (2012). Evolution, missing links and climate change: Recent advances in understanding transformational macroevolution. In A Poiani (ed.), *Pragmatic evolution: Applications of evolutionary theory* (pp. 23–36). Cambridge University Press. doi.org/10.1017/CBO9780511980381.004.

López-Gamundí, OR & Buatois, LA. (2010). *Late Paleozoic glacial events and postglacial transgressions in Gondwana*. Geological Society of America Special Paper 468. doi.org/10.1130/SPE468.

Low, T. (2014). *Where song began: Australia's birds and how they changed the world*. Melbourne.

Lowry, DC. (1970). *Geology of the Western Australian part of the Eucla Basin*. Geological Survey of Western Australia Bulletin 122.

Ludbrook, NH. (1957). Permian deposits of South Australia and their fauna. *Transactions of the Royal Society of South Australia*, 91, 65–87.

Ludbrook, NH & Lindsay, JM. (1969). Tertiary foraminiferal zones in South Australia. In P Brönnimann & HH Renz (eds), *Proceedings of the First International Conference on Planktonic Microfossils* (vol. 2, 366–374). EJ Brill.

Lukasik, J, James, NP, McGowran, B & Bone, Y. (2000). An epeiric ramp: Low-energy, cool-water carbonate facies in a Tertiary inland sea, Murray Basin, South Australia. *Sedimentology*, 47, 851–881. doi.org/10.1046/j.1365-3091.2000.00328.x.

Luterbacher, H-P & Premoli Silva, I. (1964). Biostratigrafia del limite Cretaceo-Terziarionell'Appennino centrale. *Rivista Italiana Paleontologia Stratigrafia*, 70, 67–128.

Lyell, C. (1830–33). *Principles of geology, being an attempt to explain the former changes of the earth's surface, by reference to causes now in operation.* 3 Volumes. London: John Murray. Facsimile reprint, Chicago: University of Chicago Press (1990–91).

Lyell, C. (1871). *Students' elements of geology.* John Murray. doi.org/10.5962/bhl.title.145183.

Lyle, M, Pälicke, H, Nishi, H, Raffi, I, Gamage, K, Klaus, A & the IODP Expeditions 320/321 Scientific Party. (2010). The Pacific equatorial age transect, IODP Expeditions 320 and 321: Building a 50-million-year-long environmental record of the equatorial Pacific Ocean. *Scientific Drilling*, 9, 4–15, doi.org/10.2204/iodp.sd.9.01.2010.

Macphail, M & Jordan, G. (2015). Tropical palms and arums at near-polar latitudes: Fossil pollen evidence from the Tamar ana Macquarie grabens, northern Tasmania. *Papers and Proceedings of the Royal Society of Tasmania*, 149, 23–28. doi.org/10.26749/rstpp.149.23.

Mahon, EM & Wallace, MW. (2020). Cenozoic structural history of the Gippsland Basin: Early Oligocene onset for compressional tectonics in SE Australia. *Marine and Petroleum Geology* 114, 104243. doi.org/10.1016/j.marpetgeo.2020.104243.

Mahon, EM & Wallace, MW. (2022). 3D seismic geomorphology of Early Cenozoic incised channels, Gippsland Basin, SE Australia: Evidence for submarine origin. *Sedimentary Geology* 430, 106092. doi.org/10.1016/j.sedgeo.2022.106092.

Martin, HA. (2006). Cenozoic climatic change and the development of the arid vegetation in Australia. *Journal of Arid Environments*, 66, 533–563. doi.org/10.1016/j.jaridenv.2006.01.009.

Matthew, WD. (1915). Climate and evolution. *Annals of the New York Academy of Sciences*, 24, 171–318. doi.org/10.1111/j.1749-6632.1914.tb55346.x.

Matthew WD. (1926). The evolution of the horse: A record and its interpretation. *Quarterly Review of Biology*, 1, 139–185. doi.org/10.1086/394242.

Matthew WD. (1930). The pattern of evolution. *Scientific American*, 143, 192–196. doi.org/10.1038/scientificamerican0930-192.

Matthew, WD. (1939). *Climate and evolution*. Second edition, revised and enlarged, arranged by EH Colbert. New York: Special Publications of the New York Academy of Sciences.

Matthews, KJ, Müller, RD & Sandwell, DT. (2015). Oceanic microplate formation records the onset of India–Eurasia collision. *Earth and Planetary Science Letters*, 433, 204–214. doi.org/10.1016/j.epsl.2015.10.040.

Mayr, E. (1942). *Systematics and the origin of species*. Columbia University Press.

Mayr, E. (1959). Where are we? *Cold Spring Harbor Symposia on Quantitative Biology* 24, 1–14. doi.org/10.1101/SQB.1959.024.01.003.

Mayr, E. (1972). Continental drift and the history of the Australian bird fauna. *Emu*, 72, 26–28. doi.org/10.1071/MU972022d.

Mayr, E. (1976). *Evolution and the diversity of life*. Harvard University Press.

Mayr, E. (1982). *The growth of biological thought: Diversity, evolution, and inheritance*. Cambridge: The Belknap Press of Harvard University Press.

Mayr, E. (1988). *Towards a new philosophy of biology: Observations of an evolutionist*. Cambridge: Harvard University Press.

Mayr, E. (2002). *What evolution is*. London: Orion Books.

Mayr, E & Provine, WB. (1980). *The evolutionary synthesis*. Harvard University Press. doi.org/10.4159/harvard.9780674865389.

McGowran, B. (1959). Tertiary nautiloids (*Eutrephoceras* and *Cimomia*) from South Australia. *Journal of Paleontology*, 33, 435–448.

McGowran, B. (1964). Foraminiferal evidence for the Paleocene age of the King's Park Shale (Perth Basin, Western Australia). *Journal of the Royal Society of Western Australia*, 47, 81–86.

McGowran, B. (1965). Two foraminiferal faunas from the Wangerrip Group, Pebble Point coastal section, western Victoria. *Proceedings of the Royal Society of Victoria*, 79, 9–74.

McGowran, B. (1966). Australian Paleocene *Lamarckina* and *Ceratobulimina*, with a discussion of *Cerobertina, Pseudobulimina* and the family Robertinidae. *Contributions from the Cushman Foundation for Foraminiferal Research*, 17, 77–103.

McGowran, B. (1968a). Late Cretaceous and Early Tertiary correlations in the Indo-Pacific region. In LR Rao (ed.), *Cretaceous–Tertiary formations of South India* (pp. 335–360), Geological Society of India, Memoir 2.

McGowran, B. (1968b). Reclassification of Early Tertiary *Globorotalia*. *Micropaleontology*, 14, 179–198. doi.org/10.2307/1484733.

McGowran, B. (1971). On foraminiferal taxonomy. In PA Farinacci (ed.), *Proceedings of the Second Planktonics Conference, Rome, 1970* (pp. 813–820). Edizioni Technoscienza.

McGowran, B. (1973). Rifting and drift of Australia and the migration of mammals. *Science*, 180, 759–761. doi.org/10.1126/science.180.4087.759.

McGowran, B. (1978). Stratigraphic record of Early Tertiary oceanic and continental events in the Indian Ocean region. *Marine Geology*, 26, 1–39. doi.org/10.1016/0025-3227(78)90043-9.

McGowran, B. (1979a). The Australian Tertiary: Foraminiferal overview. *Marine Micropaleontology* 4, 235–264. doi.org/10.1016/0377-8398(79)90019-7.

McGowran, B. (1979b). Comments on Early Tertiary tectonism and lateritization. *Geological Magazine*, 116, 227–230. doi.org/10.1017/S0016756800043636.

McGowran, B. (1990). Fifty million years ago. *American Scientist,* 78, 30–39.

McGowran, B. (2005a). *Biostratigraphy: Microfossils and geological time.* Cambridge University Press. doi.org/10.1017/CBO9780511610653.

McGowran, B. (2005b). Foraminifera in Cenozoic palaeoenvironments. *Journal of China University of Geosciences*, 16(3), 200–218.

McGowran, B. (2007). Preface. In B. McGowran (ed.), *Beyond the GSSP: New developments in chronostratigraphy.* Special issue of *Stratigraphy*, 4(2–3), 81–82.

McGowran, B. (2009). The Australo-Antarctic Gulf and the Auversian Facies Shift. In C Koeberl & A Montanari (eds), *The Late Eocene earth—Hothouse, icehouse, and impacts* (pp. 215–240). Geological Society of America Special Paper 452. doi.org/10.1130/2009.2452(14).

McGowran, B. (2012a). Cenozoic environmental shifts and foraminiferal evolution. In JA Talent (ed.), *Earth and life* (pp. 937–965). Springer Science + Business Media B.V. doi.org/10.1007/978-90-481-3428-1_33.

McGowran, B. (2012b). Foraminiferal micropalaeontology in Adelaide 1950–1970: Correlation and age determination in postwar mapping and subsurface exploration. *Transactions of the Royal Society of South Australia*, 136(2), 99–127. doi.org/10.1080/03721426.2012.10887166.

McGowran, B. (2013a). Martin Glaessner's foraminiferal micropalaeontology. In AJ Bowden, FJ Gregory & AS Henderson (eds), *Landmarks in foraminiferal micropalaeontology: History and development* (pp. 227–250). The Micropalaeontological Society, Special Publications. doi.org/10.1144/TMS6.18.

McGowran, B. (2013b). Organic evolution and deep time: Charles Darwin and the fossil record. *Transactions of the Royal Society of South Australia*, 137(2), 102–148. doi.org/10.1080/03721426.2013.10887188.

McGowran, B. (2013c). Scientific accomplishments of Reginald Claude Sprigg. *Transactions of The Royal Society of South Australia*, 137(1), 1–52. doi.org/10.1080/3721426.2013.10887170.

McGowran, B & Alley, NF. (2008). Chapter 2: History of the Cenozoic St Vincent Basin in South Australia. In SA Shepherd, S Bryars, I Kirkegaard, P Harbison & JT Jennings (eds), *Natural history of Gulf St Vincent* (pp. 13–28). Royal Society of South Australia, Occasional Publication No. 8.

McGowran, B & Beecroft, A. (1985). *Guembelitria* in the Early Tertiary of southern Australia and its palaeoceanographic significance. In JM Lindsay (ed.), *Stratigraphy, palaeontology, malacology: Papers in honour of Dr. Nell Ludbrook* (pp. 247–261). South Australian Department of Mines and Energy, Special Publication 5.

McGowran, B, Berggren, WA, Hilgen, FJ, Steininger, FF, Van Couvering, JA, Aubry, M-P & Lourens, LJ. (2009). Neogene and Quaternary coexisting in the geological time scale: The inclusive compromise. *Earth-Science Reviews*, 96, 249–262, doi.org/10.1016/j.earscirev.2009.06.006.

McGowran, B & Hill, RS. (2015). Cenozoic climatic shifts in southern Australia. *Transactions of the Royal Society of South Australia*, 139(1), 19–37. doi.org/10.1080/03721426.2015.1035215.

McGowran, B, Holdgate, GR, Li, Q & Gallagher, SJ. (2004). Cenozoic stratigraphic succession in southeastern Australia. *Australian Journal of Earth Sciences*, 51, 459–496. doi.org/10.1111/j.1400-0952.2004.01078.x.

McGowran, B, Lemon, NM, Preiss, WV & Olliver, JG. (2016). *Geological field excursion guide—Cenozoic Willunga Embayment: From Australo-Antarctic Gulf to Sprigg Orogeny*. Report Book 2016/00008. Department of State Development, South Australia; and Geological Society of Australia, South Australian Division.

McGowran, B & Li, Q. (1994). The Miocene oscillation in southern Australia. *Records of the South Australia Museum,* 27(2), 197–212.

McGowran, B & Li, Q. (1997). Miocene climatic oscillations recorded in the Lakes Entrance oil shaft, southeastern Australia: Reappraisal of the planktonic foraminiferal record. *Micropaleontology,* 43, 129–148. doi.org/10.2307/1485778.

McGowran, B & Li, Q. (1998). Cainozoic climatic change and its implications for understanding the Australian regolith. In RA Eggleton (ed.), *The state of the regolith: Proceedings of the Second Australian Conference on Landscape Evolution & Mineral Exploration* (pp. 86–103). Geological Society of Australia Special Publication No. 20.

McGowran, B & Li, Q. (2000). Evolutionary palaeoecology of Cainozoic foraminifera: Tethys, IndoPacific, southern Australasia. *Historical Biology,* 15, 3–28. doi.org/10.1080/10292380109380579.

McGowran, B & Li, Q. (2007). Stratigraphy: Gateway to geohistory and biohistory. In B McGowran (ed.), *Beyond the GSSP: New developments in chronostratigraphy.* Special issue of *Stratigraphy,* 4(2–3), 173–185.

McGowran, B, Li, Q, Cann, J, Padley, D, McKirdy, D & Shafik, S. (1997). Biogeographic impact of the Leeuwin Current in southern Australia since the late Middle Eocene. *Palaeogeography, Palaeoclimatology, Palaeoecology,* 136, 19–40. doi.org/10.1016/S0031-0182(97)00073-4.

McGowran, B, Li, Q & Moss, G. (1997). The Cenozoic neritic record in southern Australia: The biogeohistorical framework. In NP James & J Clarke (eds), *Coolwater carbonates in space and time* (pp. 185–203). Society of Economic Paleontologists and Mineralogists, Special Volume No. 56. doi.org/10.2110/pec.97.56.0185.

McGowran, B, Li, Q, Pledge, NS & Schmidt, R. (2009). Neogene Fossils in the Murravian Gulf (western Murray Basin) and their environmental and chronological significance. In JT Jennings (ed.), *Natural history of the Riverland & Murraylands* (pp. 178–205). Royal Society of South Australia, Occasional Publications No. 9.

McGowran, B, & Lindsay, JM. (1969). A Middle Eocene planktonic foraminiferal assemblage from the Eucla Basin. *Quarterly Geological Notes,* 30, 2–10.

McGowran, B, Lindsay, JM & Harris, WK. (1971). Attempted reconciliation of Tertiary biostratigraphic systems, Otway Basin. In H Wopfner & JG Douglas (eds), *The Otway Basin in Southeastern Australia, Adelaide and Melbourne* (pp. 273–281). Geological Surveys of South Australia and Victoria, Special Bulletin.

McInerney, FA & Wing, SL. (2011). The Paleocene–Eocene thermal maximum: A perturbation of carbon cycle, climate, and biosphere with implications for the future. *Annual Review of Earth and Planetary Sciences*, 39, 489–516. doi.org/10.1146/annurev-earth-040610-133431.

McLaren, S, Wallace, MW, Gallagher, SJ, Miranda, JA, Holdgate, GR, Gow, LJ, Snowball, I & Sandgren, P. (2011). Palaeogeographic, climatic and tectonic change in southeastern Australia: The Late Neogene evolution of the Murray Basin. *Quaternary Science Reviews* 30, 1086–1111. doi.org/10.1016/j.quascirev.2010.12.016.

Menpes, SA, Korsch, RJ & Carr, LK. (2010). Gawler Craton–Officer Basin–Musgrave Province–Amadeus Basin (GOMA) seismic survey, 08GA-OM1: Geological interpretation of the Arckaringa Basin. In RJ Korsch & N Kositcin (eds), *GOMA (Gawler Craton–Officer Basin–Musgrave Province–Amadeus Basin) seismic and MT workshop 2010: Extended abstracts* (pp.16–31). Geoscience Australia Record 2010/39.

Metzger, CA & and Retallack, GJ. (2010). Paleosol record of Neogene climate change in the Australian outback. *Australian Journal of Earth Sciences*, 57, 871–885. doi.org/10.1080/08120099.2010.510578.

Miller, AK. (1947). *Tertiary nautiloids of the Americas*. Geological Society of America, Memoir 23. doi.org/10.1130/MEM23-p1.

Miller, H. (1862). *The testimony of the rocks, or, its bearing on the two theologies, natural and revealed*. Edinburgh: Adam and Charles Black.

Milnes, AR, Bourman, R & Northcote, KH. (1985). Field relationships of ferricretes and weathered zones in southern South Australia: A contribution to 'laterite' studies in Australia. *Australian Journal of Soils Research*, 23, 441–465. doi.org/10.1071/SR9850441.

Mitchell, KJ, Llamas, B, Soubrier, J, Rawlence, N, Worthy, TJ, Wood, J, Lee, MSY & Cooper, A. (2014). Ancient DNA reveals elephant birds and kiwi are sister taxa and clarifies ratite bird evolution. *Science,* 344(6186), 898–900. doi.org/10.1126/science.1251981.

Mortimer, N, Campbell, HJ, Tulloch, AJ, King, PR, Stagpoole, VM, Wood, RA, Rattenbury, MS, Sutherland, R, Adams, CJ, Collot, J & Seton, M. (2017). Zealandia: Earth's hidden continent. *GSA Today*, 27, doi.org/10.1130/GSATG 321A.1.

Moss, G & McGowran, B. (2003). Oligocene neritic foraminifera in Southern Australia: Spatiotemporal biotic patterns reflect sequence-stratigraphic environmental patterns. In H Olson & M Leckie (eds), *Paleobiological, geochemical, and other proxies of sea level change* (pp. 117–138). SEPM Special Volume 75. doi.org/10.2110/pec.03.75.0173.

Moyle, RG, Oliveros, CH, Andersen, MJ, Hosner, PA, Benz, BW, Manthey, JD, Travers, SL, Brown, RM & Faircloth, BC. (2016). Tectonic collision and uplift of Wallacea triggered the global songbird radiation. *Nature Communications*, 7, 12709 doi.org/10.1038/ncomms12709.

Müller, RD, Gaina, C & Clark, S. (2000). Seafloor spreading around Australia. In JJ Veevers (ed.), *Billion-year history of Australia and neighbours in Gondwanaland* (pp. 18–28). Gemoc Press.

Müller, RD, Seton, M, Zahirovic, S, Williams, SE, Matthews, KJ, Wright, NM, Shephard, GE, Maloney, KT, Barnett-Moore, N, Hosseinpour, M & Bower, DJ. (2016). Ocean basin evolution and global-scale plate reorganization events since Pangea breakup. *Annual Review of Earth and Planetary Sciences*, 44, 107–138. doi.org/10.1146/annurev-earth-060115-012211.

Munz, P. (1985). *Our knowledge of the growth of knowledge: Popper or Wittgenstein?* London: Routledge & Kegan Paul.

Munz, P. (1993). *Philosophical Darwinism: On the origin of knowledge by means of natural selection*. London & New York: Routledge.

Munz, P. (2004). *Beyond Wittgenstein's poker: New light on Popper and Wittgenstein*. Aldershot: Ashgate Publishing.

Murray, J. (1912?). *The ocean: A general account of the science of the sea*. Williams & Norgate. doi.org/10.5962/bhl.title.6545.

Murray, J & Renard, AF. (1891). Report on deep-sea deposits (based on the specimens collected during the voyage of HMS Challenger in the years 1872 to 1876). *Report on the scientific results of the voyage of HMS Challenger during the years 1873–76*. John Menzies and Co.

Murray-Wallace, CV. (2002). Pleistocene coastal stratigraphy, sea-level highstands and neotectonism of the southern Australian passive continental margin—a review. *Journal of Quaternary Science*, 17, 460–489. doi.org/10.1002/jqs.717.

Murray-Wallace, CV. (2018). *Quaternary history of the Coorong coastal plain, Southern Australia: An archive of environmental and global sea-level changes*. Cham: Springer. doi.org/10.1007/978-3-319-89342-6.

Murray-Wallace, CV & Cann, JH. (2007). Quaternary history of the Coorong coastal plain, South Australia. Excursion (A6), XVII INQUA Congress, Cairns, Australia, School of Earth & Environmental Sciences, University of Wollongong.

Nagappa, Y. (1959). Foraminiferal biostratigraphy of the Cretaceous: Eocene succession in the India–Pakistan–Burma region. *Micropaleontology*, 5, 145–192. doi.org/10.2307/1484208.

Nürnberg, D, Kayode, A, Meier, KJF & Karas, C. (2022). Leeuwin Current dynamics over the last 60 kyr – relation to Australian ecosystem and Southern Ocean change. *Climate of the Past*, 18, 2483–2507. doi.org/10.5194/cp-18-2483-2022.

Nebelsick, JH, Rasser, MW & Bassi, D. (2005). Facies dynamics in Eocene to Oligocene circumalpine carbonates. *Facies*, 51, 197–216. doi.org/10.1007/s10347-005-0069-2.

Newsam, C, Bown, PR, Wade, BS & Jones, HL. (2017). Muted calcareoua nannoplankton response and the Middle/Late Eocene turnover event in the western North Atlantic Ocean. *Newsletters on Stratigraphy*, 50(3), 297–309. doi.org/10.1127/nos/2016/0306.

O'Connell, LG, James, NP & Bone, Y. (2012). The Miocene Nullarbor Limestone, southern Australia: Deposition on a vast subtropical epeiric platform. *Sedimentary Geology*, 253–254, 1–16. doi.org/10.1016/j.sedgeo.2011.12.002.

Olson, EC. (1952). The evolution of a Permian vertebrate chrono-fauna. *Evolution*, 6, 181–196. doi.org/10.1111/j.1558-5646.1952.tb01413.x.

Olsson, RK, Hemleben, C, Berggren, WA & Huber, BT. (1999). Atlas of Paleocene planktonic foraminifera. *Smithsonian Contributions to Paleobiology*, 85, 1–252. doi.org/10.5479/si.00810266.85.1.

O'Malley, MA. (2014). *Philosophy of microbiology*. Cambridge University Press. doi.org/10.1017/CBO9781139162524.

Oreskes, N. (1999). *The rejection of continental drift: Theory and method in American earth science*. New York and Oxford. doi.org/10.1093/oso/9780195117325.001.0001.

Osborn, HF. (1894). *From the Greeks to Darwin: An outline of the development of the evolution idea*. Macmillan and Co. (Second edition published 1902). doi.org/10.1037/12938-000.

Owen, R. (1848). *The archetype and homologies of the vertebrate skeleton*. London: J. van Voorst. doi.org/10.5962/bhl.title.118611.

Owen, R. (1849). *On the nature of limbs*. London: J van Voorst. doi.org/10.5962/bhl.title.50117.

Owen, R. (1860). *Palæontology, or a systematic summary of extinct animals and their geological relations*. Adam and Charles Black. doi.org/10.5962/bhl.title.153670.

Padian, K. (2008). Richard Owen's quadrophenia: The pull of opposing forces in Victorian cosmogeny. In R Owen (ed.), *On the nature of limbs: A discourse* (pp. lv–xci). University of Chicago Press.

Pagani, M, Huber, M, Liu, Z, Bohaty, SM, Henderiks, J, Sijp, W, Krishnan, S & DeConto, RM. (2011). The role of carbon dioxide during the onset of Antarctic glaciation. *Science*, 334, 1261–1263. doi.org/10.1126/science.1203909.

Pagani, M, Huber, M & Sageman, B. (2014). Greenhouse climates. *Treatise on Geochemistry* (second edition), 6, 281–304. doi.org/10.1016/B978-0-08-095975-7.01314-0.

Pagani, M, Zachos, JC, Freeman, KH, Tipple, B & Bohaty, S. (2005). Marked decline in atmospheric carbon dioxide concentrations during the Palaeogene. *Science*, 309, 600–603. doi.org/10.1126/science.1110063.

Pälike, H, Lyle, MW, Nishi, H, Raffi, I, Ridgwell, A, Gamage, K, Klaus, A, Acton, G, Anderson, L, Backman, J, Baldauf, J, Beltran, C, Bohaty, SM, Bown, P, Busch, W, Channell, JET, Chun, COJ, Delaney, M, Dewangan, P, Jones, TD, … Zeebe, RE. (2012). A Cenozoic record of the equatorial Pacific carbonate compensation depth. *Nature*, 488, 609–614. doi.org/10.1038/nature11360.

Paley, W. (1802, 1827). Natural theology. In W. Paley (1827), *The works of William Paley, D.D., Archdeacon of Carlisle* (pp. 433–554). Edinburgh: The University Press for Peter Brown and Thomas Nelson.

Parker, FL. (1948). Foraminifera of the continental shelf from the Gulf of Maine to Maryland. *Bulletin of the Museum of Comparative Zoology at Harvard College*, 100(2), 213–241.

Parr, WJ. (1938). Upper Eocene foraminifera from deep borings in King's Park, Perth, Western Australia. *Journal of the Royal Society of Western Australia*, 24, 69–101.

Parr, WJ. (1947). An Australian record of the foraminiferal genus *Hantkenina*. *Proceedings of the Royal Society of Victoria* (new series), 58, 45–47.

Parr, WJ. (1950). Foraminifera. *Reports of the B.A.N.Z. Antarctic Research Expedition 1929–1931*, series B, 5, 233–392.

Partridge, AD. (2006). Late Cretaceous–Cenozoic palynology zonations Gippsland Basin. In E Montiel (Co-ord.), *Australian Mesozoic and Cenozoic Palynology Zonations*. Geoscience Australia, Record 2006/23.

Pavord, A. (2005). *The naming of names*. New York: Bloomsbury Publishing.

Pearson, PN, John, E, Wade, BS, D'haenens, S & Lear, CH. (2022). Spine-like structures in Paleogene muricate planktonic foraminifera. *Journal of Micropalaeontology*, 41, 107–127. doi.org/10.5194/jm-41-107-2022.

Pearson, PN, Olsson, RK, Huber, BT, Hemleben, C & Berggren, WA (eds). (2006). *Atlas of Eocene planktonic foraminifera*. Cushman Foundation for Foraminiferal Research, Special Publication No 41.

Perner, K, Moros, M, De Deckker, P, Blanz, T, Wacker, L, Telford, R, Siegel, H, Schneider, R & Jansen, E. (2018). Heat export from the tropics drives Mid to Late Holocene palaeoceanographic changes offshore southern Australia. *Quaternary Science Reviews* 180, 96–110. doi.org/10.1016/j.quascirev.2017.11.033.

Phillips, J. (1860). *Life on the earth: Its origin and succession*. Macmillan. doi.org/10.5962/bhl.title.22153.

Pillans, BJ. (2002). Climate-driven weathering episodes during the last 200 Ma in southern Australia. *Geological Society of Australia Abstracts* 67, 428.

Pimm, AC, McGowran, B & Gartner, S. (1974). Early sinking history of Ninetyeast Ridge, north-eastern Indian Ocean. *Geological Society of America Bulletin*, 85, 1219–1224. doi.org/10.1130/0016-7606(1974)85<1219:ESHONR>2.0.CO;2.

Pinker, S. (2002). *The blank slate: The modern denial of human nature*. Penguin Books.

Playfair, J. (1802). *Illustrations of the Huttonian theory of the earth*. Edinburgh: William Creech. doi.org/10.5962/bhl.title.50752.

Plummer, HJ. (1936). The structure of *Ceratobulimina*. *American Midland Naturalist*, 17, 460–463. doi.org/10.2307/2419972.

Pole, MS & Macphail, MK. (1996). Eocene *Nypa* from Regatta Point, Tasmania. *Review of Palaeobotany and Palynology*, 92, 55–67. doi.org/10.1016/0034-6667(95)00099-2.

Pomar, L, Baceta, JI, Hallock, P, Mateu-Vicens, G & Basso, D. (2017). Reef building and carbonate production modes in the west-central Tethys during the Cenozoic. *Marine and Petroleum Geology*, 83, 261–304. doi.org/10.1016/j.marpetgeo.2017.03.015.

Pomar, L & Hallock, P. (2008). Carbonate factories: A conundrum in sedimentary geology. *Earth-Science Reviews*, 87, 134–169. doi.org/10.1016/j.earscirev.2007.12.002.

Popper, KR. (1957). *The poverty of historicism*. London: Routledge & Kegan Paul.

Prazeres, M & Renema, W. (2018). Evolutionary significance of the microbial assemblages of large benthic Foraminifera. *Biological Reviews*, 94, 828–848. doi.org/10.1111/brv.12482.

Preiss, WV. (2019). The tectonic history of Adelaide's scarp-forming faults. *Australian Journal of Earth Sciences*, 66(3), 305–365. doi.org/10.1080/08120 099.2018.1546228.

Preiss, WV & Cowley, WM. (2016). *Geological field excursion—rifts, reverse faults and regolith: Neoproterozoic to Cenozoic geology in the mid-north of South Australia.* South Australia Report Book 2016/00013. South Australian Department of State Development and Geological Society of Australia, South Australian Division.

Pross, J, Contreras, L, Bijl, PK, Greenwood, DR, Bohaty, SM, Schouten, S, Bendle, JA, Röhl, U, Tauxe, L, Raine, JI, Huck, CE, van de Flierdt, T, Jamieson, SSR, Stickley, CE, van de Schootbrugge, B, Escutia, C & Brinkhuis, H. (2012). Persistent near-tropical warmth on the Antarctic continent during the Early Eocene Epoch. *Nature,* 488(7409), 73–77. doi.org/10.1038/nature11300.

Pufahl, PK, James, NP, Kyser, TK, Lukasik, JK & Bone, Y. (2006). Brachiopods in epeiric seas as monitors of secular changes in ocean chemistry: A Miocene from the Murray Basin, South Australia. *Journal of Sedimentary Research,* 76, 926–941. doi.org/10.2110/jsr.2006.079.

Quilty, PG. (2013). Foraminiferology in Australia, 1843–present. In AJ Bowden, FJ Gregory & AS Henderson (eds), *Landmarks in foraminiferal micropalaeontology: History and development* (pp. 251–270). The Micropalaeontological Society, Special Publications, Geological Society of London. doi.org/10.1144/TMS6.19.

Rainger, R. (1981). The continuation of the morphological tradition: American paleontology, 1880–1910. *Journal of the History of Biology*, 14, 129–158. doi.org/10.1007/BF00127518.

Rainger, R. (1985). Paleontology and philosophy: A critique. *Journal of the History of Biology,* 18, 267–287. doi.org/10.1007/BF00120112.

Rainger, R. (1986). Just before Simpson: William Diller Matthew's understanding of evolution. *Proceedings of the American Philosophical Society*, 130, 453–474.

Reichgelt, T, Greenwood, DR, Steinig, S, Conran, JG, Hutchinson, DK, Lunt, DJ, Scriven, LJ & Zhu, J. (2022). Plant proxy evidence for high rainfall and productivity in the Eocene of Australia. *Paleoceanography and Paleoclimatology*, 37, e2022PA004418. doi.org/10.1029/2022PA004418.

Renema, W. (2015). Spatiotemporal variation in morphological evolution in the Oligocene–Recent larger benthic foraminifera genus *Cycloclypeus* reveals geographically undersampled speciation. *GeoResJ* 5, 12–22. doi.org/10.1016/j.grj.2014.11.001.

Renema, W & Cotton, L. (2015). Three dimensional reconstructions of *Nummulites* tests reveal complex chamber shapes. *PeerJ*, 3, e1072. doi.org/10.7717/peerj.1072.

Renz, O. (1936). Stratigraphische und mikropaläontologische Untersuchung der Scaglia (Obere Kreide-Tertiär) im Zentralen Appenin [Stratigraphic and micropalaeontological study of the Scaglia (Upper Cretaceous-Tertiary) in the Central Apennines]. *Eclogae Geologicae Helvetiae*, 29, 1–149.

Retallack, GJ. (2008). Cool-climate or warm-spike lateritic bauxites at high latitudes? *The Journal of Geology*, 116, 558–570. doi.org/10.1086/592387.

Reynolds, P, Holford, S, Schofield, N & Ross, A. (2017). Three-dimensional seismic imaging of ancient submarine lava flows: An example from the southern Australian margin. *Geochemistry, Geophysics, Geosystems*, 18, 3840–3853. doi.org/10.1002/2017GC007178.

Rudwick, MJS. (1978). Charles Lyell's dream of a statistical palaeontology. *Palaeontology*, 21, 225–244.

Rudwick, MJS. (1985). *The Great Devonian controversy: The shaping of scientific knowledge among gentlemanly specialists.* University of Chicago Press. doi.org/10.7208/chicago/9780226731001.001.0001.

Rudwick, MJS. (1997). *Georges Cuvier, fossil bones, and geological catastrophes.* University of Chicago Press. doi.org/10.7208/chicago/9780226731087.001.0001.

Rudwick, MJS. (2005). *Bursting the limits of time: The reconstruction of geohistory in the age of revolution.* University of Chicago Press. doi.org/10.7208/chicago/9780226731148.001.0001

Rudwick, MJS. (2008). *Worlds before Adam: The reconstruction of geohistory in the age of reform.* University of Chicago Press. doi.org/10.7208/chicago/9780226731308.001.0001.

Rudwick, MJS. (2014). *Earth's deep history: How it was discovered and why it matters.* University of Chicago Press. doi.org/10.7208/chicago/9780226204093.001. 0001.

Rupke, NA. (1983). The study of fossils in the Romantic philosophy of history and nature. *History of Science,* 21, 389–413. doi.org/10.1177/007327538302100403.

Rupke, NA. (2009). *Richard Owen: Biology without Darwin* (revised edition). University of Chicago Press. doi.org/10.7208/chicago/9780226731780.001.0001.

Russell, DA. (1971). The disappearance of the dinosaurs. *Canadian Geographical Journal,* 83, 204–215.

Russell, ES. (1916). *Form and function: A contribution to the history of animal morphology.* John Murray. doi.org/10.5962/bhl.title.3747.

Sartorio, D & Venturini, S (compilers). (1988). *Southern Tethys biofacies.* Agip (*Azienda Generale Italiana Petroli*, General Italian Oil Company), S.p.A., S. Donato Milanese.

Sauermilch, I, Whittaker, JM, Bijl, PK, Totterdell, JM & Jokat, W. (2019). Tectonic, oceanographic, and climatic controls on the Cretaceous–Cenozoic sedimentary record of the Australian–Antarctic Basin. *Journal of Geophysical Research: Solid Earth,* 124(8), 7699–7724. doi.org/10.1029/2018JB016683.

Savin, SM. (1977). The history of the Earth's surface temperature during the past 100 million years. *Annual Reviews of Earth & Planetary Sciences,* 5, 319–355. doi.org/10.1146/annurev.ea.05.050177.001535.

Schodde, R. (2005). Ernst Mayr and southwest Pacific birds: Inspiration for ideas on speciation. *Ornithological Monographs,* 58, 50–57. doi.org/10.2307/40587709.

Schodde, R. (2006). Australasia's bird fauna today—origins and evolutionary development. In JR Merrick, M Archer, GM Hickey, MSY Less (eds), *Evolution and biogeography of Australasian vertebrates* (pp. 413–458). Australian Scientific Publishing.

Schofield, A & Totterdell, JM. (2008). *Distribution, timing and origin of magmatism in the Bight and Eucla Basins.* Geoscience Australia Record 2008/24.

Schuchert, C. (1915). *A text-book of geology. Part II: Historical geology.* John Wiley & Sons.

Secord, JA. (1986). *Controversy in Victorian geology: The Cambrian–Silurian dispute.* Princeton: Princeton University Press.

Secord, JA. (1991). The discovery of a vocation: Darwin's early geology. *The British Journal for the History of Science, 24*, 133–157. doi.org/10.1017/S00070874 00027059.

Seibold, E & Berger, WH. (1993). *The sea floor: An introduction to marine geology*. Springer. (Published over 1982–2017 in four editions.)

Sepkoski, D. (2015). *Rereading the fossil record*. Chicago: Chicago University Press.

Sepkoski, D. & Ruse, M. (2009). *The paleobiological revolution: Essays on the growth of modern paleontology*. Chicago: University of Chicago Press. doi.org/10.7208/chicago/9780226748597.001.0001.

Shackleton, NJ & Kennett, JP. (1975). Paleotemperature history of the Cenozoic and the initiation of Antarctic glaciation: Oxygen and carbon isotope analyses in DSDP sites 277, 279, and 281. In JP Kennett, RE Houtz, PB Andrews, AR Edwards, VA Gostin, M Hajós, MA Hampton, DG Jenkins, SV Margolis, AT Ovenshine & K Perch-Nielsen (eds), *Initial reports of the Deep Sea Drilling Project* (Vol. 29). US Government Printing Office. doi.org/10.2973/dsdp.proc.29.117.1975.

Sharples. AGWD, Huuse, M, Hollis, C, Totterdell, JM & Taylor, PD. (2014). Giant middle Eocene bryozoan reef mounds in the Great Australian Bight. *Geology, 42*, 683–686. doi.org/10.1130/G35704.1

Shepherd, CL, Kulhanek, DK, Hollis, CJ, Morgans, HEG, Strong, CP, Pascher, KM & Zachos, JC. (2021). Calcareous nannoplankton response to Early Eocene warmth, Southwest Pacific Ocean. *Marine Micropaleontology, 165*, 101992. doi.org/10.1016/j.marmicro.2021.101992.

Simpson, GG. (1937). Patterns of phyletic evolution. *Bulletin of the Geological Society of America, 48*, 303–314. doi.org/10.1130/GSAB-48-303.

Simpson, GG. (1942). The beginnings of vertebrate paleontology in North America. *Proceedings of the American Philosophical Society, 86(1): Symposium on the Early History of Science and Learning in America (Sep. 25, 1942),* 130–188.

Simpson, GG. (1944). *Tempo and mode in evolution*. Columbia University Press.

Simpson, GG. (1960). The history of life. In S Tax (ed.), *The evolution of life*, (pp. 117–180). Chicago: The University of Chicago Press.

Simpson, GG. (1961). *Principles of animal taxonomy*. Columbia University Press. doi.org/10.7312/simp92414.

Simpson, GG. (1965). *The geography of evolution*. Capricorn Books.

Simpson, GG. (1976). The compleat palaeontologist? *Annual Review of Earth and Planetary Sciences, 4*, 1–14. doi.org/10.1146/annurev.ea.04.050176.000245.

Singleton, FA. (1940). The Tertiary geology of Australia. *Proceedings of the Royal Society of Victoria*, 53, 1–125.

Sluijs, A, Bowen, GJ, Brinkhuis, H, Lourens, LJ & Thomas, E. (2007). The Palaeocene–Eocene Thermal Maximum super greenhouse: Biotic and geochemical signatures, age models and mechanisms of global change. In M Williams, AM Haywood, FJ Gregory & DN Schmidt (eds), *Deep-time perspectives on climate change: Marrying the signal from computer models and biological proxies* (pp. 323–349). The Micropalaeontological Society, Special Publications. The Geological Society of London. doi.org/10.1144/TMS002.15.

Sluiter, IRK, Holdgate, GR, Reichgelt, T, Greenwood, DR, Kershaw, AP & Schultz, NL. (2022). A new perspective on Late Eocene and Oligocene vegetation and paleoclimates of South-eastern Australia. *Palaeogeography, Palaeoclimatology, Palaeoecology*, 596, 110985. doi.org/10.1016/j.palaeo.2022.110985.

Sprigg, RC. (1952). *The geology of the South-East Province, South Australia, with special reference to Quaternary coast-line migration and modern beach developments.* Geological Survey of South Australia Bulletin 29.

Sprigg, RC. (1979). Stranded and submerged sea-beach systems of South-East South Australia and the aeolian desert cycles. *Sedimentary Geology*, 22, 53–96. doi.org/10.1016/0037-0738(79)90022-8.

Sprigg, RC. (1989). *Geology is fun (recollections) or, The anatomy and confessions of a geological addict.* Arkaroola Pty Ltd.

Stanley, SM & Waller, TR. (1978). Aspects of the adaptive morphology and evolution of the Trigoniidae. *Philosophical Transactions of the Royal Society of London*. Series B, Biological Sciences, 284, 247–258. doi.org/10.1098/rstb.1978.0066.

Stebbins, GL. (1950). *Variation and evolution in plants*. Columbia University Press. doi.org/10.7312/steb94536.

Stilwell, JD. (2003). Macropalaeontology of the Trochocyathus–Trematotrochus band (Paleocene/Eocene boundary), Dilwyn Formation, Otway Basin, Victoria. *Alcheringa*, 27(4), 245–275. doi.org/10.1080/03115510308619107.

Stoker, MS, Holford, SP & Totterdell, JM. (2022). Stratigraphic architecture of the Cenozoic Dugong Supersequence: Implications for the late post-breakup development of the Eucla Basin, southern Australian continental margin. *Earth and Environmental Science Transactions of the Royal Society of Edinburgh*, 1–34. doi.org/10.1017/S1755691022000123.

Stott, R. (2012). *Darwin's ghosts: In search of the first evolutionists*. Bloomsbury.

Sturt, C. (1999). *Two expeditions into the interior of Southern Australia during years 1828, 1829, 1830 and 1831.* Corkwood Press. (Original work published in 1833.)

Taylor, DJ. (1975). Palaeontological report–Potoroo 1. In Shell Development (Australia), *Well completion report, Potoroo-1, Permit SA-5, Great Australian Bight Basin* (pp. 96–104). South Australian Department of Primary Industries and Resources.

Taylor, G. (1994). Landscapes of Australia: Their nature and evolution. In RS Hill (ed.), *History of the Australian vegetation: Cretaceous to recent* (pp. 60–79). Cambridge University Press. doi.org/10.20851/australian-vegetation-05.

Taylor, G & Shirtliff, G. (2003). Weathering: Cyclical or continuous? An Australian perspective. *Australian Journal of Earth Sciences,* 50, 9–18. doi.org/10.1046/j.1440-0952.2003.00970.x.

Taylor, G, Truswell, EM, McQueen, KG & Brown, MC. (1990). Early Tertiary palaeogeography, landform evolution, and palaeoclimates of the Southern Monaro, N.S.W., Australia. *Palaeogeography, Palaeoclimatology, Palaeocology,* 78, 109–134. doi.org/10.1016/0031-0182(90)90207-N.

Totterdell, JM, Blevin, JE, Struckmeyer, HIM, Bradshaw, BE, Colwell, JB & Kennard, JM. (2000). A new sequence framework for the Great Australian Bight: Starting with a clean slate. *APPEA Journal,* 40, 95–116. doi.org/10.1071/AJ99007.

Totterdell, JM & Bradshaw, BE. (2004). The structural framework and tectonic evolution of the Bight Basin. In PJ Boult, DR Johns & SC Lang (eds), *Eastern Australasian Basins Symposium II* (pp. 41–61). Petroleum Exploration Society of Australia, Special Publication.

Totterdell, JM, Bradshaw, BE & Willcox, JB. (2014). Structural and tectonic setting. *Petroleum geology of South Australia, Volume 5: Great Australian Bight.* Geoscience Australia and Department of Energy and Mining, South Australia.

Travouillon, K, Archer, M & Hand, SJ. (2012). Early to middle Miocene monsoon climate in Australia: Comment. *Geology,* 40, e273. doi.org/10.1130/G32600C.1.

Trümpy, R. (1973). The timing of orogenic events in the central Alps. In KA de Jong & R Scholten (eds), *Gravity and tectonics* (pp. 229–251). Wiley.

Truswell, EM. (1997). Palynomorph assemblages from marine Eocene sediments on the west Tasmanian continental margin and the South Tasman Rise. *Australian Journal of Earth Sciences,* 44(5), 633–654. doi.org/10.1080/08120099708728342.

Turner, D. (2011). *Paleontology: A philosophical introduction*. Cambridge: Cambridge University Press. doi.org/10.1017/CBO9780511921100.

van der Kaars, S, Miller, GH, Turney, CSM, Cook, JE, Nürnberg, D, Schönfeld, J, Kershaw, AP & Lehman, SJ. (2017). Human rather than climate the primary cause of Pleistocene megafaunal extinction in Australia. *Nature Communications*, 8, 14142. doi.org/10.1038/ncomms14142.

Veevers, JJ. (2000). Morphotectonics of the divergent margins. In JJ Veevers (ed.), *Billion-year history of Australia and neighbours in Gondwanaland* (pp. 34–50). Gemoc Press.

Vénec-Peyré, M-T. (2004). Beyond frontiers and time: The scientific and cultural heritage of Alcide d'Orbigny (1802–1857). *Marine Micropaleontology*, 50, 149–159. doi.org/10.1016/S0377-8398(03)00064-1.

Vermeij, GJ. (2009). Comparative economics: Evolution and the modern economy. *Journal of Bioeconomics*, 11, 105–134. doi.org/10.1007/s10818-009-9062-0.

Vermeij, GJ. (2011). A historical conspiracy: Competition, opportunity and the emergence of direction in history. *Cliodynamics*, 2, 187–207. doi.org/10.21237/C7CLIO21211.

Vermeij, GJ & Leigh, EG. (2011). Natural and human economies compared. *Ecosphere*, 2, 1–16. doi.org/10.1890/ES11-00004.1.

von der Borch, CC, Sclater, JG, Gartner, S, Hekinian, R, Johnson, DA, McGowran, B, Pimm, AC, Thompson, RW, Veevers, JJ & Waterman, LS. (1974). *Initial reports of the Deep Sea Drilling Project* (Vol. 22). US Government Printing Office. doi.org/10.2973/dsdp.proc.22.1974.

von Zittel, KA. (1901). *History of geology and palæontology to the end of the nineteenth century*. Walter Scott. doi.org/10.5962/bhl.title.33301.

Vrba, ES. (1980). Evolution, species and fossils: How does life evolve? *South African Journal of Science*, 76, 61–84.

Wade, BS, Pearson, PN, Berggren, WA & Pälike, H. (2011). Review and revision of Cenozoic tropical planktonic foraminiferal biostratigraphy and calibration to the geomagnetic polarity and astronomical time scale. *Earth-Science Reviews*, 104, 111–142. doi.org/10.1016/j.earscirev.2010.09.003.

Wade, M. (1964). Application of lineage concept to biostratigraphic zoning based on planktonic foraminifera. *Micropaleontology*, 10, 273–290. doi.org/10.2307/1484576.

Warnaar, J. (2006). *Climatological implications of Australian–Antarctic separation* [PhD Thesis]. Universiteit Utrecht.

Wagner, GP. (2014). *Homology, genes, and evolutionary innovation.* Princeton University Press. doi.org/10.23943/princeton/9780691156460.001.0001.

Wagner, GP. (2016). What is 'homology thinking' and what is it for? *Journal of Experimental Zoology Part B: Molecular and Developmental Evolution,* 326B, 3–8. doi.org/10.1002/jez.b.22656.

Wallace, AR. (1989). *Darwinism: An exposition of the theory of natural selection, with some of its applications.* Macmillan.

Walther, J. (1915). Laterit in Westaustralien [Laterites in Western Australia]. *Zeitschrift der Deutschen Geologischen Gesellschaft,* 67B, 113–140.

Ward, PD, Flannery, DTO, Flannery, EN & Flannery, TFF. (2016). The Paleocene cephalopod fauna from Pebble Point, Victoria (Australia)—fulcrum between two eras. *Memoirs of Museum Victoria,* 74, 391–402. doi.org/10.24199/j.mmv. 2016.74.27.

Wegener, AL. (1915). *Die Entstehung der Kontinente und Ozeane [The origin of continents and oceans].* Friedr. Viewig & Sohn.

Weidenbach, K. (2008). *The story of Reg Sprigg—an outback legend.* East Street.

Westerhold, T, Marwan, N, Drury, AJ, Liebrand, D, Agnini, C, Anagnostou, E, Barnet, JSK, Bohaty, SM. De Vleeschouwer, D, Florindo, F, Frederichs, T, Hodell, DA, Holbourn, AE, Kroon, D, Lauretano, V, Littler, K, Lourens, LJ, Lyle, M, Pälike, H, Röhl, U, Tian, J, Wilkens, RH, Wilson, PA & Zachos, JC. (2020). An astronomically dated record of Earth's climate and its predictability over the last 66 million years. *Science,* 369, 1383–1387. doi.org/10.1126/science.aba6853.

White, G. (1789). *The natural history and antiquities of Selborne.* (Compiled by Ronald Davidson-Houston.) London: White, Cochrane & Co. Illustrated edition (2004), London: Thames & Hudson.

Whittaker, JM, Müller, RD, Leitchenkov, G, Stagg, H, Sdrolias, M, Gaina, C & Goncharov, A. (2007). Major Australian–Antarctic plate reorganization at Hawaiian-Emperor Bend time. *Science,* 328, 83–86. doi.org/10.1126/science. 1143769.

Wijeratne, S, Pattiaratchi, C & Proctor, R. (2018). Estimates of surface and subsurface boundary current transport around Australia. *JGR Oceans,* 123(5), 3444–3466. doi.org/10.1029/2017JC013221.

Williams, G. (2013). *Naturalists at sea: Scientific travellers from Dampier to Darwin.* New Haven and London: Yale University Press.

Williams, SE, Whittaker, J-M, Halpin JA & Müller, RD. (2019). Australian–Antarctic breakup and seafloor spreading: Balancing geological and geophysical constraints. *Earth-Science Reviews*, 188, 41–58. doi.org/10.1016/j.earscirev.2018.10.011.

Wilson, EO. (1998). *Consilience: The unity of knowledge.* Little, Brown and Company.

Winchester, S. (2001). *The map that changed the world.* Viking Press.

Wing, SL, Sues, H-D, Potts, R, DiMichele, WA & Behrensmeyer, AK. (1992). Evolutionary paleoecology. In AK Behrensmeyer, JD Damuth, WA DiMichele, R Potts, H-D Sues & SL Wing (eds), *Terrestrial ecosystems through time: Evolutionary paleoecology of terrestrial plants and animals* (pp. 1–13). University of Chicago Press.

Woodruff, F & Savin, SM. (1991). Mid-Miocene isotope stratigraphy in the deep sea: High-resolution correlations, paleoclimatic cycles, and sedimentary preservation. *Paleoceanography*, 6, 755–801. doi.org/10.1029/91PA02561.

Woods, JE. (1862). *Geological observations in South Australia: Principally in the district of the South-East of Adelaide.* Longman, Roberts & Green.

Wopfner, H. (2020). *From Innamincka to Lake Eyre—premier's trip, 1961.* Report Book 2020/00016. South Australian Department for Energy and Mining.

Wopfner, H, Callen, R & Harris, WK. (1974). The lower Tertiary Eyre formation of the Southwestern great Artesian Basin. *Journal of the Geological Society of Australia*, 21(1), 17–51. doi.org/10.1080/00167617408728832.

Zachos, JC, Dickens, GR & Zeebe, RE. (2008). An Early Cenozoic perspective on greenhouse warming and carbon-cycle dynamics. *Nature*, 451, 279–283. doi.org/10.1038/nature06588.

Zachos, JC, Pagani, M, Sloan, L, Thomas, E & Billups, K. (2001). Trends, rhythms, and aberrations in global climate 65 Ma to present. *Science*, 292, 686–693. doi.org/10.1126/science.1059412.

Zeuner, FE. (1945). *The Pleistocene Period.* London: The Ray Society.

Zeuner, FE. (1946). *Dating the past: An introduction to geochronology.* London: Methuen.

Zhou, Y, Retallack, GJ & Huang, C. (2015). Early Eocene paleosol developed from basalt in southeastern Australia: Implications for paleoclimate. *Arab Journal of Geosciences* 8, 1281–1290. doi.org/10.1007/s12517-014-1328-8.

www.ingramcontent.com/pod-product-compliance
Lightning Source LLC
Chambersburg PA
CBHW041143230326
41599CB00039BA/7145